Edited by
Guillermo H. Kaufmann

Advances in Speckle Metrology and Related Techniques

Related Titles

Gläser, M., Kochsiek, M. (eds.)

Handbook of Metrology

2010

ISBN: 978-3-527-40666-1

Horn, A

Ultra-fast Material Metrology

2009

ISBN: 978-3-527-40887-0

Gross, H. (ed.)

Handbook of Optical Systems

Volume 1: Fundamentals of Technical Optics

2005

ISBN: 978-3-527-40377-6

Edited by
Guillermo H. Kaufmann

Advances in Speckle Metrology and Related Techniques

WILEY-VCH

WILEY-VCH Verlag GmbH & Co. KGaA

The Editor

Prof. Guillermo H. Kaufmann
Instituto de Fisica Rosario
Universidad Nacional de Rosario
Facultad de Ciencias Exactas e Ingeniería,
Department of Physics and
Centro Internacional Franco Argentino de Ciencias
de la Información y de Sistemas
Rosario, Argentina
kaufmann@ifir-conicet.gov.ar

Library of Congress Card No.: applied for

British Library Cataloguing-in-Publication Data
A catalogue record for this book is available from the British Library.

Bibliographic information published by the Deutsche Nationalbibliothek
The Deutsche Nationalbibliothek lists this publication in the Deutsche Nationalbibliografie; detailed bibliographic data are available on the Internet at http://dnb.d-nb.de.

Typesetting Thomson Digital, Noida, India
Printing and Binding Fabulous Printers Pte Ltd, Singapore
Cover Design Adam-Design, Weinheim

Printed in Singapore
Printed on acid-free paper

ISBN: 978-3-527-40957-0

To Carolina

Contents

Advances in Speckle Metrology and Related Techniques. Edited by Guillermo H. Kaufmann

ISBN: 978-3-527-40957-0

Preface

The development of laser in the early 1960s generated many new research lines in the field of physical optics. The laser is a source that produces a beam of light that is intense, collimated, and coherent. However, the greatly increased coherency of laser light also leads to the appearance of other side effects such as the speckle phenomenon. People working with lasers quickly realized that the grainy appearance of rough objects when illuminated with coherent light was caused by an interference phenomenon. During the first few years after the invention of the laser, while much effort was carried out to minimize the effects of laser speckle, little was devoted to take the more positive path of putting it to good use. Perhaps the turning point came when it was realized that the light forming each individual speckle was fully coherent and it possessed a definite phase. This ability was recognized in 1968 and was quickly applied to the development of new laser-based techniques to measure displacements, deformations, and vibrations produced by rough objects.

The first multiauthor book on the subject (*Speckle Metrology*, edited by R.K. Erf) appeared in 1978. To fill the need for a book covering new aspects of speckles and novel topics such as fringe analysis and particle image velocimetry that had been mainly developed during the next decade, a new volume appeared in 1993 also entitled *Speckle Metrology*, edited by R. S. Sirohi. In the following years, the development of video cameras with higher resolution and high-speed data acquisition systems gave great impetus to speckle interferometry and related techniques such as digital speckle photography and digital holographic interferometry. These application-oriented techniques contributed to the publication of a new volume in 2001 edited by P.K. Rastogi, entitled *Digital Speckle Pattern Interferometry and Related Techniques*.

From the year 2000, new branches in speckle metrology have appeared, several new schemes have been proposed, and also the known approaches and techniques have been revisited and improved upon. The amount and the scope of these developments are reflected in very rich material, published in the specialized journals and presented each year at various international conferences. Therefore, time has come to review and sum up the most significant of these advances, and this book is the result of such efforts.

Advances in Speckle Metrology and Related Techniques. Edited by Guillermo H. Kaufmann

ISBN: 978-3-527-40957-0

This book provides an up-to-date collection of the new material published in the field of speckle metrology and related techniques since 2000. Although there were several topics that could be included in such a book, we had to select only a few to keep its length reasonable. This means that other topics such as speckle techniques outside the visible part of the spectrum could not be treated. It is important to note that most of the selection of topics was carried out by taking into account that the book should be useful for engineers, scientists, and graduate students who are engaged in the application of speckle techniques mainly to solve specific measurement problems in optical metrology, mechanical engineering, experimental mechanics, material science, nondestructive testing, and related fields or who are contemplating its use.

The book is organized into six chapters. The measurement of radial deformations in cylinders and the evaluation of residual stresses are important engineering problems that are better understood when they are analyzed in polar coordinates. Chapter 1 describes latest digital speckle pattern interferometry (DSPI) systems, which are sensitive to polar coordinates. First, the authors present several configurations based on the use of conical mirrors mounted on piezoelectric transducers to allow the application of phase shifting algorithms for evaluating the phase distribution. A more recently developed interferometer based on the use of a diffractive optical element, which overcomes some limitations of earlier configurations, is also described. Finally, these systems are applied to the measurement of translations, mechanical stresses, and residual stresses. Application examples spread all over the chapter show the extent to which radial speckle interferometry has been developed into a powerful tool for industrial measurements.

Chapter 2 analyzes different approaches to measure internal displacement and strain fields within a weakly scattering material. These techniques have many potential applications that range from the development of failure mechanisms in different media to the detection of retinal disease. First, these authors analyze low coherent interferometry (LCI) that involves illumination of semitransparent scattering test objects with a broadband source and scanning the sample through the required depth range. After describing some limitations of the latter approach, the authors present two recently developed techniques called wavelength scanning interferometry (WSI) and tilt scanning interferometry (TSI), which have some practical advantages over LCI, the most important being an improved signal-to-noise ratio. In WSI, temporal sequences of speckle interferograms are recorded while the wavelength of the laser is tuned at a constant rate, so this technique needs a special light source. TSI is based on tilting an illuminating single wavelength beam during the acquisition of an image sequence, a procedure that provides the necessary depth-dependent phase shifts that allow the reconstruction of the object structure and its internal displacements. A theoretical framework is presented to allow the visualization of the spatial resolution and displacement component measured by any of the techniques presented in this chapter. The chapter concludes with a section on recent developments on phase unwrapping in three dimensions.

Chapter 3 presents different spatial methods to evaluate phase distributions from single-image interferograms. This chapter analyzes not only the different approaches that can be used when a spatial carrier is introduced in the interferometric data but also the more difficult task of automatic demodulation of a single interferogram containing no carrier. First, the authors review the well-known Fourier transform method to recover the phase of a single interferogram with a spatial carrier frequency and also different spatial phase shifting algorithms. They also describe various asynchronous algorithms that do not require the knowledge of the carrier frequency. In the rest of the chapter, the authors present several techniques to recover the modulating phase from a single-image interferogram without a carrier. Among the various approaches that are described, we can mention the regularized phase tracking technique and also local adaptable robust quadrature filters that do not require previous fringe normalization. They also describe single-interferogram demodulation methods based on the determination of fringe orientation, on the vortex transform, and on a general n-dimensional quadrature transform.

Chapter 4 reviews temporal speckle pattern interferometry (TSPI) and discusses several approaches that can be used to analyze the recorded data. TSPI was mainly developed during the past decade to measure the temporal evolution of low-speed dynamic deformation fields. It is based on the analysis of a time series of speckle interferograms, which codes the temporal and spatial phase changes produced during the dynamic deformation of the object. In TSPI, the optical phase distribution is extracted from the speckle intensity at each pixel independent of all the other pixels in the image, so that phase unwrapping is also performed as a function of time. As temporal phase unwrapping involves only 1D signals, this procedure is generally much easier to carry out than spatial 2D unwrapping. Until recently, the most common phase recovery technique used in TSPI was the Fourier transform due to its simplicity and short computational time. However, TSPI signals frequently present nonmodulated pixels, modulation loss and noise that affect the bias, and the modulation intensity terms of the signals to be analyzed. This chapter also describes more robust phase recovery methods that were mainly applied in the past decade to analyze TSPI signals. The numerical algorithms described in this chapter include 1D approaches based on the Fourier transform, the windowed Fourier transform, wavelet transforms, the S-transform, quadratic time–frequency distributions, and the empirical mode decomposition and the Hilbert transform. Two-dimensional and 3D approaches based on the windowed Fourier and directional wavelet transforms are also described.

One of the features that characterize a speckle field is the presence of points where both the real and the imaginary parts of the complex amplitude are equal to zero. Therefore, the intensity at this discrete number of points is also zero and their phase is not defined. Phase singularities or optical vortices can be associated with a sign or charge depending on whether the phase rotates clockwise around them or not. Chapter 5 presents some latest work on the application of optical vortices in optical metrology. These techniques are based on the fact that phase singularities are well-defined geometrical points with unique core structures and spatial configuration, which serve as unique fingerprints and present valuable information as identifiable markers. The authors also present various applications of the so-called optical vortex metrology, such as the measurement of nanometric displacements that use the information on the locations of phase singularities before and after the introduction of a displacement. Other applications presented in this chapter include the determination of rotational displacements and the use of optical vortices both for fluid mechanical investigations and for tracking the dynamics of a biological specimen.

Finally, due to the growing importance of security applications, Chapter 6 deals with the application of speckles for coding optical and digital data. In an encryption system, the information is encoded in such a way that the original information is revealed only by applying the correct key. To realize this aspect of security need, most of the optical architectures use random phase masks for coding and encryption. The phase masks made using speckle patterns work well as random phase masks, and have been used in optical and digital encoding of information. This chapter presents a broad review of various coding techniques used for optical and digital data security applications. Various speckle coding techniques are then discussed with an emphasis on the work carried out by the authors' group. This includes methods for preparation of speckle masks and techniques for their easy alignment, the use of elongated speckle patterns and various multiplexing techniques. The chapter also includes a large number of references that will be very useful for the reader interested in the implementation of these approaches.

To conclude, I would like to thank the authors of the different chapters for their contributions and cooperation. I also wish to thank Valerie Molière of Wiley-VCH Verlag GmbH for inviting me to edit this book and to Anja Tschörtner from the same editorial department for her help and support.

Last but not least, I am grateful to my wife Carolina who has tolerated me with patience over the past difficult year while I was recovering from various health problems.

Rosario, Argentina
July 2010

Guillermo H. Kaufmann

List of Contributors

Armando Albertazzi Gonçalves Jr.
Federal University of Santa Catarina
Department of Mechanical Engineering
Laboratory of Metrology and
Automation
UFSC Campus
88040-970 Florianopolis
Brazil

Julio Estrada
Centro de Investigaciones en Optica
Loma del Bosque 115
Colonia Lomas del Campestre
37150 Leon
Mexico

Alejandro Federico
Instituto Nacional de Tecnología
Industrial
Electrónica e Informática
Av. General Paz 5445
B1650KNA San Martín
Argentina

Steen G. Hanson
Technical University of Denmark
Department of Photonics Engineering
DTU Fotonik, Frederiksborgvej 399
Building 128
P.O. Box 49
4000 Roskilde
Denmark

Jonathan M. Huntley
Loughborough University
Wolfson School of Mechanical and
Manufacturing Engineering
Ashby Road
Loughborough LE11 3TU
United Kingdom

Guillermo H. Kaufmann
Instituto de Fisica Rosario
Blvd. 27 de Febrero 210 bis
S2000EZP Rosario
Argentina

Arvind Kumar
Indian Institute of Technology Delhi
Department of Physics
Hauz Khas
New Delhi 110016
India

Antonio Quiroga
Universidad Complutense de Madrid
Facultad de Ciencias Fisicas
Departamento de Optica
Cuidad Universitaria
28040 Madrid
Spain

Advances in Speckle Metrology and Related Techniques. Edited by Guillermo H. Kaufmann

ISBN: 978-3-527-40957-0

Pablo D. Ruiz
Loughborough University
Wolfson School of Mechanical and Manufacturing Engineering
Ashby Road
Loughborough LE11 3TU
United Kingdom

Manuel Servin
Centro de Investigaciones en Optica
Loma del Bosque 115
Colonia Lomas del Campestre
37150 Leon
Mexico

Kehar Singh
Indian Institute of Technology Delhi
Department of Physics
Hauz Khas
New Delhi 110016
India

Madan Singh
Instruments Design, Development and Facility Centre
Staff Road
Ambala 133001
India

Mitsuo Takeda
The University of Electro-Communications
Department of Information and Communication Engineering
Laboratory for Information Photonics and Wave Signal Processing
1-5-1, Chofugaoka, Chofu
182-8585 Tokyo
Japan

Matías R. Viotti
Federal University of Santa Catarina
Department of Mechanical Engineering
Laboratory of Metrology and Automation
UFSC Campus
88040-970 Florianopolis
Brazil

Wei Wang
Heriot-Watt University
School of Engineering and Physical Sciences
Department of Mechanical Engineering
Edinburgh EH14 4AS
United Kingdom

1
Radial Speckle Interferometry and Applications

Armando Albertazzi Gonçalves Jr. and Matías R. Viotti

1.1
Introduction

The invention of laser in the 1960s led to the development of sources of light with a high degree of coherence and allowed to see a new effect with a grainy aspect, which appeared when optically rough surfaces were illuminated with a laser light. This effect was called *speckle effect* characterized by a random distribution of the scattered light.

After the advent of laser sources, this effect was considered a mere nuisance, mainly for holography techniques. Nevertheless, important research efforts began in the late 1960s and early 1970s, focusing on the development of new methods for performing high-sensitivity measurements on diffusely reflecting surfaces. These efforts paved the way for the development of *electronic speckle pattern interferometry* (ESPI), the basic principle of which was to combine speckle interferometry with electronic detection and processing. ESPI avoided the awkward and high time-consuming need for film processing, thus allowing real-time measurement of the object. However, first results were a bit discouraging due to low detector resolution, low sensitivity, and high signal-to-noise ratio.

Constant advances in technology, particularly with respect to high resolution and speed data acquisition systems, and software development for data processing allowed linking, first, vacuum-tube television cameras or, until today, CCD or CMOS cameras to a host computer in order to acquire a digital image of the surface illuminated with laser light. Advances in data transmission enabled to directly link cameras to the computer (IEEE-1394 interface) and transmit digital images without extra elements to digitize the acquired image (such as the well-known *frame grabbers*). Because of the use of both digital images and processing techniques, ESPI was called DSPI (digital speckle pattern interferometry).

Nowadays, there are a large number of interferometric systems that allow to monitor a large variety of physical parameters. They can be mainly grouped in two families: (i) interferometers with sensitivity to *out-of-plane* displacements and (ii) interferometers with *in-plane* sensitivity. Several approaches can be put in these two

Advances in Speckle Metrology and Related Techniques. Edited by Guillermo H. Kaufmann

ISBN: 978-3-527-40957-0

families. Among them, radial interferometers can be highlighted, a special class of interferometers that are able to measure in polar or cylindrical coordinates. *Radial out-of-plane* interferometers are very convenient for some engineering applications dealing with measurement of deformations in pipes, bearings, and other cylinders. Since *radial in-plane* interferometers can be made in a robust and compact way, they are also of great engineering interest as they allow small interferometers to perform measurements outside the laboratory. This kind of interferometers will be discussed in the following sections. Section 1.2 will describe radial out-of-plane interferometers to measure internal and external cylinders. In-plane interferometers will be discussed in Section 1.3, showing two different configurations. Finally, Section 1.4 will show some applications of in-plane radial interferometers.

1.2
Out-of-Plane Radial Measurement

Perhaps the simplest way to measure the out-of-plane displacement component on a surface is by illuminating it and viewing it in the normal direction. Figure 1.1 shows a possible configuration for out-of-plane measurement in Cartesian coordinates. The laser light is expanded and collimated by the lens and is directed to a partial plane mirror that splits the laser light into two beams. Part of the light is deflected to the right and illuminates the rough surface to be measured, which scatters the light forming a speckle pattern. The other part is transmitted through the partial plane mirror and illuminates a rough surface that produces another speckle pattern, which is taken as a reference. The camera captures both images of the measured surface, viewed through the partial plane mirror, and the image of the reference surface reflected by the partial plane mirror. The resulting image shows the coherent

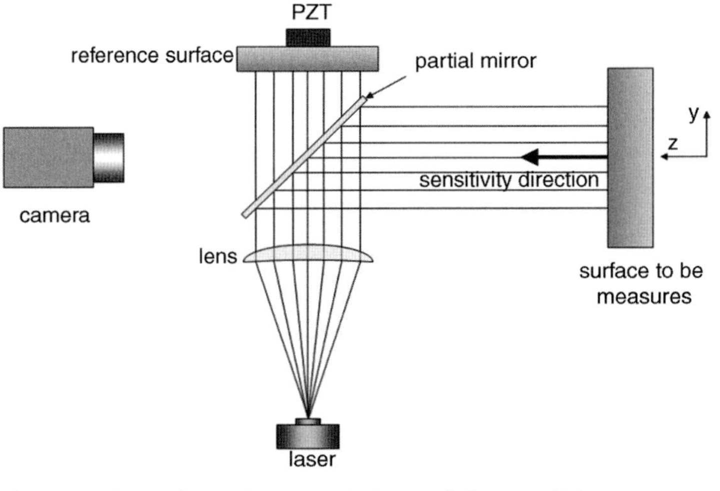

Figure 1.1 A typical optical setup to obtain out-of-plane sensitivity.

interference of the two speckle patterns emerging from both surfaces. A piezo translator (PZT) is used to move the reference surface in a submicrometric range to produce controlled phase shifts.

The sensitivity direction of this configuration is represented by the vector drawn on the surface to be measured. It is computed by the vector addition of two unitary vectors pointing to the illumination source and to the camera pupil center, respectively. In this case, since both are practically aligned with the z-axis, the sensitivity vector is also almost aligned with the z-axis and its magnitude is very close to 2.0. For the case of illumination with collimated light and imaging through telecentric lenses, the sensitivity vector is equal to 2.0 and perfectly parallel to the z-axis. Therefore, in this case the sensitivity vector has a component only along the z-axis and it is given by Equation 1.1:

$$k_z = \frac{4\pi}{\lambda}. \tag{1.1}$$

The out-of-plane displacement component w along the z-axis between two object states can be computed from the measured phase difference $\Delta\varphi$ by Equation 1.2:

$$w = \frac{\Delta\varphi}{k_z} = \frac{\lambda}{4\pi}\Delta\varphi. \tag{1.2}$$

In some cases where noncollimated illumination is used or nontelecentric imaging is involved, Equation 1.1 has to be modified to accomplish for a small amount of in-plane sensitivity. Those cases are discussed in Refs [1, 2].

The meaning of *radial out-of-plane measurement* here is related to the measurement of the displacement component normal to a cylindrical surface or, in other words, in the direction of the radius of the cylinder. As usual in cylindrical coordinates, a positive radial out-of-plane displacement increases the value of the radius. Radial out-of-plane displacement components are very important in engineering applications. They are responsible for the diameter and form deviations of cylindrical surfaces, which are very closely connected to the technical performance of cylindrical parts. Therefore, sometimes they are referred to as *radial out-of-plane deformations*. Since the measured quantity is the displacement field between two object states, the expression *radial out-of-plane displacement* is preferred in this chapter.

Pure radial out-of-plane displacement measurement can be accomplished only by DSPI using special optics. The main idea is to use optical elements to promote illumination and viewing directions that result in radial sensitivity. This section presents possible configurations for three application classes: short internal cylinders, long internal cylinders, and external cylinders.

1.2.1 Radial Deformation Measurement of Short Internal Cylinders

To measure the radial out-of-plane displacement component, special optical elements are required. Ideally, it should optically transform Cartesian coordinates into cylindrical ones. In 1991, Gilbert and Matthys [2, 3] used two panoramic annular lenses to

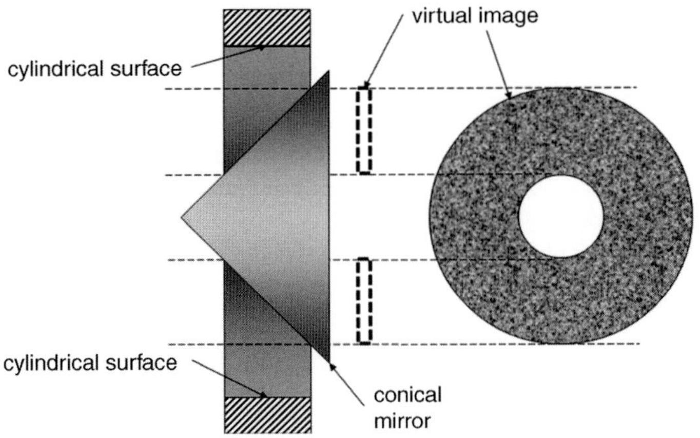

Figure 1.2 Optical transformation produced by a conical mirror placed inside a cylindrical surface.

obtain out-of-plane radial sensitivity. This special lens produces a 360° panoramic view of the scene. When introduced inside a cylinder, such lenses image the inner surface of the cylinder from a near-radial direction. They used two lenses: one panoramic annular lens to illuminate the inner surface of the cylinder in a near-radial direction and another one in the opposite side for imaging. The measurement was possible in a cylindrical ring region between both lenses.

Another possibility to produce radial sensitivity is by using conical mirrors. Figure 1.2 shows the very interesting optical transformation produced by a 45° conical mirror when it is introduced inside an inner cylindrical surface and is aligned with the cylinder axis. When viewed from left to right, the inner surface of the cylinder is reflected on the conical mirror surface all the way around 360°, producing a panoramic image. If the observer is far enough, the inner cylindrical surface is optically transformed into a *virtual flat disk*. Therefore, the out-of plane displacement component of this *virtual flat disk* corresponds to the radial out-of-plane displacement component.

Figure 1.3 shows a possible optical setup to measure the radial out-of-plane displacement component of an inner cylinder. A 45° conical mirror is placed inside the internal cylindrical surface to be measured and is aligned to the cylinder axis. Laser light is collimated and split by a partial mirror into two beams: the active and the reference beams. The active beam is deflected toward the conical mirror. The light that reaches the conical mirror is deflected toward the internal surface of the inner cylinder and reaches it orthogonally, producing a speckle field. The light coming back from the speckle field of the cylindrical surface is reflected back by the conical mirror, goes through the partial plane mirror, and is imaged by the camera lens. The reference beam reaches the reference surface, produces a speckle field, and is reflected back to the partial plane mirror and imaged by the camera lens at the same time. The two speckle fields imaged by the camera lens interfere coherently, and the resulting intensities are grabbed by the camera and digitally processed. A

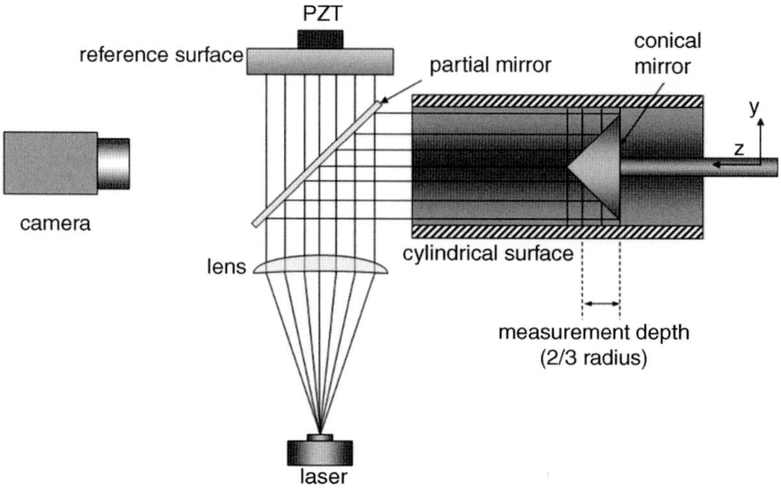

Figure 1.3 Basic configuration for radial out-of-plane displacement measurements of short cylinders using a 45° conical mirror.

piezoelectric translator is placed behind the reference surface to displace it and apply phase shifting to improve image processing capabilities.

If collimated light is used for illumination and a telecentric imaging system is used, or the camera is far enough, the sensitivity vector is always radial and with constant magnitude equal to 2.0. The radial out-of-plane displacement component u_r between two object states is computed for each point on the measured region from the phase difference $\Delta\varphi$ by Equation 1.3:

$$u_r = \frac{\Delta\varphi}{k_r} = \frac{\lambda}{4\pi}\Delta\varphi. \tag{1.3}$$

The measurement depth along the cylinder axis is limited by the conical mirror dimensions. Since the conical mirror angle is 45°, its radius cannot be greater than the inner cylinder radius, which makes the maximum theoretically possible measurement depth to be equal to the conical mirror radius.

In practice, the measurement depth along the cylinder axis is smaller. The image reflected by the conical mirror becomes very compressed near the conical mirror vertex, which reduces the lateral resolution of the reflected image by an unacceptable level. Therefore, the practical measuring limit is about two-thirds of the conical mirror radius. The inner third of the image of the virtual flat disk is not used at all.

In order to reconstruct the radial out-of-plane displacement field on the cylindrical surface, and to present the results in an appropriate way, a numerical mapping can be applied. Figure 1.4a represents the camera view. The gray area corresponds to the measurement region on the cylindrical surface. A point P in such image corresponds to a defined position in the cylindrical surface, as shown in Figure 1.4b. The geometrical mapping is straightforward and can be done by the set of Equation 1.4:

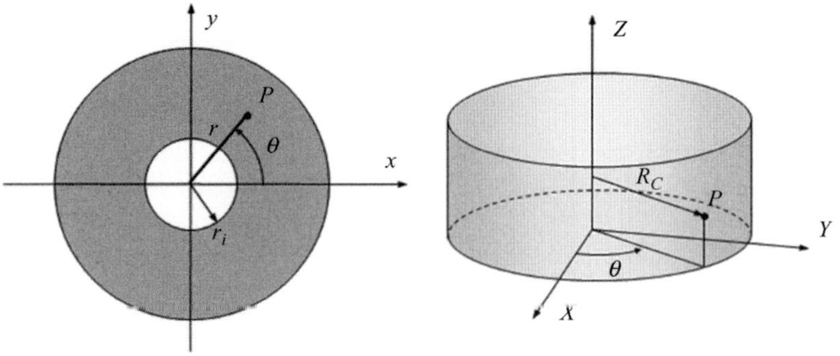

Figure 1.4 Relationship between the virtual flat disk (a) and the cylindrical surface (b).

$$
\begin{aligned}
X &= R_C \cos(\theta), \\
Y &= R_C \sin(\theta), \\
Z &= M(r - r_i),
\end{aligned}
\tag{1.4}
$$

where X, Y, and Z are Cartesian coordinates of points on the cylindrical surface, R_C is the reconstructed cylinder radius, x and y are Cartesian coordinates in the image plane, $r = \sqrt{x^2 + y^2}$ is the polar radius in the image plane, $\theta = \tan^{-1}(y/x)$ is the polar angle in both image plane and cylindrical coordinates, r_i is the inner radius of the region of interest in the image plane, and M is a calibration constant related to image magnification.

In most engineering applications, only the radial deformation of the cylindrical surface is of interest since it produces form deviations. However, in practice, it is almost inevitable that some amounts of rigid body motion – translations and rotations – are superimposed onto the radial deformation component. That comes from the limited stiffness of the mechanical fixture that is unable to keep the conical mirror and/or the cylindrical part to be measured unchanged in the exact place. Fortunately, it is possible to compensate small translations and tilts with the help of software.

A small amount Δx of lateral translation in the X-direction in the cylinder to be measured will produce radial displacement components δr that are not constant in all directions, but depend upon the cosine of the polar angle θ. It is given by

$$\delta r(r, \theta) = \Delta x \cos(\theta), \tag{1.5}$$

where δr is the radial displacement, Δx is the amount of lateral displacement in the X-direction, r is the radius, and θ is the polar angle. Note that δr depends on $\cos(\theta)$ and that the coefficient of $\cos(\theta)$ is the translation amount Δx.

The amount of rigid body translations in both X- and Y-axes in a given cross section can be determined from the Fourier series coefficients. To do that, the radial displacement field u_r must be determined all the way around 360° along a circle that corresponds to such section as a function of the polar angle θ. The amount of translation can be computed by the first-order Fourier coefficients

$$\Delta x = \int_0^{2\pi} u_r(\theta)\cos(\theta)\mathrm{d}\,\theta, \qquad \Delta y = \int_0^{2\pi} u_r(\theta)\sin(\theta)\mathrm{d}\,\theta, \tag{1.6}$$

where Δx and Δy are the rigid body translation components in the X- and Y-axes, respectively, and $u_r(\theta)$ is the radial displacement component all the way around this section.

The above procedure can be repeated for each section of the conical mirror. It is then possible to compute the mean translation components for each different section along the cylindrical surface. Here, if all translations have the same value and direction, it means that only rigid body translation is present. If not, a relative rotation between the mirror and the cylinder axis happens and/or there is a kind of bending of the cylinder axis due to deformation.

If it is possible to connect all different rigid body translation vector ends of each cylinder section by the same straight line, it means that a rigid body rotation is present. In order to quantify the amount of rotation, one can apply linear regression for all Δx and another linear regression for Δy for all sections. The obtained slope is related to the rotation components of xz and yz planes. Then, these rotation values can be used to mathematically compensate that undesirable effect. It is important to make it clear that even if the rotation is superimposed onto any other kind of displacement pattern, this procedure can quantify and remove only the rigid body rotations and displacement components, without affecting or distorting the remaining displacement field.

1.2.2
Radial Deformation Measurement of Long Internal Cylinders

There are a large number of practical applications where longer cylinders have to be measured. For these cases, the configuration present in the previous section is limited by the maximum measurement depth of two-thirds of the conical mirror radius. One possibility would be to measure the cylinder deformations in a piecewise manner. The idea is to divide the cylinder in few virtual sections and measure each of them sequentially. The data are separately processed and then stitched together to produce the total results. However, this approach requires an excellent loading repeatability, very stable experimental conditions, and is an intensively time-consuming procedure. Consequently, this piecewise approach is not practical.

In most engineering applications, the deformation of cylindrical surfaces does not need to be known for each point on the surface. It could be good enough to measure the deformation field in few separate measurement rings, each one in a different section. Therefore, the idea of a piecewise measurement comes back, but it must be done simultaneously.

A special design of a stepped 45° conical mirror can be used to make possible the simultaneous measurement of radial out-of-plane displacements of long inner cylinders [4]. The main idea is presented in Figure 1.5. The continuous 45° conical mirror is replaced for a stepped version. In this figure, four conical sections are

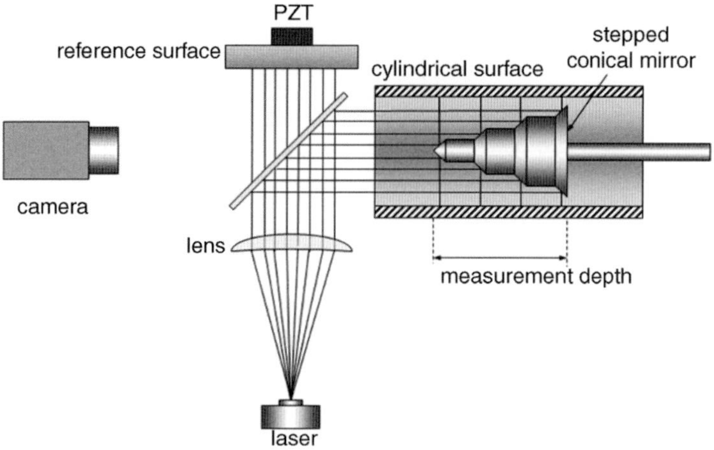

Figure 1.5 Basic configuration for radial out-of-plane displacement measurements of long cylinders using a stepped 45° conical mirror.

separated apart by three cylindrical connecting rods. Each conical section of the stepped mirror reflects the collimated light and forms a measurement ring where the radial out-of-plane measurement is done. The gap between each conical section of the stepped conical mirror is not measured at all. In practice, the lack of this information is not important in most applications where the radial deformations fields are quite smooth. In these cases, the information in few equally spaced sections is sufficient to describe the main behavior of the cylindrical part from the engineering point of view. Only four measurement zones are represented in the figure for simplicity. In practice, a larger number of measurement zones can be achieved.

Figure 1.6 shows an example of an actual stepped conical mirror with seven measuring zones. It was designed for a specific application, which required the

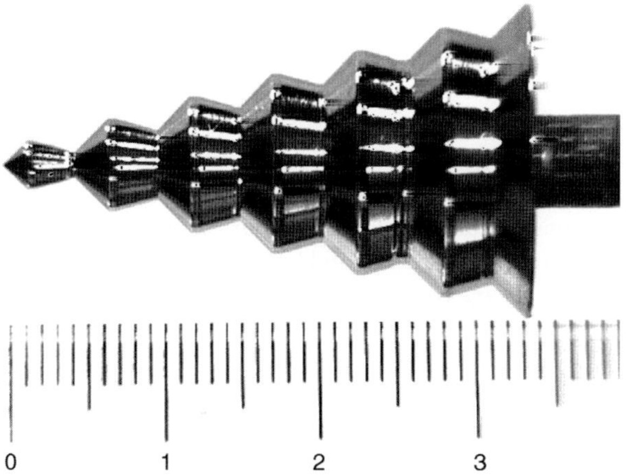

Figure 1.6 Actual view of the seven sections of a stepped 45° conical mirror.

length to be about 34 mm. It was machined in copper in a high-precision diamond turning machine and a layer of titanium was applied to increase the reflectivity and to protect the reflecting surface against mechanical damages. The reflecting areas are oriented at 45° with respect to the mirror axis. The regions in between the reflecting areas have a negative conical angle due to the geometry constraints of the available diamond tool used in the machining process. In practice, it is not possible to use the first conical section for measurement since it is too small and the lateral resolution of the image reflected on that area is unacceptably poor.

The stepped conical mirror of Figure 1.6 was used in a configuration similar to Figure 1.5 to measure the deformations of an inner cylinder of a hermetic gas compressor used in domestic refrigerators. The goal was to study the effects of tightening the four clamping bolts, shown in Figure 1.7 under the four vertical arrows, on the shape of the inner cylinder of the compressor. A set of four 90° phase-shifted images was acquired with an equal initial torque level applied to all bolts. The corresponding phase pattern was stored as the reference phase pattern. After that, the final torque level was applied to the four bolts and another sequence of four 90° phase-shifted images was acquired and the loaded phase pattern was computed and stored.

The resulting phase difference can be seen in Figure 1.8. The top left side of the figure shows the natural image. Seven annular regions can be distinguished, each one corresponding to each conical mirror section and to the radial displacement field of a different section in the inner cylinder. Fringe discontinuities can be present between neighbor annular regions since there is no surface continuity between them. A polar to Cartesian mapping was first applied to extract data. The resulting image is shown on the right-hand side of the figure. The horizontal axis corresponds to the polar angle. The vertical axis is related to the radius, which is connected to the axial

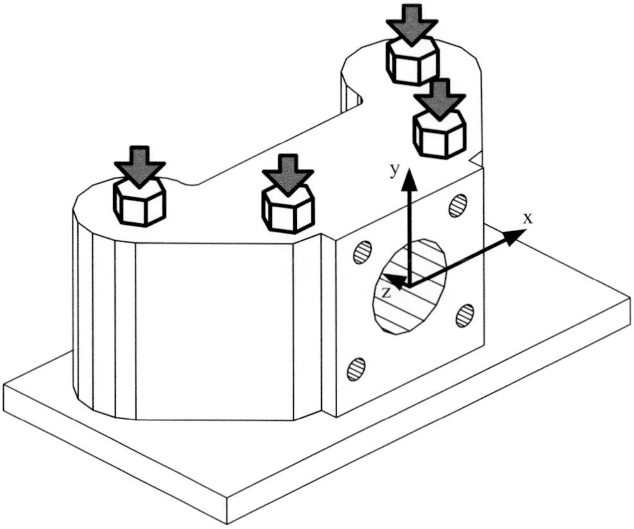

Figure 1.7 Deformation of the inner cylinder of a hermetic gas compressor was measured after tightening four bolts.

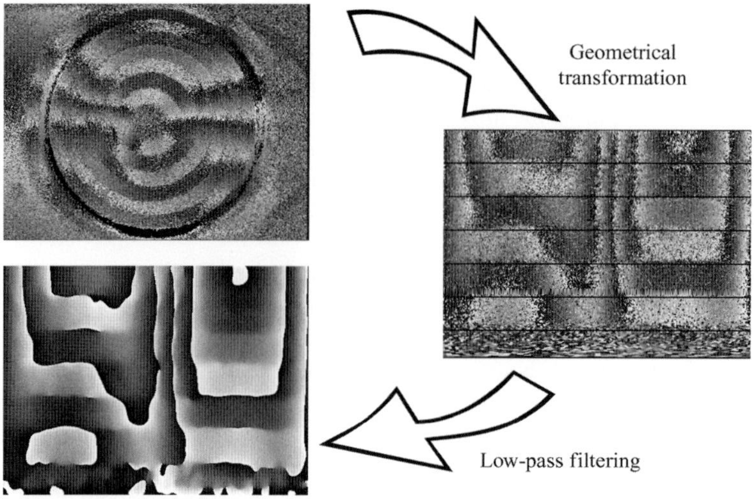

Figure 1.8 Phase difference pattern on the stepped conical mirror surface. *Top left*: the original image. *Right*: after a polar to Cartesian mapping. *Bottom left*: low-pass filtered version.

position of the measured ring. Seven horizontal stripes are visible in this image. The first one in the bottom corresponds to the first section on the nose of the conical mirror. The poor lateral resolution of this stripe is evident in this image. Finally, the bottom left image is the low-pass filtered version of the previous image.

One line was extracted from the center of each stripe and processed. The radial displacement field for six sections was computed. The results are shown in Figure 1.9. Figure 1.9a shows a polar diagram of all sections. The scale division is 1.0 μm. A 3D representation of the deformed cylinder is presented in Figure 1.9b on a much exaggerated scale. This analysis is very useful in engineering for understanding the optimization of the design for stiffness of high-precision cylindrical surfaces.

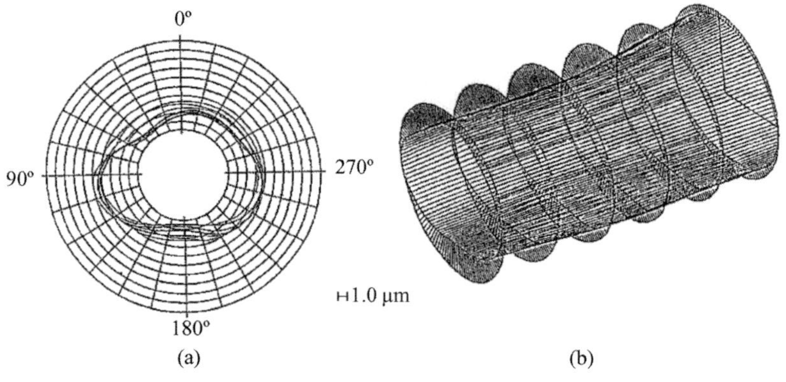

Figure 1.9 (a and b) Measurement results for the deformation of the inner cylindrical surface.

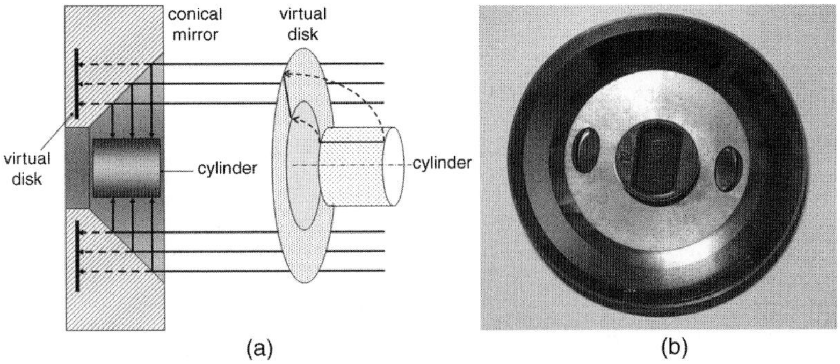

Figure 1.10 (a and b) Optical transformation produced in a cylindrical surface due to an internal 45° conical mirror.

1.2.3 Radial Deformation Measurement of External Cylinders

Radial out-of-plane displacement components can also be measured on external cylindrical surfaces by DSPI. The main idea is represented in Figure 1.10: an internal 45° conical mirror produces an appropriate optical transformation that maps the external cylindrical surface into a flat virtual disk. The ray diagram in Figure 1.10a makes it clear that parallel rays are reflected by the conical mirror and are transformed in radial rays. Figure 1.10b shows an example of a small piston inside a 45° inner conical mirror. The central part shows the upper part (top) of the piston. The cylindrical surface is reflected on the conical mirror and is transformed into a flat disk. The two lateral circular bearings (pinholes) are also visible on the virtual disk area and are distorted due to the reflection on the conical mirror surface.

The DSPI interferometer to measure the radial out-of-plane displacement component is schematically shown in Figure 1.11. The part to be measured is placed and aligned in a 45° external conical mirror. To measure only the radial out-of-plane component, the angle of the conical mirror should be 45° and both illumination source and viewing directions must come from infinity. That can be obtained with collimated illumination and a telecentric imaging system. However, if the diameter of the conical mirror is quite large, collimated illumination and telecentric imaging costs become prohibitive. For these cases, the configuration of Figure 1.12 is feasible since some degree of axial sensitivity is tolerated. Alternatively, to obtain pure radial sensitivity to measure large cylinders, the 45° conical mirror of Figure 1.12 can be replaced by a quasi-conical mirror with curved reflecting surface calculated in such a way to reflect the diverging light coming from a point source like it was a collimated (plane) wavefront and to generate radial illumination and viewing on the cylindrical surface. However, the manufacturing of such special curved mirror can be very expensive.

The configuration of Figure 1.12 was used to measure the thermal deformation of an automotive engine piston [37]. It is made of aluminum and has some steel inserts used to control the thermal deformation and the shape of the engine piston at high

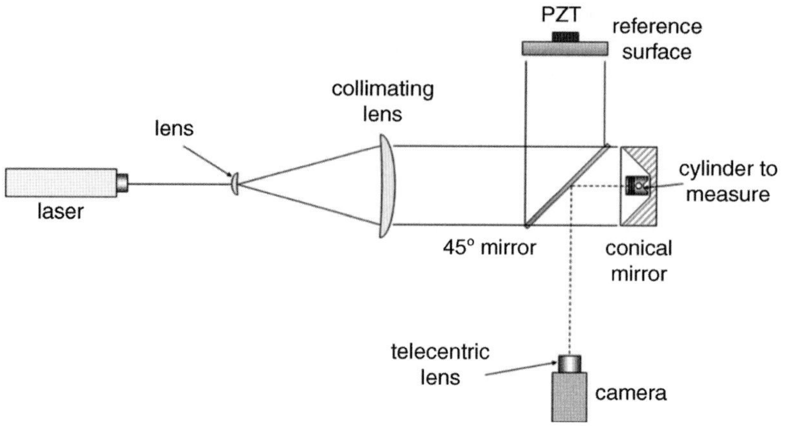

Figure 1.11 Basic configuration for pure radial out-of-plane displacement measurements of external cylinders using a 45° internal conical mirror, collimated light, and telecentric imaging.

temperatures. The way both materials interact and the resulting deformation mechanism were of interest in this investigation.

A large stainless steel conical mirror was used and the engine piston was mounted inside it. Electrical wires were wrapped in the groove of the first piston ring for heating the piston close to its crown. Controlled current levels were applied for heating the piston incrementally. Figure 1.13a shows the camera view of the piston inside the conical mirror. The groove of the first ring was filled with heating wires and covered with thermal paste. The next two grooves are clearly visible as darker circular lines near the maximum diameter. The pinhole of the piston looks distorted due to

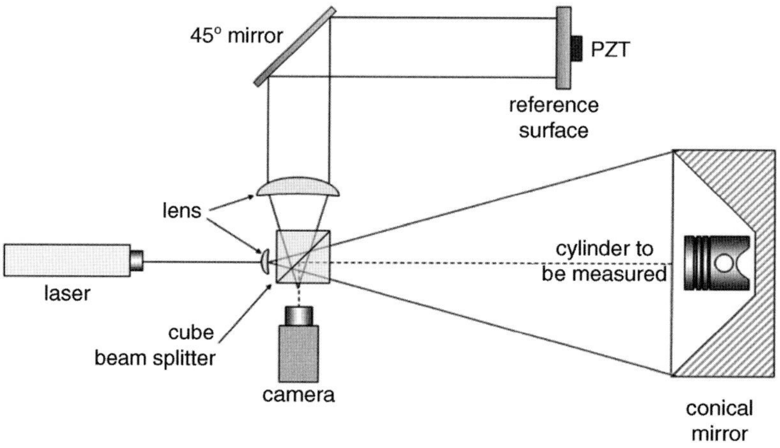

Figure 1.12 Basic configuration for quasi-radial out-of-plane displacement measurements of large external cylinders using a 45° internal conical mirror.

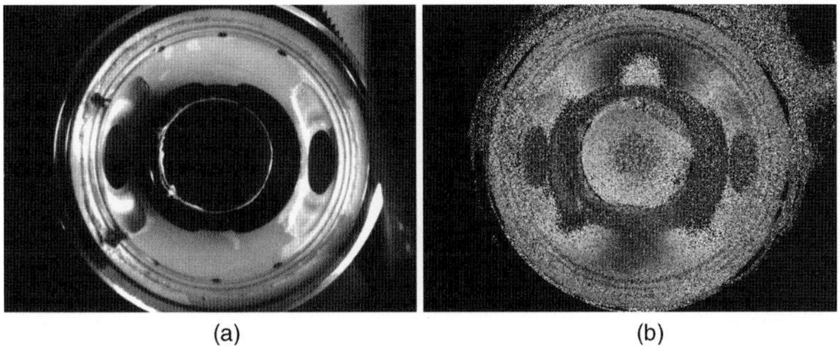

Figure 1.13 (a) Camera view of the engine piston reflected by the conical mirror. (b) The phase difference pattern after heating the engine piston.

reflection in the conical mirror. A set of four 90° phase-shifted images was first acquired and the reference phase pattern was computed and stored. A controlled current was applied in order to increase the piston temperature to about 1 K. After the temperature stabilized, another series of four 90° phase-shifted images were acquired and another phase pattern computed and stored. The phase difference pattern is shown in Figure 1.13b. From the phase difference pattern, it is possible to see that the shape deviation is much stronger in the central part of the image, which corresponds to the bottom of the piston, and less intense near the crown. This happens due to the presence of the steel inserts located somewhere between the crown and the bottom of the piston. This effect can be clearly seen after extracting and analyzing the behavior of the four sections represented in Figure 1.14. The section represented in polar coordinates in Figure 1.14a was extracted from the bottom of the piston, where strong shape deformations are present. The sections in Figure 1.14b–d are located closer to the piston crown, where the shape deformations are smaller. Finally, a 3D plot of the deformed piston is represented on a much exaggerated scale in Figure 1.15. The piston crown is located in the left part of the figure.

1.3 In-Plane Measurement

Optical configurations for measuring in-plane displacements are usually based on the two-beam illumination arrangement first described by Leendertz in 1970 [5]. These interferometers are generally capable of measuring the displacement component, which is coincident with the in-plane direction.

Figure 1.16 shows the basic setup for this kind of interferometer. Two expanded and eventually collimated beams illuminate the object surface forming two angles with the direction of illumination, namely, β_1 and β_2. Thus, two speckle distributions coming from the object surface, with their respective sensitivity vectors $\mathbf{k}_{i1}$ and $\mathbf{k}_{i2}$, interfere in the imaging plane of the camera. The change in the speckle phase will be [1]

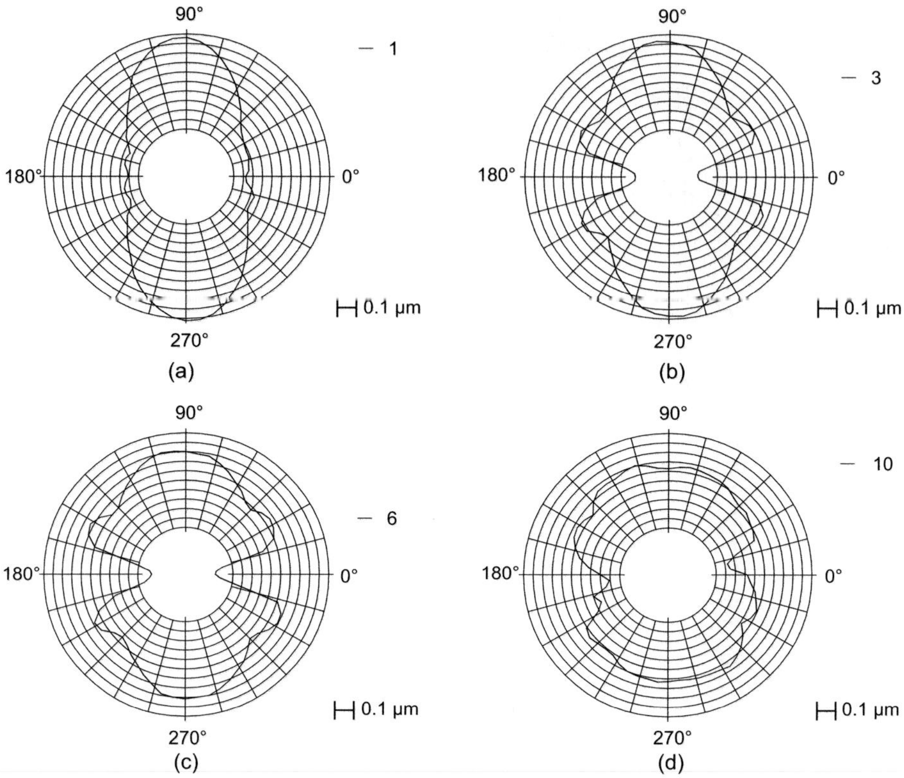

Figure 1.14 (a–d) Polar graphics of the thermal deformations of four sections of the engine piston after heating.

$$\Delta\varphi = (\mathbf{k}_{i1} - \mathbf{k}_{i2}) \cdot \mathbf{d} = \mathbf{k} \cdot \mathbf{d}, \tag{1.7}$$

where $\mathbf{k}$ represents the resultant sensitivity vector obtained from the subtraction between the sensitivity vectors from every beam and it becomes perpendicular to the z-direction of observation when $\beta_1 = \beta_2 = \beta$. In this case, if the illumination vectors are in the xy plane, the net sensitivity can be expressed as [1]

$$k_x = \frac{-4\pi}{\lambda} \sin\beta, \tag{1.8}$$

where k_x is the component of the sensitivity vector along the x-direction and λ is the wavelength of light source. According to this equation, it is noted that β can be changed in order to adjust the sensitivity of the interferometer from zero (illumination perpendicular to the object surface) to a maximum limit value of $-4\pi/\lambda$ (illumination parallel to the object surface).

To obtain the phase difference for two object states, Equation 1.8 should be substituted into Equation 1.7:

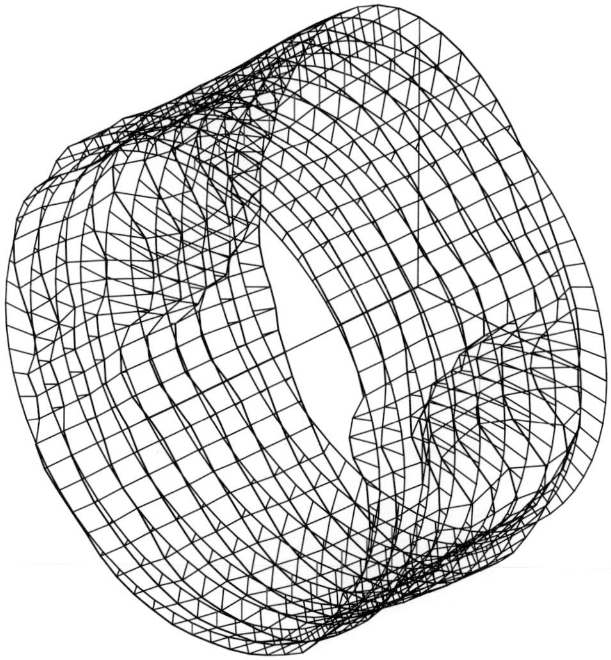

Figure 1.15 3D representation of the thermal deformation of the engine piston after heating.

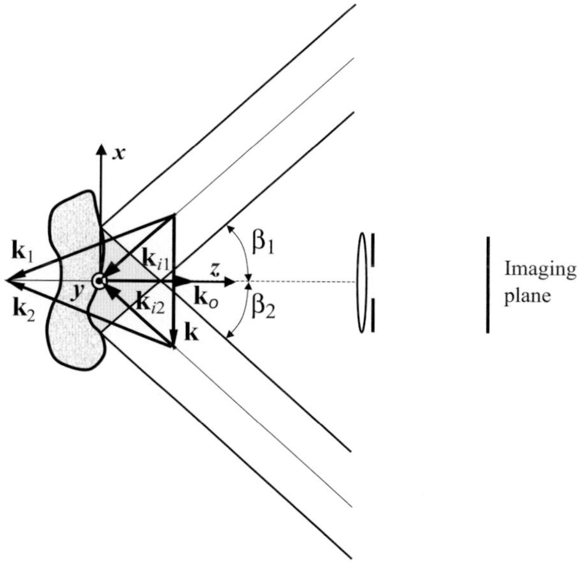

Figure 1.16 Optical setup to obtain in-plane sensitivity.

$$\Delta\varphi = \mathbf{k} \cdot \mathbf{d} = k_x u = \frac{-4\pi}{\lambda} u \sin\beta, \tag{1.9}$$

where u is the component of the displacement field along the x-direction.

For this kind of interferometer, maximum visibility of subtraction fringes will be obtained when the optical system correctly resolves every speckle produced by the scattering surface and the ratio between both illumination beams intensities is equal to 1 [6].

Figure 1.17 shows a drawing of a conventional in-plane digital speckle pattern interferometer with symmetrical dual beam illumination. According to this figure, two expanders are used to illuminate the object. As the distance between the object and the expander lens is a hundred times larger than the measurement region, the variation in the sensitivity vector across the field of view can be considered negligible.

In practical situations, three-dimensional displacement fields are frequently separated in one component normal to the surface to be measured and two components along the tangential direction. For a plane or smooth surface, the former will be known as the out-of-plane displacement component and latter ones as in-plane components. In-plane displacements are more interesting mainly for engineering applications where the main task is to determine strain and stress fields applied in mechanical parts when their integrity has to be evaluated. Nowadays, electrical strain gauges are the most widely used devices in industrial and academic laboratories to monitor strain and stress fields [7]. Even though portability, robustness, accuracy, and range of measurement of strain gauges have been firmly

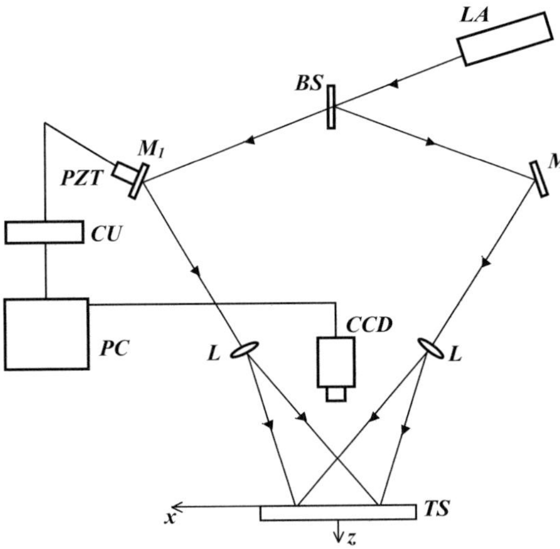

Figure 1.17 Dual-beam illumination interferometer. LA, He–Ne laser; BM, beam splitter; M1 and M2, mirrors; PZT, piezoelectric-driven mirror; L, lens; CCD, camera; CU, control unit; PC, personal computer; TS, test specimen.

established, their installation is time consuming and requires skills and aptitude of a well-trained technician.

The interferometer shown in Figure 1.17 presents sensitivity in only one direction (1D sensitivity). An important requirement in many engineering measurements is to simultaneously compute both in-plane components [1] necessary to measure in two determined directions (2D sensitivity). These systems are made of two interferometers sensitive to two orthogonal displacement directions and are based on polarization discrimination methods by using a polarizing beam splitter that splits the laser beam into two orthogonal linearly polarized beams [8, 9]. Thus, it is possible to simultaneously measure both displacement components. Two drawbacks can be found for this approach, namely, (i) test surface can appreciably depolarize the two orthogonal polarized dual-beam illumination sets causing cross interference between them and (ii) optical setup becomes more bulky and complex. References [10, 11] have managed to deal with these limitations by developing a novel double illumination DSPI system. This interferometer presents an optical arrangement that gives radial in-plane sensitivity and its first version will be described in detail in the following section.

1.3.1
Configuration Using Conical Mirrors

Figure 1.18 shows a cross section of the interferometer used to obtain radial in-plane sensitivity [10–12]. The most important component is a conical mirror that is positioned close to the specimen surface. This figure also displays two particular light rays chosen from the collimated illumination source. Each light ray is reflected by the conical mirror surface toward a point P over the specimen surface, reaching it with the same incidence angle. The illumination directions are indicated by the unitary vectors $\mathbf{n}_A$ and $\mathbf{n}_B$ and the sensitivity direction is given by the vector $\mathbf{k}$ obtained

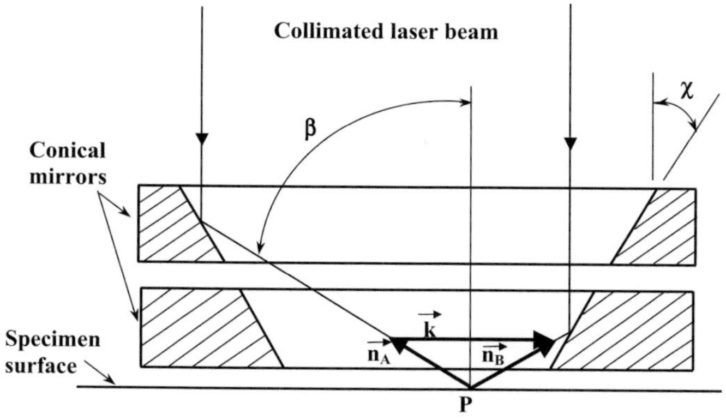

Figure 1.18 Cross section of the upper and lower parts of the conical mirror to show the radial in-plane sensitivity of the interferometer.

from the subtraction of both unitary vectors. As the angle is the same for both light rays, in-plane sensitivity is reached at point P. Over the same cross section and for any other point over the specimen surface, it can be verified that there is only one couple of light rays that merge at that point. Also, in the cross section shown in Figure 1.18, the incidence angle is always the same for every point over the specimen surface and symmetric with respect to the mirror axis. By taking into account unitary vectors and by comparing Figures 1.16 and 1.18, the reader can note similarities in both configurations. As a consequence, if the direction of the normal of the specimen surface and the axis of the conical mirror are parallel to each other, then $\mathbf{n}_A$ and $\mathbf{n}_B$ will have the same angle. Therefore, the sensitivity vector $\mathbf{k}$ will be parallel to the specimen surface and in-plane sensitivity will be obtained.

The above description can be extended to any other cross sections of the conical mirror. If the central point is kept out from this analysis, it can be demonstrated that each point of the specimen surface is illuminated by only one pair of the light rays. As both rays are coplanar with the mirror axis and symmetrically oriented to it, a full 360° radial in-plane sensitivity is obtained for a circular region over the specimen.

A practical configuration of the radial in-plane interferometer is shown in Figure 1.19. The light from a diode laser is expanded and collimated via two convergent lenses and the collimated beam is reflected toward the conical mirror by a mirror that forms a 45° angle with the axis of the conical mirror. The central hole placed on this mirror prevents the laser light from directly reaching the sample surface having triple illumination and provides a viewing window for the CCD camera.

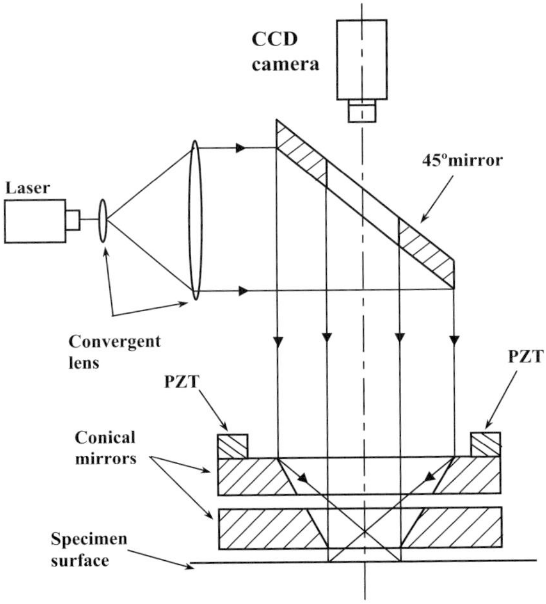

Figure 1.19 Optical arrangement of the radial in-plane interferometer.

The intensity of the light is not constant over the whole circular illuminated area on the specimen surface and it is particularly higher at the central point because it receives light contribution from all cross sections. As a result, a very bright spot will be visible in the central part of the circular measurement region and consequently fringe quality will be reduced. To reduce this effect, the conical mirror is formed by two parts with a small gap between them. The distance of this gap is adapted in such a way that the light rays reflected at the center are blocked. Thus, a small circular shadow is created in the center of the illuminated area and fringe blurring is avoided.

As can be seen from Figure 1.19, for each point over the specimen the two rays of the double illumination originate from the reflection of the upper and lower parts of the conical mirror. A piezo translator was used to join the upper part of the conical mirror, so that its lower part is fixed while the upper part is mobile. As a consequence, the PZT moves the upper part of the conical mirror along its axial direction and the gap between both parts is increased. Then, a small optical path change between both light rays that intersect on each point is produced and the PZT device allows the introduction of a phase shift to evaluate the optical phase distribution by means of any phase shifting algorithm [13].

Due to the use of collimated light, it can be verified that the optical path change is exactly the same for each point of the illuminated surface. The relation between the displacement ΔPZT of the piezoelectric transducer and the optical path change ΔOPC is given by the following equation [12, 14]:

$$\Delta OPC = [1 - \cos(2\chi)]\Delta PZT, \tag{1.10}$$

where χ is the angle between the conical mirror axis and its surface in any cross section.

Finally, the radial in-plane displacement field $u_r(r, \theta)$ can be calculated from the optical phase distribution $\phi(r, \theta)$ [1]:

$$u_r(r, \theta) = \frac{\phi(r, \theta)\lambda}{4\pi \sin \beta}, \tag{1.11}$$

where λ is the wavelength of the laser and β is the angle between the illumination direction and the normal direction of the specimen surface.

1.3.2
Configuration Using a Diffractive Optical Element

Two main drawbacks can be identified in the setup shown in Figure 1.19: (i) it uses a high-quality conical mirror that is quite expensive and (ii) it requires wavelength stabilization of the laser used as light source, which cannot be easily achieved for a compact and cheap diode laser. As a consequence, applications outside the laboratory can be difficult or even unfeasible.

As it is well known, diffractive structures can separate white light into its spectrum of colors. However, if the incident light is monochromatic, the grating will generate

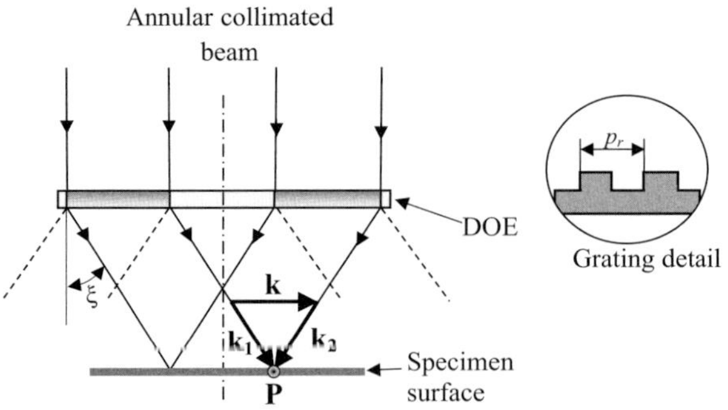

Figure 1.20 Cross section of the diffractive optical element showing radial in-plane sensitivity.

an array of regularly spaced beams in order to split and shape the wavefront beam [15]. The diffraction angle ξ of the spaced beams is given by the well-known grating equation [15, 16]

$$p_r \sin\xi = m\lambda \Rightarrow \sin\xi = \frac{m\lambda}{p_r}, \tag{1.12}$$

where p_r is the period of the grating structure and ξ is the diffraction angle for the order m. From this equation, it is clear that the orders -1 and $+1$ have symmetrical angles with the incident rays.

The recent development of microlithography manufacturing allowed the production of diffractive optical elements (DOEs). The ability to manufacture diffraction gratings with a large variety of geometries and configurations made possible the development of a new and flexible family of optical elements with tailor-made functions. Diffractive lenses, beam splitters, and diffractive shaping optics are some examples of the many possibilities. A special diffractive optical element can be designed to achieve radial in-plane sensitivity with DSPI. It is made as a circular diffraction grating with a binary profile and a constant pitch p_r as shown in Figure 1.20. Its geometry is like a disk with a clear aperture in the center.

If an axis-symmetric circular binary DOE (see Figure 1.20) is used instead of conical mirrors, a double illuminated circular area with radial in-plane sensitivity will also be achieved [17, 18]. The symmetry of the orders -1 and $+1$ will produce double illumination with symmetrical angles, which produces radial in-plane sensitivity. Some advantages can be found by comparing DOE and conical mirror usage: (i) due to advances in microlithography techniques, DOE manufacturing has reached a certain maturity that makes it less expensive than special fabricated conical mirrors, and (ii) because of dual-beam illumination setup, interferometer sensitivity is independent of the wavelength of the laser used as the light source, which will be discussed next.

By considering Equation 1.11, the corresponding fringe equation is as follows:

$$u_r(r,\theta) = \frac{\lambda}{2\sin\beta}. \tag{1.13}$$

According to Equation 1.13, sensitivity of the method would change if angle β or the wavelength of the light source is modified. For example, if angle β is increased, sensitivity would also increase.

By observing Figure 1.20, it is evident that the diffraction angle ξ and the angle between the direction of illumination and the normal to the specimen surface (β) have the same magnitude. Thus, $\sin\xi = \sin\beta$. By substituting Equation 1.12 in Equation 1.11 and by considering the first-order diffraction ($m = 1$)

$$u_r(r,\theta) = \frac{\phi(r,\theta)\lambda}{4\pi(\lambda/p_r)} = \frac{\phi(r,\theta)p_r}{4\pi}. \tag{1.14}$$

In the same way, the corresponding fringe equation will be

$$u_r(r,\theta) = \frac{p_r}{2}. \tag{1.15}$$

Equations 1.14 and 1.15 show that the relationship between the displacement field and the optical phase distribution depends only on the period of the grating of the DOE and not on laser wavelength. This particular and curious effect can be understood through the following explanation: when wavelength of the illumination source increases/decreases, sine function of the diffraction angle decreases/increases by the same amount (see Equation 1.13). As λ is divided by $\sin\beta$ in Equation 1.11, the ratio between them will be constant.

Reference [18] compares the influence on the sensitivity of the interferometer when a DOE is used instead of conical mirror. According to Viotti *et al*, when the setup shown in Figure 1.17 is used with a red light source or with a green one, phase maps obtained with the green laser had approximately 1.5 more fringes compared to those obtained with the red laser. Figure 1.21a shows a phase map obtained for a red light source and Figure 1.21b shows a phase map obtained for green light for the same displacement field.

On the other hand, Figure 1.22a and b shows the phase maps for the same displacement field obtained by using the diffractive optical element instead of the conical mirror. As the figure shows, it can be noted that fringe amounts are the same for both. Thus, Figure 1.22a and b clearly confirms the result obtained in Equation 1.14.

As shown in Figure 1.19, a similar optical arrangement can be built in order to integrate the diffractive optical element. This new practical configuration of the radial in-plane interferometer is shown in Figure 1.23. The light from a diode laser (L) is expanded by a plane concave lens (E). Then, it passes through the elliptical hole of the mirror M_1, which forms a 45° angle with the axis of the DOE, illuminating mirrors M_2 and M_3 and being reflected back to the mirror M_1. Thus, the central hole placed at M_1 allows that the light coming from the laser source reaches mirrors M_2 and M_3. In addition, this hole has other functions, namely, (i) to prevent the laser light from directly reaching the specimen surface having triple illumination and (ii) to provide a viewing window for the CCD camera. Mirror M_1 directs the expanded laser light to the lens (CL) in order to obtain an annular collimated beam. Finally, the light is diffracted by the DOE mainly in the first diffraction order toward the specimen

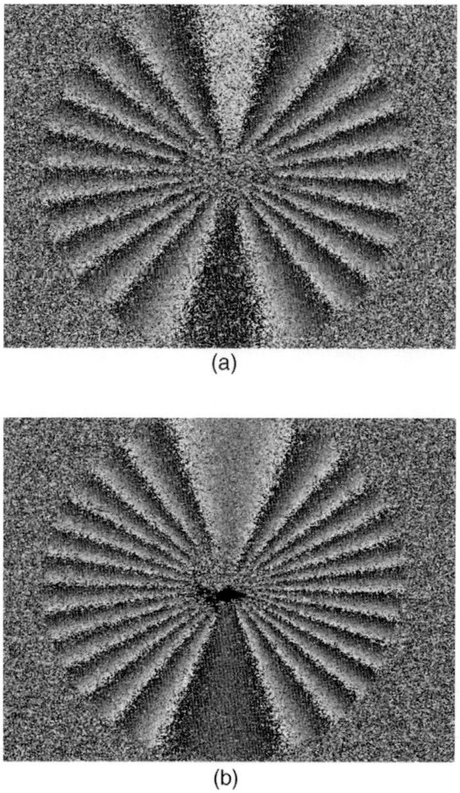

Figure 1.21 Phase maps obtained by using the radial in-plane interferometer with conical mirror for wavelength light source of (a) 658 nm and (b) 532 nm [18].

surface. Residual nondiffracted light or light from higher diffraction is not considered a problem since this kind of light is not directed to the central measuring area on the specimen surface.

M_2 and M_3 are two special circular mirrors. The former is joined to a piezoelectric actuator (PZT) and the later has a circular hole with a diameter slightly larger than the diameter of M_2. Mirror M_3 is fixed while M_2 is mobile. The PZT actuator moves the mirror M_2 along its axial direction generating a relative phase difference between the beam reflected by M_2 (central beam) and the one reflected by M_3 (external beam). The boundary between both beams is indicated in Figure 1.23 with dashed lines. According to this figure, it is possible to see that every point over the illuminated area receives one ray coming from M_2 and other one from M_3. Thus, PZT enables the introduction of a phase shift to calculate the optical phase distribution by means of phase shifting algorithms.

As stated before, the intensity of light is not constant over the whole circular illuminated area on the specimen surface and it is particularly higher at the central point because it receives light contribution from all cross sections. As a result, a very bright spot will be visible in the central part of the circular measurement region and

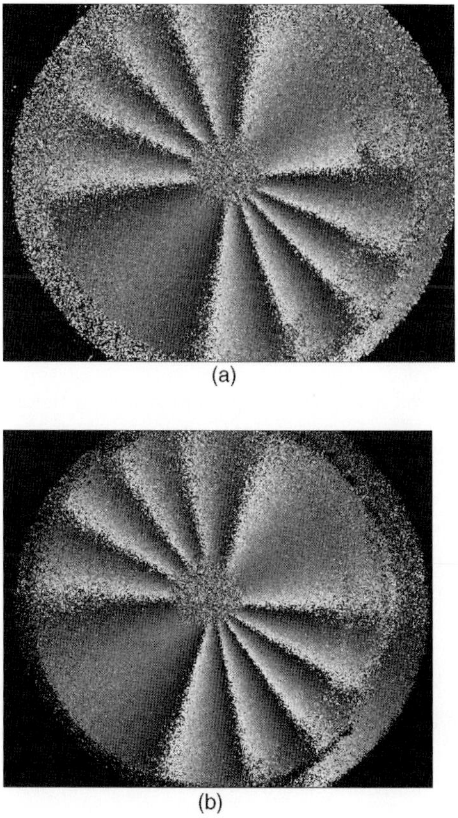

Figure 1.22 Phase maps obtained by using the radial in-plane interferometer with DOE for wavelength light source of (a) 658 nm and (b) 532 nm [18].

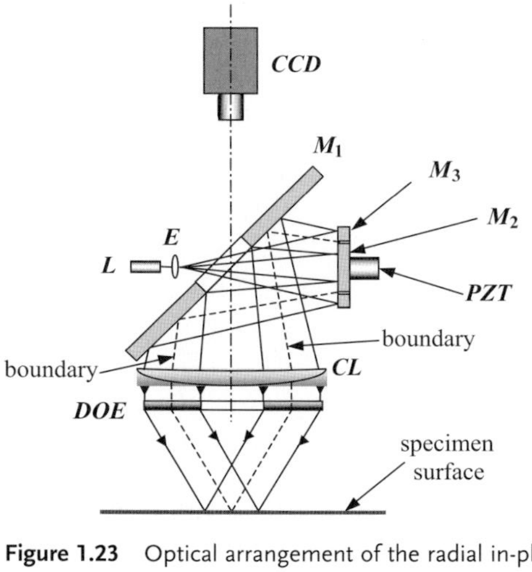

Figure 1.23 Optical arrangement of the radial in-plane interferometer with DOE.

consequently fringe quality will be reduced. For this reason, the outlier diameter of mirror M_2 and the diameter of central hole of M_3 are computed obtaining a gap of about 1.0 mm and blocking the light rays reflected to the center of the measurement area.

1.4 Applications

1.4.1 Translation and Mechanical Stress Measurements

The polar radial displacement field measured in a circular region provides sufficient information to characterize the mean level of both rigid body translations and strains or stresses that occur in that region. For uniform displacement, strain, or stress fields, the complete determination of the associated parameters is almost a straightforward process [19, 20]. In this section, rigid body computation will be analyzed. Mechanical stress field computation will be considered in the next section.

If a uniform in-plane translation is applied on the specimen surface, the following radial displacement field is developed:

$$u_r(r,\theta) = u_t \cos(\theta - \alpha), \tag{1.16}$$

where u_r is the radial component of the in-plane displacement, u_t is the amount of uniform translation, α is the angle that defines the translation direction, and r and θ are polar coordinates. Readers can note that the displacement field does not depend on the radius r at all.

When a uniform stress field is applied to the measured region, the radial in-plane displacement field can be derived from the linear strain–displacement or stress–displacement relations. Usually x and y Cartesian coordinates are used to describe strain or stress states. Since the radial in-plane speckle interferometer measures polar coordinates, the strain and stress states are better described in terms of the principal axes 1 and 2, where the strains and stresses assume the maximum and minimum values, respectively. If η is the angle that the principal axis 1 forms with the x-axis, the in-plane radial displacement field is related to the principal strain and stress components by the following equations [21]:

$$u_r(r,\theta) = \frac{r}{2}[(\varepsilon_1 + \varepsilon_2) + (\varepsilon_1 - \varepsilon_2)\cos(2\theta - 2\eta)], \tag{1.17}$$

$$u_r(r,\theta) = \frac{r}{2E}[(1-\upsilon)(\sigma_1 + \sigma_2) + (1+\nu)(\sigma_1 - \sigma_2)\cos(2\theta - 2\eta)], \tag{1.18}$$

where ε_1 and ε_2 are the principal strains, σ_1 and σ_2 are the principal stresses, E and υ are the material's Young modulus and Poisson ratio, respectively, and η is the principal angle.

Figure 1.24 shows two examples of interferograms obtained with the radial in-plane speckle interferometer. The phase difference patterns correspond to the radial

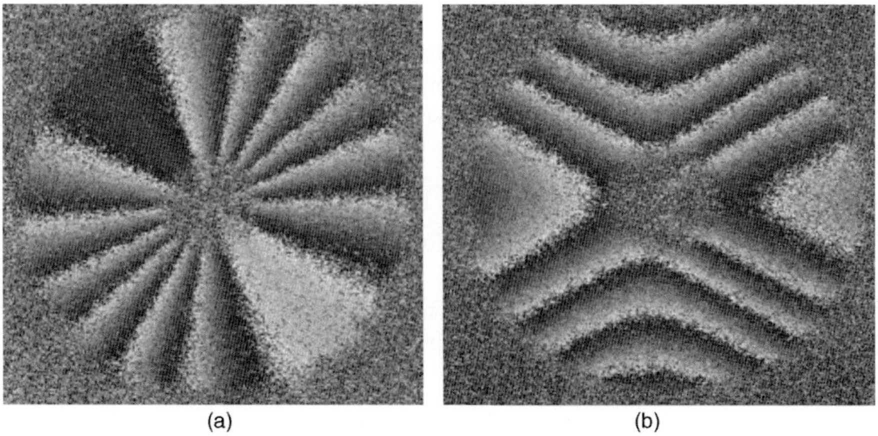

Figure 1.24 Two wrapped phase maps obtained with the radial in-plane speckle interferometer: (a) is due to pure translation and (b) is due to a uniaxial stress field applied in the vertical direction.

displacement component. Figure 1.24a corresponds to a displacement pattern of pure translation of about $u_t = 1.5\ \mu m$ in the direction of $\alpha = 120°$ with the horizontal axis. Note that the fringes caused by pure translation are straight lines pointing to the polar origin. This behavior is predicted by Equation 1.16 since the radial displacement component is independent of the radius r. The phase difference pattern of Figure 1.24b is due to a single stress state of about 40 MPa applied in a steel specimen in the vertical direction. Note that, due to Poisson's effect, the number of fringes in the vertical axis is about three times larger than that in the horizontal one.

In order to quantify the rigid body translations or mechanical stress fields from the measured radial in-plane displacement field, two approaches can be used, namely, (i) the Fourier approach or (ii) the least squares one.

The former uses data of a single sampling circle, concentric with the polar origin, and the latter uses the whole image.

For the Fourier approach, a finite number of regularly spaced sampling points can be extracted from the same circular line all the way around 360°. From this data set, the first three Fourier series coefficients are computed by Equation 1.19. To determine the amount of translation u_t, it is necessary to compute the sine and cosine components and the total magnitude of the first Fourier series coefficient by [21]

$$\begin{aligned} H_{nS}(r_s) &= \int_0^{2\pi} u_r(r_s, \theta)\sin(n\theta)\mathrm{d}\,\theta, \\ H_{nC}(r_s) &= \int_0^{2\pi} u_r(r_s, \theta)\cos(n\theta)\mathrm{d}\,\theta, \\ H_n(r_s) &= \sqrt{H_{nS}^2(r_s) + H_{nC}^2(r_s)}, \end{aligned} \tag{1.19}$$

where r_s is the sampling radius, $H_{nS}(r_s)$ and $H_{nC}(r_s)$ are, respectively, the sine and cosine component of the nth Fourier series coefficient, and $H_{nS}(r_s)$ is the total magnitude of the nth harmonic. As a singular case, readers can note that if $n=0$, components $H_{0S}(r_s)=0$ and $H_{0C}(r_s)=H_0(r_s)$ will be equal to the mean value of $u_r(r_s, \theta)$ along the sampling radius r_s.

To compute the translation component u_t, Equation 1.16 can be expanded to

$$u_r(r,\theta) = u_t\cos(\alpha)\cos(\theta) + u_t\sin(\alpha)\sin(\theta). \tag{1.20}$$

In this case, only the first harmonic is present. The translation amount u_t and its direction α can be computed from the first Fourier series coefficient by

$$\begin{aligned} u_t &= H_1(r_s), \\ \alpha &= \tan^{-1}\left(\frac{H_{1S}(r_s)}{H_{1C}(r_s)}\right). \end{aligned} \tag{1.21}$$

In the same way, the cos term of Equation 1.18 can be expanded to obtain

$$\begin{aligned} u_r(r,\theta) = {} & \frac{r(1-\upsilon)}{2E}(\sigma_1+\sigma_2) + \frac{r(1+\upsilon)}{2E}(\sigma_1-\sigma_2)\cos 2\theta\cos 2\eta \\ & + \frac{r(1+\upsilon)}{2E}(\sigma_1-\sigma_2)\sin 2\theta\sin 2\eta. \end{aligned} \tag{1.22}$$

As stated before, it is possible to verify that the principal stresses and direction can be determined from the zero- and second-order Fourier coefficients by

$$\begin{aligned} \sigma_1 &= \frac{E}{r_s}\left(\frac{H_0(r_s)}{1-\nu} + \frac{H_2(r_s)}{1+\nu}\right), \\ \sigma_2 &= \frac{E}{r_s}\left(\frac{H_0(r_s)}{1-\nu} - \frac{H_2(r_s)}{1+\nu}\right), \\ \eta &= \frac{1}{2}\mathrm{tg}^{-1}\left(\frac{H_{2S}(r_s)}{H_{2C}(r_s)}\right). \end{aligned} \tag{1.23}$$

In practical situations, it is very usual that both stresses and rigid body translations appear mixed up in the same interferogram. They can be measured simultaneously and computed independently since different Fourier series coefficients are involved and the terms of a Fourier series are mutually orthogonal.

The other approach is based on the least squares method. In this approach, a set of experimental data is sampled from the measured displacement field. No particular sampling strategy is required, but it is a good practice to select sampling points regularly distributed over all measured region. The sampled data are fitted to a mathematical model by least squares. An appropriate mathematical model can be obtained by adding and rewriting Equations 1.20 and 1.22:

$$u_r(r,\theta) = K_{OR}r + K_{1C}\cos(\theta) + K_{1S}\sin(\theta) + K_{2C}r\cos(2\theta) + K_{2S}r\sin(2\theta) + K_0. \tag{1.24}$$

Terms K_{OR}, K_{1C}, K_{1S}, K_{2C}, and K_{2S} are easily identified by comparison with Equations 1.21 and 1.23. K_0 is an additional term that was introduced only to take into account a constant bias in the phase pattern that can be occasionally caused by a thermal drift.

At least six measured points are necessary to determine all the six coefficients. Usually, few tens of thousands measured points are used and the coefficients are computed by the least squares method. Since the coefficients are all linear, the least squares can be carried out in a straightforward way using a multilinear fitting procedure. The displacement and stress components can be computed from the fitted coefficients by the following set of equations:

$$\begin{aligned}
u_t &= \sqrt{K_{1C}^2 + K_{1S}^2},\\
\alpha &= \tan^{-1}\left(\frac{K_{1S}}{K_{1C}}\right),\\
\sigma_1 &= E\left(\frac{K_{OR}}{1+\nu} + \sqrt{K_{2C}^2 + K_{2S}^2}\right),\\
\sigma_2 &= E\left(\frac{K_{OR}}{1+\nu} - \sqrt{K_{2C}^2 + K_{2S}^2}\right),\\
\eta &= \frac{1}{2}\text{tg}^{-1}\left(\frac{K_{2S}}{K_{2C}}\right).
\end{aligned} \tag{1.25}$$

1.4.2
Residual Stress Measurement

The stress field that exits in the bulk of some materials without application of external loads or other stress sources is known as residual stress [22, 23]. Many service failures of structural or mechanical components are caused by a combination of residual stress fields present in the material and mechanical stresses produced by applied loads. As a consequence, accurate residual stress measurement becomes a valuable task when the structure integrity must be evaluated. Although recent advances in finite element-based analyses have improved predictions of residual stress distributions, it is essential to accurately know the history of the structure of the mechanical part, which can be done in a few experimental cases. For this reason, nowadays, experimental methods cannot be fully replaced to determine magnitude and principal direction of residual stresses, not only in raw materials but also in components under operating conditions.

There are several methods to characterize residual stresses in engineering materials. Among them, the hole drilling technique is the most widely used for industrial and laboratory applications [24, 25]. This method involves the measurement of in-plane strains generated by relieved stresses when a small hole is drilled into the stressed material, either in a single pass or using multiple increments. Despite strains being usually monitored with specialized three-element strain gauge rosettes, the combined hole drilling strain gauge method presents some practical and economical drawbacks, for example, (i) the specimen surface has to be flat and smooth to bond the rosettes, (ii) the hole has to be drilled exactly in the center of the rosette in order to avoid eccentricity errors, and (iii) the significant cost and time associated with installation of rosettes, which can exceed 1 h for each measurement [24, 26, 27].

Due to these disadvantages, several optical techniques have been developed in the past decades [28]. Among them, digital speckle pattern interferometry is a very attractive technique because of its noncontacting nature and its high relative speed of inspection procedure. Application of digital techniques allows the automation of the data analysis process, which is usually based on the extraction of the optical phase distribution encoded by correlation fringes [13]. Diaz *et al.* [29] presented a hole drilling and DSPI combined system with automated data analysis to measure uniaxial residual stress fields whose direction was coincident with the direction of the in-plane illumination. For this system, the main residual stress direction should be known before starting the measurement in order to adequately orient the in-plane illumination. Some experimental applications have shown that unwanted rigid body displacements can be introduced when hole drilling is performed with this combined system. For this reason, Dolinko and Kaufmann [30] have developed a least squares method to cancel rigid body motion by computing correction parameters determined from two evaluation lines located near the edge of the phase map.

As was clearly explained in Section 1.3, DSPI systems based on two sets of dual-beam illumination arrangements can be used to separately determine both orthogonal components. Thus, measurement of residual stress fields whose principal direction is unknown becomes possible. As previously explained, these polarization systems present some practical drawbacks making difficult their application outside the laboratory.

In order to perform successful measurements outside the laboratory, a set of requirements should be fulfilled by the interferometer [31]:

- **Robust**: The interferometer must be able to successfully work in places with environmental demands. It must be tightly clamped to the specimen surface and stiff enough to be able to keep negligible internal and external relative motions produced by mechanical vibrations. It must be able to handle both environmental temperature variations and voltage oscillations or be battery operated. It also must have some protection against dust, moisture, and daylight.
- **Flexible**: The interferometer must be attachable and adjustable to a variety of specimen geometries and materials. Relative positioning and alignment requirements must be handled in a very flexible way. It should be possible to place the measuring device flexibly and precisely in a given point of interest on the specimen surface and in several positions.

- **Compact**: The device has to be as small as possible. That makes it easy to transport and increases the chances to fit the interferometer in small places. A compact device is an important issue to keep it stiff and robust against mechanical vibrations and relative motions.
- **Stable**: The interferometer must keep stable its metrological performance. No temperature or time dependence of the calibration is desirable. It must be trustworthy everywhere and every time.
- **Friendly**: Frequently, there is not enough time or working conditions for complicated adjustments in out-of-laboratory applications. Therefore, the interferometer must be easy to install, easy to adjust, and easy to operate. In addition, it is important to present clear results on demand for the cases where decisions must be taken in-field.

The practical configuration shown in Figure 1.23 can be used to measure residual stress fields when combined with a hole drilling device. Thus, a portable measurement device can be built having a modular configuration with three parts: (i) a universal base (UB), (ii) a measurement module (MM), and (iii) a hole drilling module (HM) [32]. The universal base is rigidly clamped to the specimen surface by four adjustable and strong magnetic legs and three feet with sharp conical tips to reduce the relative motion between the base and the specimen surface.

The measurement module implements the radial in-plane interferometer shown in Figure 1.23. A 50 mW diode laser with a wavelength $\lambda = 658$ nm was used as a light source. The angle β between the directions of illumination and the normal to the specimen surface was chosen as 30°. The test specimen surface was monitored live by a CCD camera, whose output was digitized by a frame grabber with a resolution of 1280×1024 pixels and 256 gray levels (8 bits). This camera provided a field of view that included the illuminated area of 10 mm in diameter over the specimen.

The hole drilling module is based on an air turbine with a tungsten end mill of 1.6 mm in diameter that is moved by means of a manual micrometric screw. The air turbine has a specified speed of about 320 000 rpm generating minimal induced residual stress during its operation [33].

The measurement and the hole milling modules are fixed to the universal base by an interface that allows a fast and accurate reposition of the modules. The interface is shown in Figure 1.25. Both modules have three spheres (Sph) of steel positioned at 120° and a set of nine strong magnets (Mg2) is fixed rigidly to them. The interface has three pairs of cylindrical supports (Cyl) positioned at 120°, another similar set of nine magnets (Mg1) is fixed rigidly to it, and also a mobile steel plate (Pl). When the measurement or the hole drilling modules are placed over the universal base, the three spheres are precisely positioned on each pair of cylindrical supports forming a kinematic mounting. The magnet sets are aligned in such a way that a light repulsion force is present between the movable module and the clamping base. That avoids mechanical shocks. After positioning the measurement or hole drilling modules on the base, the plate (Pl) is laterally displaced to be located between both sets of magnets (Mg1 and Mg2). In this way, the light repulsion force is smoothly changed to a strong attraction force, which keeps both modules rigidly fixed to the universal base. Using

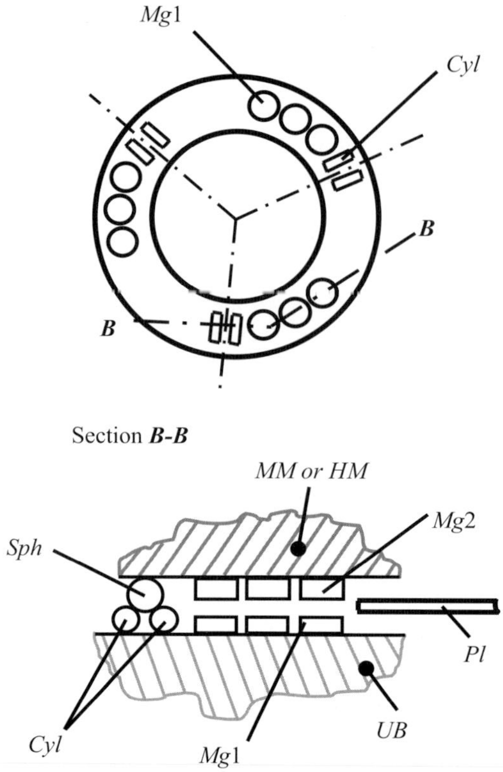

Figure 1.25 Scheme of the kinematic interface of the universal base.

an unloaded specimen, it was tested that the measurement module can be repositioned in the universal base with an error much lower than $\lambda/4$ [12]. Figure 1.26 shows a photograph of the portable system.

To perform the measurements with the portable system, the following procedure is applied. First, the universal base is positioned over the surface to be measured and the measurement module is fixed using the kinematic interface. After that, a set of phase-shifted speckle interferograms is acquired and the reference phase distribution is computed and stored in the portable computer. Then, the measurement module is taken off the universal base and replaced by the hole drilling module. A blind hole is drilled with a depth of about 2 mm. After waiting some seconds for the measurement region to cool down, a second set of phase-shifted speckle interferograms is acquired and a new phase distribution is calculated and stored. Finally, the wrapped phase difference map is evaluated and the continuous phase distribution is obtained by applying a flood-fill phase unwrapping algorithm [34]. Figure 1.27 gives a typical wrapped phase difference pattern. By applying Equation 1.14, the radial in-plane displacement field generated around the hole is calculated from the optical phase distribution. The last step involves the computation of the principal residual stresses and their direction that is accomplished by using the numerical solution developed by

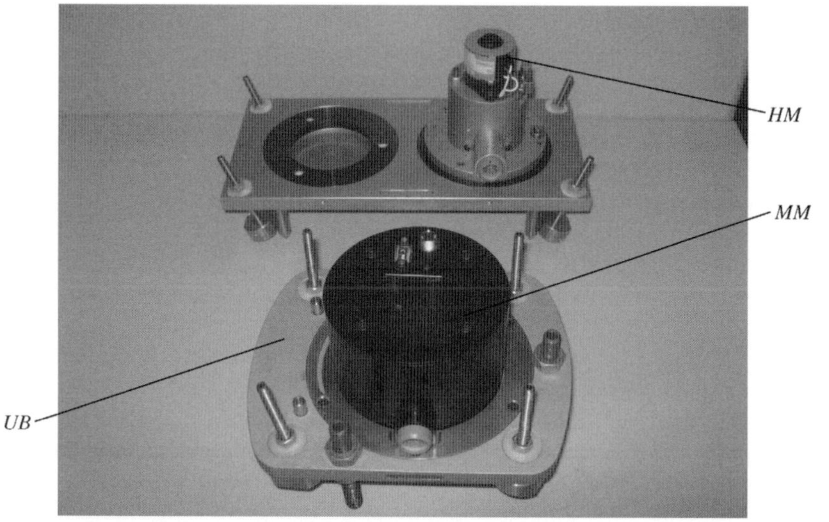

Figure 1.26 Photograph of the portable device. UB, universal base; HM, hole drilling module; MM, measuring module with the radial in-plane interferometer.

Makino and Nelson [35] or the ASTM solution [25] both obtained from the analytical Kirsch's solution [36]. As a consequence, relieved residual stresses were computed from the radial in-plane displacement field, developed by the introduction of the hole with Equation 1.26.

$$u_r(r,\theta) = A(\sigma_{R1} + \sigma_{R2}) + B(\sigma_{R1} - \sigma_{R2})\cos(2\theta - 2\eta), \tag{1.26}$$

where σ_{R1} and σ_{R2} are the principal residual stresses, η is the angle of the principal directions, and r and θ are polar coordinates. A and B are constants given by the

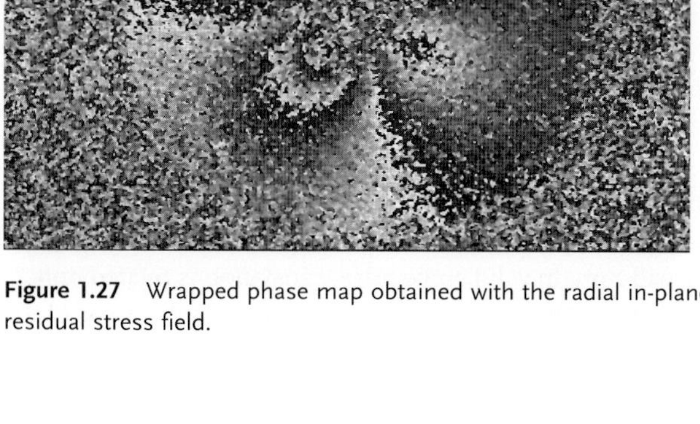

Figure 1.27 Wrapped phase map obtained with the radial in-plane speckle interferometer for a residual stress field.

following equations:

$$A = \frac{r_0}{2E}(1+\upsilon)\varrho, \qquad B = \frac{r_0}{2E}\left[4\varrho-(1+\upsilon)\varrho^3\right], \tag{1.27}$$

where E is the modulus of elasticity (Young's modulus), υ is the Poisson's ratio, and $\varrho = r_0/r$ is the ratio of the hole radius to the radial coordinate.

By replacing the defined values of these constants into Equation 1.26:

$$u_r(r,\theta) = \frac{r_0}{2E}(1+\nu)\varrho(\sigma_{R1}+\sigma_{R2}) + \frac{r_0}{2E}\left[4\varrho-(1+\nu)\varrho^3\right](\sigma_{R1}-\sigma_{R2})\cos(2\theta-2\eta). \tag{1.28}$$

Also here, the principal residual stresses can be determined using two approaches: the Fourier approach or the least squares approach.

After computing the radial in-plane field, a finite number of regularly spaced sampling points can be extracted from a single sampling circle radius r_s for the Fourier approach. It is important to highlight that the sampling circle should be simultaneously concentric with the interferometer axis and the center of the drilled hole. Finally, in order to be in accordance with the ASTM E837 [25], a good practice is to use the sampling radius value given by [21]

$$r_s = 3.25\, d_t, \tag{1.29}$$

where d_t is the diameter of the end milling tool. From this data set, the first three Fourier series coefficients are computed by Equation 1.19.

Equation 1.28 is formed by two additive terms. The first one does not depend on θ at all being associated with the zero-order Fourier coefficient (H_0). The second term depends on $\cos(2\theta)$; therefore, it is connected with the second Fourier coefficient (H_2). Thus, the following relations can be written:

$$\begin{aligned} H_0(\varrho_s) &= A(\sigma_{R1}+\sigma_{R2}), \\ H_2(\varrho_s) &= B(\sigma_{R1}-\sigma_{R2}). \end{aligned} \tag{1.30}$$

Equation 1.30 can be solved in terms of the principal stress components (σ_{R1}, σ_{R2}) and the principal direction (η):

$$\begin{aligned} \sigma_{R1} &= \frac{1}{2}\left(\frac{H_0(\varrho_S)}{A} + \frac{H_2(\varrho_S)}{B}\right) \\ \sigma_{R2} &= \frac{1}{2}\left(\frac{H_0(\varrho_S)}{A} - \frac{H_2(\varrho_S)}{B}\right) \\ \eta &= \tan^{-1}\left(\frac{H_{2S}(\varrho_S)}{H_{2C}(\varrho_S)}\right), \end{aligned} \tag{1.31}$$

where $\varrho_S = r_0/r_S$ is the ratio of the hole radius (r_0) to the polar coordinate r_S of the sampling circle. If some amount of pure translation is mixed up with this residual stress signal, Equation 1.31 will remain valid since pure translation is related only to the first Fourier harmonic. However, if the hole is not drilled in the optical axis of the radial in-plane speckle interferometer, the second Fourier harmonic becomes influenced by the pure translation component, bringing errors to the residual stress measurement.

For the least squares approach, a set of experimental data is sampled from the unwrapped phase difference pattern. A good practice is to sample regularly spaced data in a circular region, concentric with the interferometer optical axis and the center of the drilled hole. The region very close to the edge of the hole and points very far from it should not be taken into account during computation. According to the ASTM E837 [25], the minimum and maximum sampling radius should be given by

$$\begin{aligned} r_{\mathrm{s\,min}} &= 2.25 d_{\mathrm{t}}, \\ r_{\mathrm{s\,max}} &= 4.25 d_{\mathrm{t}}. \end{aligned} \tag{1.32}$$

As mentioned before, d_t is the diameter of the end milling tool. The sampled data are fitted to the mathematical model described by

$$\begin{aligned} u_{\mathrm{r}}(r,\theta) = {} & K_{OR}(1+\upsilon)\varrho + K_{1C}\cos(\theta) + K_{1S}\sin(\theta) + K_{2C}[4\varrho-(1+\upsilon)\varrho^3]\cos(2\theta) \\ & + K_{2S}[4\varrho-(1+\upsilon)\varrho^3]\sin(2\theta) + K_0. \end{aligned} \tag{1.33}$$

The uniform translation u_t and its direction α as well as the principal residual stresses σ_{R1} and σ_{R2} and their principal direction η can be computed by

$$\begin{aligned} u_{\mathrm{t}} &= \sqrt{K_{1C}^2 + K_{1S}^2}, \\ \alpha &= \tan^{-1}\left(\frac{K_{1S}}{K_{1C}}\right), \\ \sigma_{R1} &= \frac{E}{r_0}\left(K_{OR} + \sqrt{K_{2C}^2 + K_{2S}^2}\right), \\ \sigma_{R2} &= \frac{E}{r_0}\left(K_{OR} - \sqrt{K_{2C}^2 + K_{2S}^2}\right), \\ \eta &= \frac{1}{2}\,tg^{-1}\left(\frac{K_{2S}}{K_{2C}}\right). \end{aligned} \tag{1.34}$$

1.5 Conclusions

The choice of a coordinate system has a strong impact on how easily and efficiently a physical problem can be solved. Cartesian coordinates are widely used in a variety of engineering problems. However, polar and cylindrical coordinates are much better choices for certain classes of problems, especially when axial symmetry is involved. The calculations become much more straightforward and the outputs are more naturally connected with the physical phenomena.

This chapter presented special configurations of digital speckle pattern interferometers that can be used to measure in polar or cylindrical coordinates. The radial in-plane interferometer measures in polar coordinates, which is very appropriate for determining the mechanical strain, stress, and residual stress states on the surface of elastic and isotropic materials. The concept of principal strains, stresses, and principal directions is very conveniently handled with polar coordinates since both the principal directions and the principal values are naturally determined.

The measurement of both inner and outer cylindrical surface deformations by digital speckle pattern interferometers using conical mirror is a very appropriate and natural way. Shape deviations caused by mechanical or thermal deformations of bearings, shafts, pistons, and cylinders are of very great engineering interest. This chapter shows that DSPI can be successfully used for that.

References

1 Rastogi, P.K. (2001) Digital speckle pattern interferometry and related techniques, in *Digital Speckle Pattern Interferometry and Related Techniques* (ed. P.K. Rastogi), John Wiley & Sons, Inc., New York.

2 Gilbert, J.A., Greguss, P., and Kransteuber, A.S. (1991) Holo-interferometric patterns recorded through a panoramic annular lens. Three-Dimensional Holography: Science, Culture, Education (eds T.H. Jeong and V.B. Markov), *Proc. SPIE*, **1238**, 412–420.

3 Matthys, D.R., Gilbert, J.A., and Greguss, P. (1991) Endoscopic measurement using radial metrology with digital correlation. *Opt. Eng.*, **30** (10), 1455–1460.

4 Albertazzi Gonçalves, A. Jr., Coutinho, G., Manzolli, A., Kapp, W., and Santos, D. (1997) Measurement of radial deformation of a long internal cylinder using ESPI. Seminario Internacional de Metrologia para Controle da Qualidade-SI/MpCQ97, Florianópolis, SC, Brazil, pp. 22–32.

5 Leendertz, J.A. (1970) Interferometric displacement measurement on scattering surfaces utilizing speckle effect. *J. Phys. E*, **3**, 214–218.

6 Jones, R. and Wykes, K. (1989) *Holographic and Speckle Interferometry*, 2nd edn, Cambridge University Press, Cambridge.

7 Dally, J.W. and Riley, W.F. (1991) *Experimental Stress Analysis*, 3rd edn, McGraw-Hill, Inc., New York.

8 Moore, A.J. and Tyrer, J.R. (1990) An electronic speckle pattern interferometer for complete in-plane measurement. *Meas. Sci. Technol.*, **1**, 1024–1030.

9 Moore, A.J. and Tyrer, J.R. (1996) Two-dimensional strain measurement with ESPI. *Opt. Lasers Eng.*, **24**, 381–402.

10 Albertazzi, A. Jr., Borges, M.R., and Kanda, C. (2000) A radial in-plane interferometer for residual stresses measurement using ESPI. Proceedings of SEM IX International Congress on Experimental Mechanics, Society for Experimental Mechanics, Bethel, pp. 108–111.

11 Viotti, M.R., Albertazzi, A. Jr., and Kaufmann, G.H. (2005) Measurement of residual stresses using local heating and a radial in-plane speckle interferometer. *Opt. Eng.*, **44** (9), 093606.

12 Viotti, M.R., Sutério, R., Albertazzi, A., and Kaufmann, G.H. (2004) Residual stress measurement using a radial in-plane speckle interferometer and laser annealing: preliminary results. *Opt. Lasers Eng.*, **42**, 71–84.

13 Huntley, J.M. (2001) Automated analysis of speckle interferograms, in *Digital Speckle Pattern Interferometry and Related*

Techniques (ed. P.K. Rastogi), John Wiley & Sons, Inc., New York, pp. 59–139.

14 Boettger, J. (1998) Development of a portable field residual stresses measurement device using ESPI and hole drilling method (in Portuguese). Master in Mechanical Engineering Dissertation, Federal University of Santa Catarina, Brazil.

15 O'Shea, D.C., Suleski, T.J., Kathman, A.D., and Prather, D.W. (2003) *Diffractive Optics: Design Fabrication and Test*, Tutorial Texts in Optical Engineering, **TT62**, SPIE, Washington, DC

16 Hecht, E. and Zajac, A. (1974) *Optics*, Addison-Wesley Publishing Company.

17 Viotti, M.R., Albertazzi, A. Jr., and Kapp, W. (2008) Experimental comparison between a portable DSPI device with diffractive optical element and a hole drilling strain gage combined system. *Opt. Lasers Eng.*, **46**, 835–841.

18 Viotti, M.R., Kapp, W., and Albertazzi, A. Jr. (2009) Achromatic digital speckle pattern interferometer with constant radial in-plane sensitivity by using a diffractive optical element. *Appl. Opt.*, **48**, 2275–2281.

19 Albertazzi, A. Jr., Willemann, D.P., and Veiga, C.L.N. (2002) Preliminary evaluation of the optical rosette: a portable device for stress measurement with ESPI. Proceedings of the 7th International Symposium on Laser Metrology Applied to Science, Industry and Everyday Life, Novosibirsk, Russia, pp. 1073–1084.

20 Albertazzi, A., Veiga, C.L.N., and Willemann, D.P. (2003) Evaluation of the optical rosette for translation, stresses, and stress gradients measurement. Optical Measurement Systems for Industrial Inspection III, Munich, *Proc. SPIE*, **5144**, 533–544.

21 Kaufmann, G.H. and Albertazzi, A. (2008) Speckle interferometry for the measurement of residual stresses, in *New Directions in Holography and Speckle* (eds H.J. Caulfield and C.S. Vikram), American Scientific Publishers, California, USA.

22 Lu, J. (1996) *Handbook of Measurement of Residual Stresses*, The Fairmont Press, Lilburn, GA.

23 Lu, J. (2005) *Handbook on Residual Stress*, Society of Experimental Mechanics.

24 Rendler, N.J. and Vigness, I. (1966) Hole-drilling strain-gage method of measuring residual stresses. *Exp. Mech.*, **6**, 577–586.

25 ASTM E837-08e1 (2008) *Standard Test Method for Determining Residual Stresses by the Hole-Drilling Strain-Gage Method*, Annual Book of ASTM Standards, American Society for Testing and Materials.

26 Wang, H.P. (1979) The alignment error of the hole-drilling method. *Exp. Mech.*, **19**, 23–27.

27 Viotti, M.R. and Kaufmann, G.H. (2004) Accuracy and sensitivity of a hole drilling and digital speckle pattern interferometry combined technique to measure residual stresses. *Opt. Lasers Eng.*, **41**, 297–305.

28 Furgiuele, F.M., Pagnotta, L., and Poggialini, A. (1991) Measuring residual stresses by the hole drilling and coherent optics techniques: a numerical calibration. *J. Eng. Mater. Technol.*, **113**, 41–50.

29 Diaz, F.V., Kaufmann, G.H., and Galizzi, G.E. (2000) Determination of residual stresses using the hole drilling and digital speckle pattern interferometry with automated data analysis. *Opt. Lasers Eng.*, **33**, 39–48.

30 Dolinko, A.E. and Kaufmann, G.H. (2006) A least-squares method to cancel rigid body displacements in a hole drilling and DSPI system for measuring residual stresses. *Opt. Lasers Eng.*, **44**, 1336–1347.

31 Viotti, M.R. and Albertazzi, A. (2009) Industrial inspections by speckle interferometry: general requirements and a case study. Optical Measurement Systems for Industrial Inspection (ed. P.H. Lehmann), *Proc. SPIE*, 73890G.

32 Sutério, R., Albertazzi, A., and Cavaco, M.A.M. (2003) Preliminary evaluation: the indentation method combined with a radial interferometer for residual stress measurement. SEM Annual Conference and Exposition on Experimental and Applied Mechanics, Society of Experimental Mechanics.

33 Flaman, M.T. (1982) Brief investigation of induced drilling stresses in the centre-hole method of residual stress measurement. *Exp. Mech.*, **22**, 26–30.

34 Asundi, A. and Wenzen, Z. (1998) Fast phase-unwrapping algorithm based on a gray-scale mask and flood fill. *Appl. Opt.*, **37**, 5416–5420.

35 Makino, A. and Nelson, D. (1994) Residual-stress determination by single-axis holographic interferometry and hole drilling. Part I: theory. *Exp. Mech.*, **34**, 66–78.

36 Mathar, J. (1934) Determination of initial stresses by measuring the deformations around drilled holes. *Trans. ASME*, **56**, 249–254.

37 Albertazzi Gonçalves, A. Jr., Melao, I., and Devece, E. (1998) Measurement of thermal deformation of an engine piston using a conical mirror and ESPI. Laser Interferometry IX: Applications, (eds R.J. Pryputniewicz, G.M. Brown, and W.P. Jueptner), *Proc. SPIE*, **3479**, 274–283.

2
Depth-Resolved Displacement Field Measurement

Jonathan M. Huntley and Pablo D. Ruiz

2.1
Introduction

Following its initial development in the 1970s, electronic speckle pattern interferometry (ESPI) and its successor digital speckle pattern interferometry (DSPI) have become powerful tools for the measurement of displacement fields on objects with rough surfaces. Part of their success stems from the fact that they provide dense data coverage over a complete two-dimensional surface. This contrasts with the more restricted pointwise or linear coverage provided by techniques such as the strain gauge or Bragg grating fiber sensor, which risk missing important phenomena as the sample is loaded.

In view of their success, it is natural to consider how one may extend the measurement process to encompass volumes rather than just surfaces. A family of interferometric techniques has been developed over the past few years to achieve this aim for weakly scattering materials. Several non optical techniques have also been developed such as phase contrast magnetic resonance imaging (MRI) [1] and volumetric digital image correlation (DIC) combined with X-ray computed tomography [2, 3]. The latter two techniques work well for opaque objects, where the interferometric techniques are unsuitable, but have in turn their own limitations such as cost and weight of the experimental facility as well as typically rather long acquisition times. The purpose of this chapter is to present an overview of the current state of the art of volumetric displacement field imaging using optical interferometric techniques.

Quite apart from the possibility of observing internal phenomena that might otherwise be invisible to a surface-only technique, there are in fact several distinct reasons for choosing a 3D technique. Several postprocessing steps, for example, phase unwrapping, become more reliable in three dimensions than in two. Reconstruction of 3D elastic modulus distributions becomes a tractable proposition with 3D data [4], whereas in general it is not possible with only 2D surface data without significant simplifying assumptions (see Ref. [5] for an example involving delamination cracks in composites). A further benefit provided by the 3D interferometric

Advances in Speckle Metrology and Related Techniques. Edited by Guillermo H. Kaufmann

ISBN: 978-3-527-40957-0

techniques is that one of the basic assumptions of standard DSPI – that the object is sufficiently opaque for the scattering of the incoming light to be confined to a narrow layer at the sample surface – can be relaxed. As the scope of DSPI is extended beyond traditional engineering materials to weakly scattering materials such as biological tissues, backscattering from within the sample can become an important issue. In such a case, contributions to the recorded speckle patterns are received from throughout the sample volume and as a result the patterns decorrelate strongly with only very small deformations of the sample. Traditional "tricks of the trade," such as coating the sample surface with developer powder, are not viable options when dealing, for example, with the human retina or cornea. The ability to separate signal from different depths within the measurement volume can in such cases allow measurements to be made where previously none would be feasible.

Analysis of traditional 2D speckle interferograms normally follows the flow diagram shown in Figure 1 of Ref. [6]. The interferogram is specified by the digitized intensity distribution $I(m, n, t)$, where m and n are spatial indices, and t is a time index. The first step is extraction of the phase, $\phi(m, n, t)$, which is carried out by measuring the fringe or speckle intensity as known phase shifts are introduced between the two interfering waves. The phase shifts can be a function either of time or of position in the image, known, respectively, as temporal and spatial phase shifting. The next step is calculation of the phase change, $\Delta\phi(m, n, t)$, relative to some initial reference interferogram, which might correspond, for example, to the undeformed state of the object. The calculated phase change values obtained from the first two steps lie in the range $(-\pi, \pi)$. Phase unwrapping refers to the process of adding integral multiples of 2π to the phase change values so as to remove the resulting 2π phase discontinuities. The removal can be carried out either along one or more of the spatial axes, or along the time axis, known, respectively, as spatial and temporal phase unwrapping. The last step in the analysis is conversion of the unwrapped phase map $\Delta\phi_u(m, n, t)$ to the parameter of interest in the coordinate system of the object.

The situation for depth-resolved measurements is similar; however, as shown in Figure 2.1, the time variation now often refers to changes in some quantity such as wavenumber or angle of the object beam while the sample is kept stationary. As we will see, a changing wavenumber or illumination direction introduces into the interference signal a temporal carrier with a frequency that depends on depth. The carrier thus has a twofold function of providing (i) the required temporal phase shifts and (ii) a mechanism to separate the signal into separate depth-resolved bands through Fourier transformation along the time axis. The resulting three-dimensional phase distribution can be used to measure depth-resolved displacement fields by loading the sample, performing a second scan of the deformed state, and then calculating the difference between the two phase volumes. The wrapped phase maps can then be unwrapped by a 1D, a 2D, or a full 3D unwrapping algorithm to produce the required 3D displacement field.

Historically, the first interferometric depth-resolving displacement-sensing technique to be developed was known as low-coherence speckle pattern interferometry and this is therefore described first (Section 2.2). Much of the chapter is devoted to the two main techniques, namely, wavelength scanning interferometry (WSI;

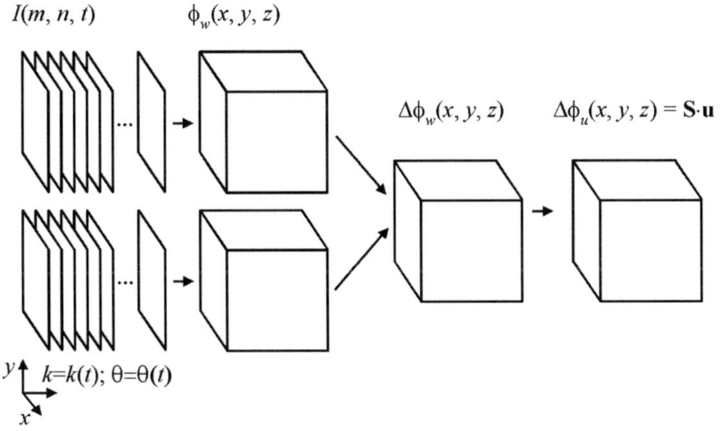

Figure 2.1 Main steps in the analysis of data from depth-resolving displacement-sensing interferometric techniques.

Section 2.3) and tilt scanning interferometry (TSI; Section 2.5), which refer, respectively, to the cases of time-varying wavenumber and time-varying illumination direction. A family of related techniques has also been developed based on the concept of wavelength scanning interferometry, but where the temporal scan has been replaced with a spatial separation of the different wavenumbers. These techniques offer the important benefit of improved temporal resolution, albeit at the expense of some reduction in spatial resolution, and are described in Section 2.4. A theoretical framework based on the Ewald sphere is presented in Section 2.6 that, by means of a simple geometrical construction, allows one to visualize the spatial resolution and displacement component measured by any of the techniques presented in this chapter. The chapter then concludes in Section 2.7 with a discussion on recent developments on phase unwrapping in three dimensions.

2.2 Low-Coherence Electronic Speckle Pattern Interferometry

With any type of interferometer, it is necessary to match the path lengths of object and reference beams within the coherence length l_c of the light ($l_c \sim \lambda^2/\Delta\lambda$, where λ is the center wavelength and $\Delta\lambda$ is the bandwidth) in order to observe interference fringes. The invention of the laser in the 1960s increased l_c from the tens of micrometers typically achievable with thermal light sources to several meters or more, thereby making interferometers much easier to set up and align. However, in certain situations the fact that the whole sample is coherently illuminated can be a drawback, and a low coherence light source may be chosen deliberately in order to localize the interference signal to a region within $l_c/2$ of the surface of zero optical path difference. This region is sometimes known as the "coherence gate."

Two widespread recent applications of low-coherence interferometry are high-resolution surface profilometry, using scanning white light interferometry (SWLI) [7–10], and depth-resolved measurement of tissue structure using time-domain optical coherence tomography (OCT). Several groups contributed to the early development of OCT (see, for example, Refs [11–14]), which has now expanded into a large and sophisticated family of related techniques. Refs [15, 16] are useful review papers, and Ref. [17] is a comprehensive textbook, describing these techniques and some of the medical and nonmedical applications of OCT.

Both SWLI and time-domain OCT involve sweeping the surface of zero optical path difference through the measurement volume, typically using some form of mechanical actuators. The similarities between the two approaches is clear from the optical setups drawn schematically in Figure 2.2a and b. As scattering points fall within the coherence gate, a high-frequency interference signal modulated by an envelope given by the Fourier transform of the source spectrum is detected, as shown in Figure 2.3.

The width of the envelope (l_c) determines the depth resolution of OCT and ranges from typically a few to tens of micrometers. In SWLI, the surface is normally assumed to be opaque so that scattering occurs only from a thin zone at the sample surface. In this case, the position of the peak in the envelope can be determined to much higher precision than the width of the envelope. Subnanometer depth resolution becomes possible with this assumption.

The other main difference between the two techniques is that time-domain OCT normally works with pointwise illumination and detection optical setups (see Figure 2.2b), whereas SWLI normally works with two-dimensional arrays

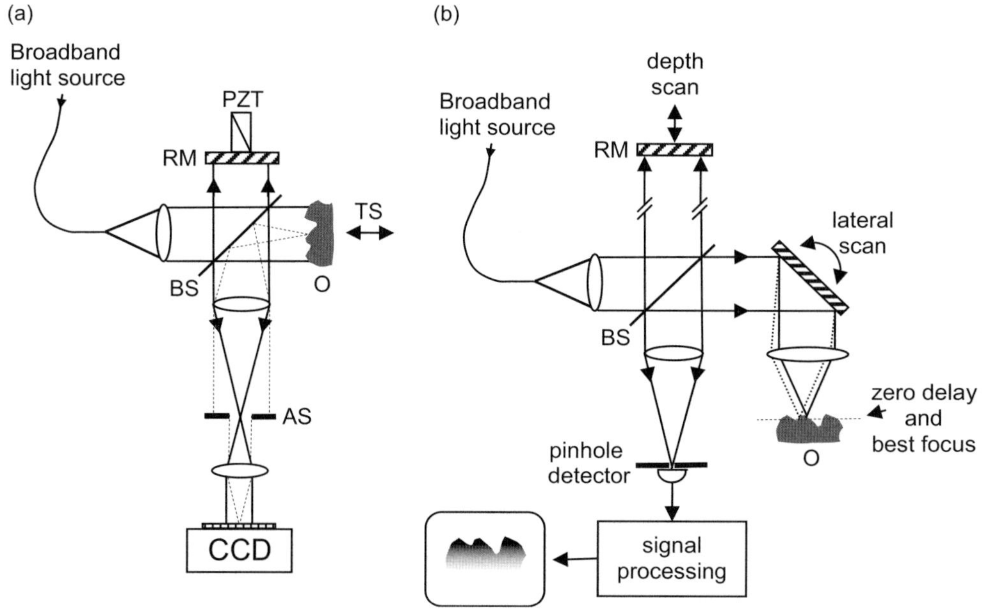

Figure 2.2 Optical setups for (a) scanning white light interferometry and (b) time-domain OCT, showing beam splitter (BS), translation stage (TS), object (O), reference mirror (RM), piezo-electric transducer (PZT), aperture stop (AS) and camera with charged coupled device (CCD).

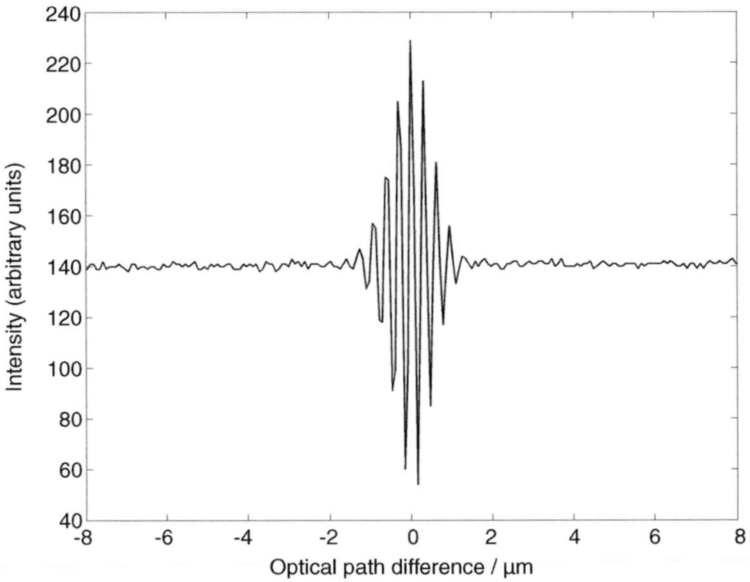

Figure 2.3 Measured intensity at one pixel of a SWLI (Zygo New View 500, with tungsten halogen light source, and numerical aperture = 0.55) as a function of optical path difference between the object and reference waves.

(sometimes called "parallel detection" in the OCT literature), as seen in Figure 2.2a. There are several reasons for preferring pointwise detection schemes for OCT, which are summarized in Ref. [15]. These include the ability to use AC detection (thereby increasing the effective dynamic range of the analogue-to-digital converter) and to operate at higher frequencies, thereby reducing the influence of $1/f$ noise. On the other hand, OCT uses almost exclusively the magnitude of the interference signal. Although the magnitude distributions can be used for subsurface displacement field measurement, for example, by applying DIC techniques [18], the measurement accuracy is limited by the depth resolution of the OCT system. High-sensitivity displacement field measurement requires the *phase* of the interference signal, and in particular *changes* to this phase over time. The accuracy of the phase difference measurement is highly sensitive to the repeatability of the scanning mechanisms, so that parallel detection – which has only one scanning mechanism in place of three – may well offer the best approach for depth-resolved displacement field measurement.

The first to describe the use of low-coherence interferometry for the measurement of subsurface displacement fields in weakly scattering materials were, to the best of our knowledge, Kraft *et al.* [19]. In this method, a standard fiber-based ESPI setup is used, but with a superluminescent laser diode as the light source in place of the usual laser (see Figure 2.4). Only the scattered light from points lying within the coherence gate coherently interferes with the reference wave. An off-axis reference is chosen so as to introduce a high-frequency carrier to the image-plane speckle pattern formed by these scattering points. All the other scattering points contribute to a speckle pattern that is limited in spatial frequency by the size of the imaging lens aperture. The signal from

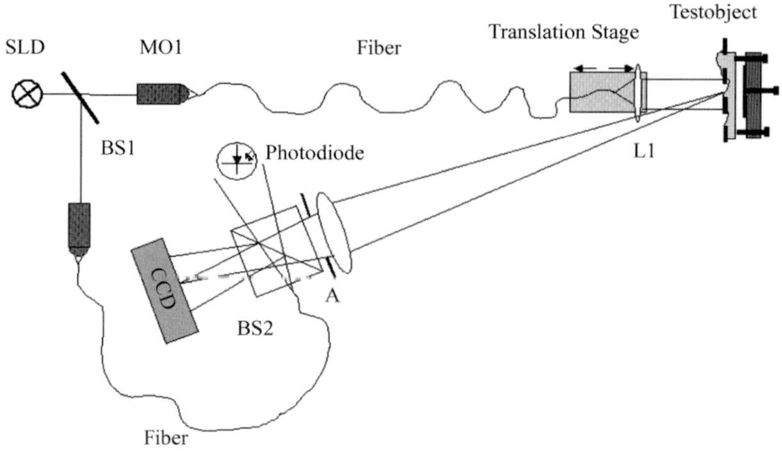

Figure 2.4 Experimental arrangement for low-coherence speckle interferometry. MO: microscope objective; BS: beam splitter; A: aperture. (Reprinted with permission from Ref. [19].)

the scatterers in the coherence gate can therefore be separated from the others by 2D Fourier techniques, already commonly used in digital holography [6]. By repeating the measurement before and after the sample deformation, a phase difference map can then be computed that is proportional to the out-of-plane displacement component for the slice within the coherence gate. By varying the position of the translation stage, the coherence gate can be placed at different locations within the sample, thereby allowing the full 3D out-of-plane displacement field to be evaluated.

Examples of applications of LCSI include the measurement of hygral deformations of subsurface paint layers on ancient terracotta soldiers [20] and displacement fields around adhesively bonded joints in multilayer structures [21]. Figure 2.5 shows one

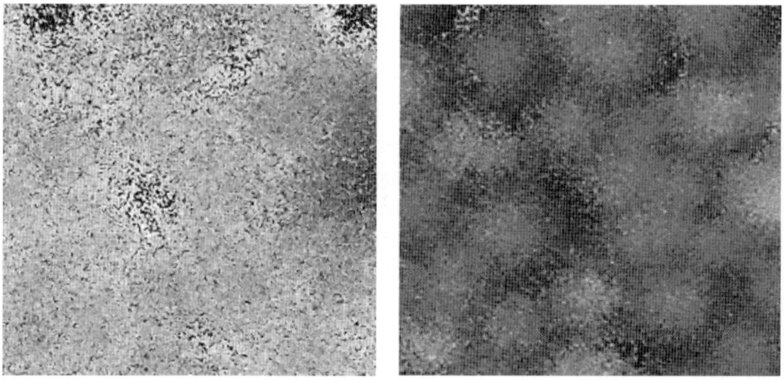

(a) (b)

Figure 2.5 Depth-resolved displacement field measurement on painted fragment from ancient terracotta soldier due to humidity change from 68 to 73% RH. (a) Lower and (b) upper paint layers. Field of view $\sim 1 \times 1\,\text{mm}^2$. (Reprinted with permission from Ref. [20].)

example from the former application. An overview of recent developments of the technique, including dual wavelength measurements, temporal phase shifting, and optimization of the interferometer, is given in Ref. [22].

One of the main attractive features of LCSI is that it provides depth-resolved displacement field measurements with a relatively simple experimental setup. If one is interested in displacements on only one slice within a relatively thin sample, it may well be the technique of choice. On the other hand, there are some drawbacks such as limited modulation depth arising from the incoherent background and fringe visibility reductions due to sample-induced dispersion in the object wave. These are discussed in more detail in Section 2.3.

2.3 Wavelength Scanning Interferometry

The LCSI technique, described in the previous section, provides depth-resolving capability by simultaneously illuminating the sample with light of multiple wavelengths. WSI, on the other hand, while also involving the use of multiple wavelengths, is based on *sequential* illumination with light from a tunable source. By tuning the wavenumber linearly with time, an interference signal is generated whose temporal frequency is proportional to the range. Like LCSI, WSI provides decoupling of the depth resolution and displacement sensitivity, but with a few additional benefits, including removal of the need for any mechanical scanning device, better signal-to-noise ratio, and greater immunity from dispersion artifacts.

Several terms have been used in the literature to describe essentially the same technique. For example, optical frequency domain reflectometry was developed to locate defects in optical fibers [23, 24]; in the OCT community, it is commonly called wavelength tuning interferometry or "swept-source" OCT [25, 26]. Variations on the basic theme, called frequency scanning interferometry in Ref. [27], have also been developed for large-volume metrology applications. Parallel versions of WSI were developed in the mid-1990s for profilometry applications [28, 29] in which each pixel of a 2D array is turned into an independent range sensor. More recently, de Groot described a wavelength scanning interferometer able to separate the interferograms from two different optical surfaces on a common optical path [30]. A phase shifting formula tuned to the temporal frequency of the surface of interest effectively eliminated the signal from the unwanted surface, allowing an accurate phase map of the first surface to be obtained.

The use of phase information to measure 2D depth-resolved displacement fields takes this idea a stage further. Proof-of-principle experiments on a pair of specularly reflecting surfaces [31] and diffusely scattering surfaces [32] showed how independent displacement fields associated with rigid body rotation of the two surfaces could be separated. In these cases, a windowed Fourier transform was used to separate the signal from the different surfaces and to extract phase maps. Phase difference maps then encode the out-of-plane displacement field.

In this section, the basic concepts behind WSI are described and the key length scales of the technique are given: the gauge volume, depth range, and displacement sensitivity. For simplicity, Section 2.3.1 starts with the case of a single scattering surface, before extending to the more general case of a scattering volume in Section 2.3.2. Finally in Section 2.3.3, some of the error sources are discussed, and the advantages and disadvantages of WSI over LCSI for depth-resolved displacement field mapping are described.

2.3.1
WSI with a Single Scattering Surface

Consider the surface of an opaque sample that is illuminated and observed using an interferometer of the type shown in Figure 2.2a. A given scattering point with coordinates (x, y, z), where the illumination and observation directions are parallel to z-axis, and where z measures distance from the zero optical path difference surface, is imaged onto a pixel of the photodetector array with indices (m, n). The optical path difference between the backscattered light and the reference wave is therefore $\Lambda = 2n_0z(x, y)$, where n_0 is the refractive index of the surrounding medium.

If the amplitude of the object and reference waves at this pixel are denoted A_o and A_r, respectively, the measured intensity can be written as [6]

$$\begin{aligned} I(m,n,t) &= |A_r(m,n,t) + A_o(m,n,t)\exp[i\phi(m,n,t)]|^2 \\ &= I_0(m,n,t) + I_M(m,n,t)\cos[\phi(m,n,t)], \end{aligned} \tag{2.1}$$

where the DC and modulation terms, I_0 and I_M, respectively, are given by

$$I_0 = A_o^2 + A_r^2, \tag{2.2}$$

$$I_M = 2A_oA_r, \tag{2.3}$$

and where ϕ is the phase shift between the waves. The spatial indices m and n take the values $m = 0, 1, 2, \ldots, N_m - 1$; $n = 0, 1, 2, \ldots, N_n - 1$; and t is a nondimensional time variable defined as the true time normalized by the camera interframe time. ϕ can be represented, in turn, as a sum of two contributions:

$$\phi(m,n,t) = k(t)\Lambda(m,n) + \phi_s(m,n), \tag{2.4}$$

where k is the wavenumber $2\pi/\lambda$ with λ the wavelength in vacuum, and ϕ_s is a phase shift that may arise even at zero nominal path difference due to, for example, a phase change on reflection or due to the microscopically random arrangements of scatterers contributing to the amplitude at pixel (m, n).

Wavelength scanning interferometry involves changing k with time over a total range Δk, while image sequences are recorded using the 2D photodetector array. Ideally, one aims for a linear variation in k with t, that is,

$$k(t) = k_c + \delta k t, \tag{2.5}$$

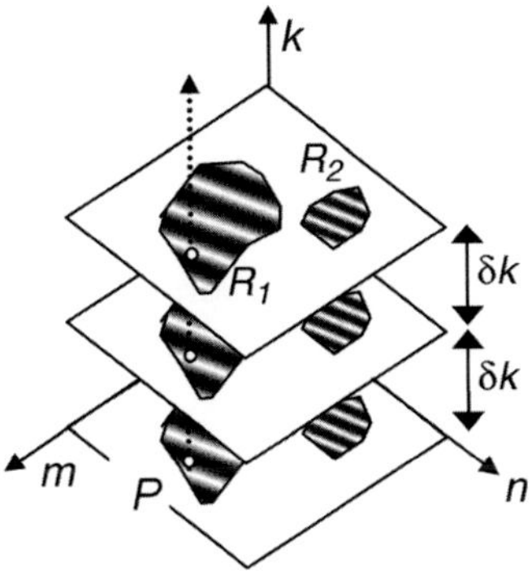

Figure 2.6 Schematic time-varying intensity distribution from a single scattering surface measured during a wavelength scan.

where k_c is the central wavenumber, δk is the increment in wavenumber between successive frames and t ranges from $-\Delta k/2\delta k$ to $+\Delta k/2\delta k$.

The resulting stack of images forms a 3D intensity distribution, as schematically shown in Figure 2.6. Subsequent data analysis takes place on a pixelwise basis, shown in Figure 2.6 for an arbitrary point P on the sample, by extracting the intensity signal along lines parallel to the k-axis. Such paths avoid the geometrical boundaries of the object (in this case two separated regions R_1 and R_2) that will cause problems for conventional spatial fringe analysis techniques.

Substitution of Equation 2.4 into Equation 2.5 leads to the result

$$\phi(m,n,t) = k_c\Lambda(m,n) + \phi_s(m,n) + \delta k\Lambda(m,n)t. \tag{2.6}$$

Two important consequences follow from Equation 2.6. First, the linear time variation causes the intensity to modulate at a temporal frequency $f = f_\Lambda$, where

$$f_\Lambda = \delta k\Lambda(m,n)/2\pi, \tag{2.7}$$

that is, at a frequency proportional to the optical path difference. Note that the units of f are "cycles per frame" because of the definition of nondimensional time t. It is convenient to also define an alternative frequency $\hat{f}$ as

$$\hat{f} = f\Delta k/\delta k, \tag{2.8}$$

which has units of "cycles per scan duration." Thus, whether through f or $\hat{f}$, measuring the modulation frequency on a pixelwise basis provides a direct measure of $\Lambda(m,n)$. Second, if two scans are performed before and after sample deformation, then – provided ϕ_s does not change between the scans – the phase change $\Delta\phi(m,n,0)$ provides a direct measure of the change in optical path length $\Delta\Lambda(m,n)$ due to the deformation:

$$\Delta\phi(m,n,0) = k_c\Delta\Lambda(m,n). \tag{2.9}$$

The calculation of both the modulation frequency and the phase change between two successive scans can be achieved by means of the 1D Fourier transform, which thus has a dual purpose function that is described further in the following two sections.

2.3.1.1 Fourier Transform for Measurement of Optical Path Length

The analysis given below and in the next section is an extension of the approach outlined in Ref. [6] that, in turn, follows the earlier work of de Groot and Surrel [33–35].

The Fourier transform of a set of N_t sampled intensity values $I(t)$ can be written

$$\tilde{I}(\hat{f}) = \int_{-\infty}^{\infty} I(\breve{t}) W(\breve{t}) \left[\sum_{t'=-\infty}^{\infty} \delta(\breve{t}-t') \right] \exp(-2\pi i \hat{f}\, \breve{t}/N_t) \mathrm{d}\breve{t}, \tag{2.10}$$

where δ is the Dirac delta function, $\breve{t}$ is a continuous version of the discrete nondimensional time variable t, $W(\breve{t})$ is a continuous window function assumed to be even and nonzero only for $|\breve{t}| \leq N_t/2$, and $\hat{f}$ is a (continuous) nondimensional temporal frequency. The right-hand side of Equation 2.10 can be interpreted as the Fourier transform of either (i) a sampled intensity distribution, multiplied by a continuous window function, or (ii) a continuous intensity distribution, multiplied by a sampled window function, $W_S(\breve{t})$.

If we consider the first interpretation, then by the convolution theorem we have

$$\tilde{I}(\hat{f}) = \tilde{W}(\hat{f}) * \sum_{j=-\infty}^{\infty} \left[I_0 \delta(\hat{f}-jN_t) + \frac{I_M}{2} \exp(i\phi_0) \delta(\hat{f}-jN_t-\hat{f}_\Lambda) \right.$$
$$\left. + \frac{I_M}{2} \exp(-i\phi_0) \delta(\hat{f}-jN_t+\hat{f}_\Lambda) \right], \tag{2.11}$$

where $\tilde{W}(\hat{f})$ is the Fourier transform of $W(\breve{t})$, $*$ denotes convolution, and $\phi_0 = \phi(m, n, 0)$. The three terms within the summation corresponding to $j=0$ are shown schematically in Figure 2.7. These occur at $\hat{f} = -\hat{f}_\Lambda$, 0 and $+\hat{f}_\Lambda$ and represent the continuous Fourier transform of a signal consisting of a cosine wave plus a DC offset. These three peaks are repeated along the $\hat{f}$ axis at integral multiples of N_t (corresponding to nonzero values of j in the sum above) due to the sampling of the original intensity signal. The vertical dashed lines at $\hat{f} = \pm N_t/2$ represent the Nyquist frequency. Aliasing will occur if the signal frequency falls outside the range from $\hat{f} = -N_t/2$ to $\hat{f} = +N_t/2$.

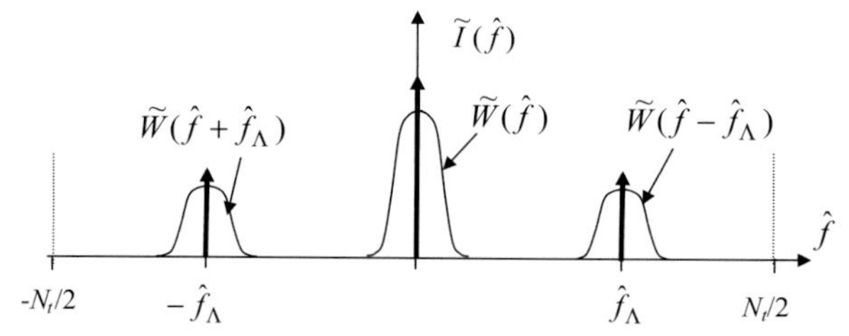

Figure 2.7 1D Fourier transform $\tilde{I}(\hat{f})$ of the time-varying intensity at one pixel during a wavelength scan of a sample consisting of a single scattering surface.

Convolution with the transform of the window function is equivalent to placing a copy of $\tilde{W}(\hat{f})$ centered on each delta function. As a result of the even symmetry assumption, $\tilde{W}(\hat{f})$ also possesses even symmetry and so the positions of the peaks in $\tilde{I}(\hat{f})$ are also located at $\hat{f} = 0$ and $\pm\hat{f}_\Lambda$. Thus, the value of $\hat{f}$ that maximizes $\tilde{I}(\hat{f})$ over the region $\hat{f} > 0$ of the function $\tilde{I}(\hat{f})$ is an unbiased estimator of the optical path Λ, provided the leakage of the signal from the other peaks in the spectrum can be neglected. The maximum value will in general lie between the integral values of $\hat{f}$ that are provided by the discrete Fourier transform; however, the true peak position can be located to high accuracy by searching the magnitude of the continuous transform for the location of the maximum (nonintegral) value of $\hat{f}$ [36]. It should also be pointed out that in the definition of the discrete Fourier transform, which reads as follows

$$\tilde{I}(\hat{f}) = \sum_{t=0}^{N_t-1} I(t)\,W(t)\exp\left(-2\pi i\hat{f}t/N_t\right), \tag{2.12}$$

the windowed signal has been shifted by $N_t/2$ along the $+\,t$-axis compared to that in the continuous Fourier transform (Equation 2.10). However, by the Fourier shift theorem this introduces only a phase factor that disappears when computing the magnitude of the Fourier transform.

With the approach to 3D profilometry outlined above, WSI is therefore being used for its depth-resolving capabilities and not for its ability to measure phase. In the next section, we consider WSI from the opposite point of view.

2.3.1.2 Fourier Transform for Calculation of Interference Phase

At the point $\hat{f} = \hat{f}_\Lambda$, $\tilde{I}(\hat{f})$ from Equation 2.11 takes the complex value

$$\tilde{I}(\hat{f}_\Lambda) = \frac{I_M}{2}\tilde{W}(0)\exp\left(i\phi_0\right), \tag{2.13}$$

so that ϕ_0 can be calculated from the real and imaginary parts as

$$\phi(0) = \tan^{-1}\left\{\frac{\mathrm{Im}\left[\tilde{I}(\hat{f}_\Lambda)\right]}{\mathrm{Re}\left[\tilde{I}(\hat{f}_\Lambda)\right]}\right\}. \tag{2.14}$$

In writing Equation 2.13, the spectral peaks are assumed to be sufficiently far apart so that the signal leakage between them can be neglected. Substituting $\tilde{I}(\hat{f}_\Lambda)$ from Equation 2.12 into Equation 2.14 results in the standard temporal phase shifting equation

$$\phi(0) = \tan^{-1}\left\{\frac{\sum_{t=0}^{N_t-1} I(t)b(t)}{\sum_{t=0}^{N_t-1} I(t)a(t)}\right\}, \tag{2.15}$$

where the a and b coefficients are obtained by comparing real and imaginary parts of the two sides of the following equation [33, 37]:

$$a(t) + ib(t) = W(t)\exp\left(-2\pi i \hat{f}_\Lambda t/N_t\right). \tag{2.16}$$

Equations 2.15 and 2.16 allow one to produce a phase shifting formula that is customized to any particular optical path difference Λ and wavenumber step δk. The relationship between a given displacement vector **u** at position vector **r**, and the resulting change in Λ, depends only on the geometry of the optical setup as described in Section 2.3.1.4. Thus, quite apart from its depth-sensing capabilities, wavelength-scanning interferometry can be used as a temporal phase shifting technique in its own right for the measurement of deformation fields on opaque surfaces. One such example is given in Ref. [38] where phase shifting formulae corresponding to Equation 2.15 with up to 114 frames were used to measure the out-of-plane displacement fields on a single reflective surface to subnanometer accuracy. So far, the measurement of optical path difference distributions and of phase distributions has been considered separately. As we will see in Section 2.3.2, the ability of WSI to perform two distinct roles in one experiment is exploited for measurements of volume scatterers – first to localize a slice of interest within the volume and second to measure one component of the displacement field on that slice.

2.3.1.3 Range and Resolution of Optical Path Difference Measurement

The *maximum unambiguous range* that the optical path difference may take is given by the Shannon sampling theorem, which states that in order to ensure adequate sampling of the $I(m, n, t)$ signal, the term $\delta k \Lambda(m,n)t$ in Equation 2.6 should not change by more than π between successive t values. This leads to a maximum allowed optical path difference $\Lambda = \Lambda_M$, where

$$\Lambda_M = \frac{\pi}{\delta k}. \tag{2.17}$$

Any larger Λ value will be aliased onto a lower Λ value thus creating an undersampling artifact. For the case of a sample illuminated and viewed parallel to the z-axis, one can define a corresponding depth range $z_M = \Lambda_M/2n_0$.

The minimum allowed value of Λ is, on the other hand, $\Lambda = 0$ because negative Λ values cause modulation of Equation 2.1 at the same frequency and cannot, therefore, be distinguished from positive ones. The allowable optical path difference range is therefore

$$0 \le \Lambda \le \Lambda_M. \tag{2.18}$$

The *resolution* to which the optical path length may be measured can be characterized by the width of the peak in the 1D Fourier domain. A usual resolution criterion is a frequency difference between two neighboring peaks of at least twice the distance from their centers to their first zero. The peak width is the width of the Fourier transform $\tilde{W}(\hat{f})$ of the window function $W(\breve{t})$. The scaling of the transform from nondimensional $\hat{f}$ to optical path difference Λ follows from the fact that the discrete Fourier transform $\tilde{I}(m, n, \hat{f})$ contains $N_t/2$ positive frequency components, with a separation between sample points of $2\pi/N_t\delta k$. The width of the spectral peak (i.e., distance between zero crossing points) is $4\pi/N_t\delta k$ for the case of a rectangular window function of width $N_t\delta k$. In general, we can write the width of the spectral peak, and hence the optical path resolution, as

$$\delta\Lambda = \gamma \frac{2\pi}{N_t \delta k}, \tag{2.19}$$

where the constant γ takes the value 2 for a rectangular window and 4 for a Hanning window, for example. Although the rectangular window has superior resolution, this is accompanied by the undesirable presence of large secondary lobes in $\tilde{W}(\hat{f})$ that may strongly interfere with other peaks, leading to phase errors. This effect is significantly reduced if a Hanning window is used instead. The relationship between $\delta\Lambda$ and the corresponding depth resolution δz depends on the geometry of the illumination and observation directions. For the simple geometry considered in Section 2.3.1, $\delta z = \delta\Lambda/2n_0$. Using Equation 2.19, and the approximate relationship $|\Delta k| \approx 2\pi\Delta\lambda/\lambda^2$, this reduces to $\delta z = \gamma\pi/n_0\Delta k \approx \gamma\lambda^2/2n_0\Delta\lambda$. It can be shown that this is the same as the depth resolution of LCSI for the same source bandwidth if the window function in WSI is equated with the envelope of the source spectrum in LCSI. Analysis of a more general optical geometry is considered in Section 2.3.1.5.

The precision to which Λ may be determined may be much better than the value given by Equation 2.19 for the case of scattering surfaces (as opposed to scattering volumes). Just as SWLI can be used to measure surface profiles to much finer precision than the optical depth resolution predicted by time-domain OCT, so WSI can measure profiles of single scattering surfaces to much finer precision than $\delta\Lambda$. This is possible because of the assumption that the thickness of the scattering layer is very small compared to the depth resolution. As a result, the relevant portion of the Fourier domain contains a single isolated peak, and the position of the peak maximum can be determined to an accuracy that is much smaller than the width of the peak (see, for example, Ref. [39]). The ultimate factor limiting the accuracy of such an approach is the noise in the intensity signal, rather than the resolution defined by Equation 2.19.

2.3.1.4 Determination of Scattering Point Location

There are several steps required to convert the measured quantities (Λ and $\Delta\phi$) to the location and displacement component of the scattering points in the coordinate system of the specimen. This section deals with the calculation of location, while Section 2.3.1.5 considers the relationship between the phase change distribution and the displacement field.

In many instances involving a single scattering layer, almost the entire optical path will be in air with a refractive index close to unity. In this case, the equations outlined in Ref. [6] can be adapted fairly easily to the current problem. In this section, however, we consider the more general case of two media with independent refractive indices n_0 and n_1, separated by a planar interface lying normal to the z-axis as shown in Figure 2.8. This will be useful when turning to the problem of depth localization within volume scatterers in Section 2.3.2. The simpler case of a surface scattering object in air can also be derived easily from the following equations by substituting the values $n_0 = n_1 \approx 1$.

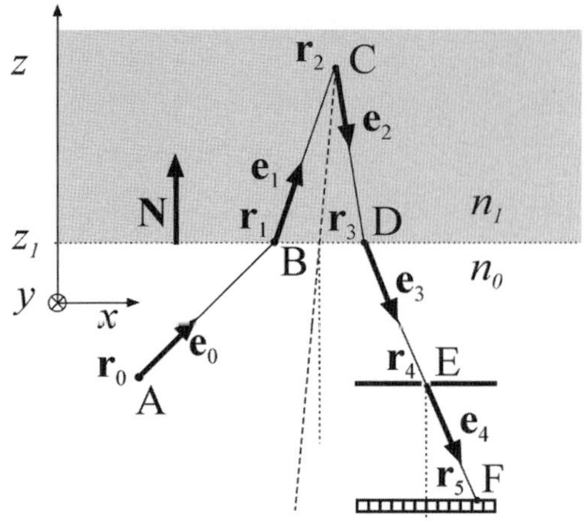

Figure 2.8 Ray diagram used to compute the location of a scattering point from a measured optical path difference Λ.

The optical geometry shown in Figure 2.8 consists of a point source of light for the object beam at position vector $\mathbf{r}_0$, an arbitrary scattering point at $\mathbf{r}_2$, and the center of the entrance pupil at $\mathbf{r}_4$. To simplify the equations, a pinhole camera model is assumed here, although additional radial and tangential distortion parameters could be introduced relatively easily to improve the accuracy if required [40]. The position vectors $\mathbf{r}_1$ and $\mathbf{r}_3$ define the points of intersection of the incoming and outgoing beams, respectively, with the interface between the two media. Finally, $\mathbf{r}_5$ is the position vector of the pixel with indices (m, n) in the photodetector array, which is assumed to be lying with its normal parallel to the z-axis. If one writes the position vector $\mathbf{r}_j$ in terms of its Cartesian coordinates $(x_j, y_j, z_j)^T$, where superscript T denotes the transpose operator, then the pixel coordinates can be expressed in terms of the pinhole location $\mathbf{r}_4 = (x_4, y_4, z_4)^T$ as

$$x_5 = x_4 + (m - m_c)p_x,$$

$$y_5 = y_4 + (n - n_c)p_y, \tag{2.20}$$

where (m_c, n_c) are the indices of the pixel lying closest to the pinhole and p_x and p_y are the pitches of the pixels in the photodetector array along the x- and y-axes, respectively.

The unit vectors $\mathbf{e}_0, \mathbf{e}_1, \ldots, \mathbf{e}_4$ shown in Figure 2.8 are defined as follows:

$$\mathbf{e}_j = \frac{\mathbf{r}_{j+1} - \mathbf{r}_j}{|\mathbf{r}_{j+1} - \mathbf{r}_j|}. \tag{2.21}$$

The object wave is regarded as originating from the point source ($\mathbf{r}_0$) and the reference wave, as with conventional ESPI, is assumed to come from the camera pinhole at position vector $\mathbf{r}_4$. If we denote by Λ_{ro} the difference in optical path between the reference wave as it passes from the beam splitter to $\mathbf{r}_4$ and the object illumination wave as it passes from the beam splitter to $\mathbf{r}_0$, then the total optical path difference between reference and object waves at pixel (m, n) may be written as

$$\Lambda = n_0|\mathbf{r}_1 - \mathbf{r}_0| + n_1|\mathbf{r}_2 - \mathbf{r}_1| + n_1|\mathbf{r}_3 - \mathbf{r}_2| + n_0|\mathbf{r}_4 - \mathbf{r}_3| - \Lambda_{\mathrm{ro}}. \tag{2.22}$$

The objective of the analysis in simple terms is to deduce the coordinates of the scattering point, $\mathbf{r}_2$, from a measured Λ value. All the other terms in Equation 2.22 are in principle obtainable from the optical geometry ($\mathbf{r}_0$, $\mathbf{r}_3$, $\mathbf{r}_4$, and Λ_{ro}), are measurable material properties (n_0 and n_1), or are a known function of $\mathbf{r}_2$ and $\mathbf{r}_0$ ($\mathbf{r}_1$).

It is convenient to analyze the problem by working backward from the pixel of interest. $\mathbf{e}_4$ can be written in terms of the pixel indices and the known lens-photodetector array separation $z_4 - z_5$ using Equations 2.20 and 2.21. The vectors $\mathbf{e}_3$ and $\mathbf{e}_4$ are identical, so that the position vector $\mathbf{r}_3$ can be determined by tracing the ray back from the pinhole according to the equation

$$\mathbf{r}_3 = \mathbf{r}_4 + \frac{z_3 - z_4}{z_4 - z_5}(\mathbf{r}_4 - \mathbf{r}_5). \tag{2.23}$$

The vector $\mathbf{e}_2$ can be computed from Snell's law in vector form [41] from which

$$n_1(\mathbf{N} \times \mathbf{e}_2) = n_0(\mathbf{N} \times \mathbf{e}_3), \tag{2.24}$$

where $\mathbf{N}$ is the normal vector to the interface. Substituting the vectors $\mathbf{N} = (0, 0, 1)^T$ and $\mathbf{e}_2 = (\alpha_2, \beta_2, \gamma_2)^T$ into Equation 2.24 allows one to derive the components of $\mathbf{e}_2$ as follows:

$$\begin{aligned} \alpha_2 &= \frac{n_0}{n_1}\frac{x_5 - x_4}{|\mathbf{r}_5 - \mathbf{r}_4|}, \\ \beta_2 &= \frac{n_0}{n_1}\frac{y_5 - y_4}{|\mathbf{r}_5 - \mathbf{r}_4|}, \\ \gamma_2 &= -\sqrt{1 - \alpha_2{}^2 - \beta_2{}^2}. \end{aligned} \tag{2.25}$$

The scattering point that is imaged onto pixel (m, n) must lie on the line defined by the equation

$$\mathbf{r}_2 = \mathbf{r}_3 + \mu\mathbf{e}_2, \tag{2.26}$$

where μ is a (negative) scalar. For any choice of μ, $\mathbf{r}_2$ is defined by Equation 2.26, and the x- and y-coordinates of $\mathbf{r}_1$ may therefore be calculated once again using Snell's law as follows:

$$\begin{aligned} x_1 &= \frac{x_2 n_1|\mathbf{r}_1 - \mathbf{r}_0| + x_0 n_0|\mathbf{r}_2 - \mathbf{r}_1|}{n_1|\mathbf{r}_1 - \mathbf{r}_0| + n_0|\mathbf{r}_2 - \mathbf{r}_1|}, \\ y_1 &= \frac{y_2 n_1|\mathbf{r}_1 - \mathbf{r}_0| + y_0 n_0|\mathbf{r}_2 - \mathbf{r}_1|}{n_1|\mathbf{r}_1 - \mathbf{r}_0| + n_0|\mathbf{r}_2 - \mathbf{r}_1|}. \end{aligned} \tag{2.27}$$

The right-hand side of Equation 2.27 also contains x_1 and y_1, so the equation needs to be solved iteratively.

The optical path length Λ for any μ may therefore be calculated from Equations 2.20 to 2.27. By creating a $\mu - \Lambda$ "look-up" table, the required position vector $\mathbf{r}_2$ of the scattering point can thus be computed from any given measured Λ value.

2.3.1.5 Gauge Volume and Displacement Sensitivity

Figure 2.9 is an enlargement of Figure 2.8 showing the region around the scattering point C with position vector $\mathbf{r}_2$. We denote the gauge volume AA′B′B by G, which is the region in space centered on point C that contributes signal at a given temporal frequency to the pixel of interest. Ω_1 represents the set of wavefronts from the source that illuminates G. Likewise, Ω_2 represents the set of wavefronts that are scattered from within the gauge volume and that can pass through the camera pinhole to the pixel. Both sets can be regarded as approximately planar over the length scale of G provided the source and the pinhole are sufficiently far away.

In two dimensions, the gauge volume can be represented as a parallelogram. Lines AB and A′B′, which run approximately parallel to the observation direction, are the edges of the point spread function of the imaging system. For a point detector and diffraction-limited imaging system, these are the loci of the zero amplitude of the Airy's disk diffraction pattern, giving rise to a lateral resolution (defined as the distance between the zero amplitude points)

$$\delta x' = 1.22\lambda / N_A, \tag{2.28}$$

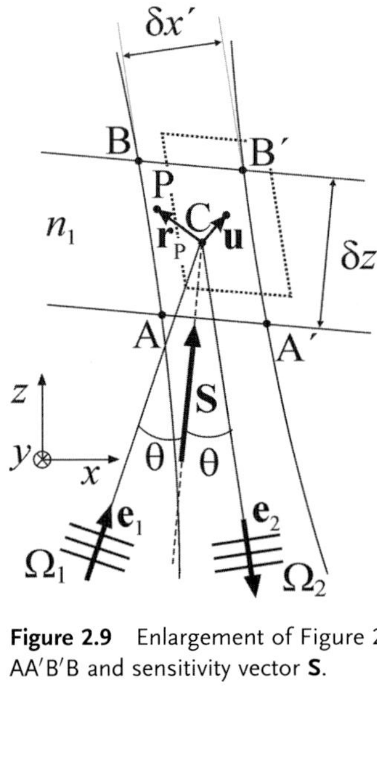

Figure 2.9 Enlargement of Figure 2.8 near the scattering point C, showing the gauge volume AA′B′B and sensitivity vector **S**.

where the numerical aperture N_A is given by

$$N_A = n_1 \sin\alpha, \qquad (2.29)$$

and where α is the half-angle subtended by the aperture at a point in the object.

The other dimension, $\delta z'$, of the gauge volume depends on the optical path length resolution, $\delta\Lambda$, which is given by Equation 2.19. Consider a point P with position vector $\mathbf{r}_P$ relative to the center C of the gauge volume. The difference in optical path between the light scattered by point P and the light scattered by point C has two contributions, which are related, respectively, to the change in distance perpendicular to the incoming waves, Ω_1, and to the outgoing waves, Ω_2. Adding these two contributions allows the change in optical path at P relative to C to be expressed as

$$\delta\Lambda_P = n_1(\mathbf{e}_1 - \mathbf{e}_2)\cdot\mathbf{r}_P. \qquad (2.30)$$

This is the equation of a plane $d = \hat{\mathbf{n}}\cdot\mathbf{r}_P$, with unit normal $\hat{\mathbf{n}} = (\mathbf{e}_1 - \mathbf{e}_2)/|\mathbf{e}_1 - \mathbf{e}_2|$ and with a perpendicular distance

$$d = \delta\Lambda_P/n_1|\mathbf{e}_1 - \mathbf{e}_2| \qquad (2.31)$$

from C. Lines of constant $\delta\Lambda_P$, therefore, lie perpendicular to $\hat{\mathbf{n}}$, that is, to the bisector of the illumination and observation directions. The lines AA′ and BB′ in Figure 2.9, which lie a distance $\delta z'/2$ from C, represent the locus of points for which $\delta\Lambda_P = \pm\delta\Lambda/2$. Substitution of the value $d = \pm\delta z'/2$ and $\delta\Lambda_P = \pm\delta\Lambda/2$ into Equation 2.31 results in the following equation for the depth resolution:

$$\delta z' = \gamma\frac{\pi}{n_1\cos\theta N_t\delta k}, \qquad (2.32)$$

where 2θ is the angle between the illumination and the observation directions. An alternative method for calculating the shape of the gauge volume and depth resolution, based on the Ewald sphere construction, is given in Section 2.6.

Suppose the sample now undergoes a deformation so that the scattering points initially inside the gauge volume centered at $\mathbf{r}_2$ move to a new position $\mathbf{r}_2 + \mathbf{u}$, indicated with a dashed line parallelogram in Figure 2.9, where $\mathbf{u}$ is the displacement $\mathbf{u} = (u_x, u_y, u_z)^T$. G is assumed to be small enough for the variation in $\mathbf{u}$ across G to be considered negligible. Provided $|\mathbf{u}|$ is, in turn, smaller than $\delta x'$ and $\delta z'$, some of the points originally within G remain within it after the displacement. The common scattering points are shown in Figure 2.9 as the region of overlap between the two parallelograms.

The arguments for the calculation of $\delta\Lambda$ earlier in the section can then be used to show that the change in phase of the scattered light at the pixel of interest is

$$\Delta\phi = n_1 k(\mathbf{e}_1 - \mathbf{e}_2)\cdot\mathbf{u}. \qquad (2.33)$$

Equation 2.33 implicitly assumes that the geometry of the interface between the two regions of refractive index n_0 and n_1, and the values of refractive index themselves, are unaffected by the loading of the sample. While this is a reasonable assumption for the case of a single scattering surface, it may be less valid in the case of a volume scattering material. The surface profile, and hence the shape of the interface, will in general change with load, and the refractive index may depend on stress. In such cases, $\Delta\phi$ will include additional terms to those in Equation 2.33. Their effects can be minimized through the use of a liquid cell with index matching fluid and through the use of materials with low stress-optical coefficients.

The quantity

$$\mathbf{S} = n_1 k(\mathbf{e}_1 - \mathbf{e}_2) \tag{2.34}$$

is known as the sensitivity vector. Equations 2.33 and 2.34 show that the interferometer measures the displacement component parallel to the sensitivity vector, that is, along the bisector of the illumination and observation directions [42]. For the simple on-axis case where both the point source and the lens are positioned at $x_0 = y_0 = x_4 = y_4 = 0$, and $z_0 = z_4 \rightarrow -\infty$, Equation 2.33 reduces to the standard out-of-plane interferometer formula, allowing one to estimate the component u_z from the measured unwrapped phase map $\Delta\phi_u$ as

$$u_z(x, y, z) = \frac{\lambda \Delta\phi_u(m, n, \hat{f})}{4\pi n_1}. \tag{2.35}$$

The displacement resolution σ_z (sometimes called the sensitivity of the technique) is decoupled from the depth resolution δz and depends only on the central wavelength of the tunable laser and – for a rough surface – the degree of speckle decorrelation. In the case of an interferometer sensitive to out-of-plane displacements, with a smooth reference wave, σ_z is typically better than $\lambda_c/30$.

2.3.2
WSI with Volume Scatterers

When the scattering points lie within a volume rather than on a surface, each camera pixel will receive light with a range of optical path differences so that the simple two-beam interference pattern of Equation 2.1 is no longer applicable. It is convenient to model the object as a set of N_s discrete thin scattering layers, each with its own distribution of scattered amplitude A_j, intensity $I_j = |A_j|^2$, optical path difference Λ_j, and phase offset $\phi_{s,j}$ relative to the reference wave. The model is illustrated in the top graph of Figure 2.10. If the sample is weakly scattering so that multiple reflections may be neglected, the measured intensity can be written as

$$I(m, n, t) = \left| A_r(m, n, t) + \sum_{j=1}^{N_s} A_j(m, n, t) \exp\left[i\phi_j(m, n, t)\right] \right|^2. \tag{2.36}$$

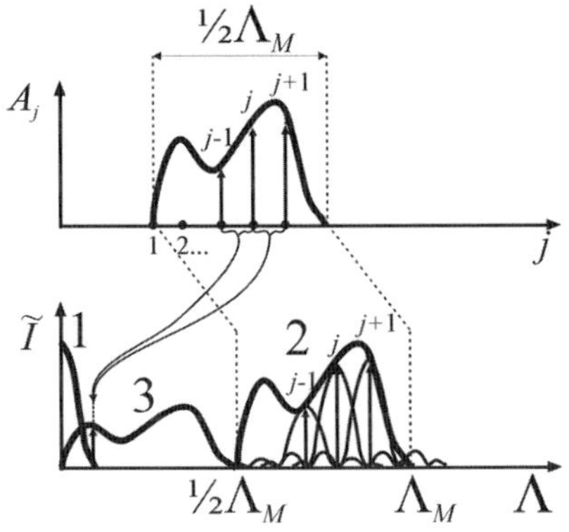

Figure 2.10 Scattered light amplitude distribution for N_s scattering layers (top) and its reconstruction (below) from the Fourier transform of the intensity measured at a single camera pixel. The labels 1, 2, and 3 refer to the three terms on the right-hand side of Equation 2.37.

Expanding Equation 2.36 and dropping the (m, n, t) dependence in all the intensity and phase variables for clarity gives

$$I(m,n,t) = \left(A_r^2 + \sum_{j=1}^{N_s} A_j^2\right) + 2A_r \sum_{j=1}^{N_s} A_j \cos\phi_j + 2\sum_{j=1}^{N_s}\sum_{k=j+1}^{N_s} A_j A_k \cos(\phi_j - \phi_k). \tag{2.37}$$

The right-hand side of Equation 2.37 consists of three terms, of which the first is a slowly varying function of time due both to the envelope of the source power spectrum and to the wavelength-dependence of the scattered amplitude. The second and third terms are the more interesting interference signals. Term 2 represents interference between the scattered light from the jth layer and the reference wave, whereas term 3 represents interference between the light from the jth and the light from the kth layers, and is sometimes called the autocorrelation term. During a wavelength scan, the phase ϕ_j changes with time according to Equation 2.6:

$$\phi_j(m,n,t) = k_c\Lambda_j(m,n) + \phi_{s,j}(m,n) + \delta k \Lambda_j(m,n)t. \tag{2.38}$$

As for the case of the single scattering layer, the $\cos\phi_j$ term in term 2 of the right-hand side of Equation 2.37 therefore modulates at a temporal frequency

$$f_{\Lambda,j} = \delta k \Lambda_j(m,n)/2\pi, \tag{2.39}$$

that is, at a frequency proportional to the optical path difference between the jth layer and the reference wave. The contributions to term 2 from the N_s scattering layers, each of which has its own unique Λ_j, are therefore separated by the Fourier transform. On the other hand, the contributions to term 3 cannot be separated easily because many distinct j–k pairs of scattering layers can have the same difference in optical path $\Lambda_j - \Lambda_k$, and are therefore still superimposed after Fourier transformation. The three terms are illustrated schematically in the lower graph of Figure 2.10.

Fortunately, it is easy to ensure that terms 2 and 3 occupy different regions of the frequency domain simply by positioning the reference surface sufficiently far away from the sample. The highest temporal frequency from term 3 is $f_{\Lambda,N_s} - f_{\Lambda,1} = \delta k(\Lambda_{N_s} - \Lambda_1)/2\pi$, whereas the lowest frequency from term 2 is $f_{\Lambda,1} = \delta k \Lambda_1/2\pi$. Provided the condition $\Lambda_{N_s} - \Lambda_1 < \Lambda_1$, that is,

$$\Lambda_1 > \Lambda_{N_s}/2 \tag{2.40}$$

is satisfied, the required term 2 can be obtained from the high-frequency portion of the Fourier transform of $I(m, n, t)$. Λ_{N_s} must not exceed Λ_M to avoid aliasing of the signal. The inequality (2.18) for the case of the single scattering surface is therefore modified for the case of volume scatterers to

$$\Lambda_M/2 < \Lambda \le \Lambda_M, \tag{2.41}$$

which means that the available depth range is reduced by a factor of 2. Alternatively, phase shifting can be used to eliminate the low-frequency autocorrelation terms without loss of the depth range [43].

The goal of the measurement process is to reconstruct the amplitude, $A_j(m, n)$, and phase, $\phi_j(m, n)$, distributions of the scattered light from each of the N_s scattering layers. The amplitude depends on the "scattering potential" that, in turn, measures refractive index and absorption variations throughout the sample, as discussed in Section 2.6. Conventional OCT normally provides images only of the scattering potential and ignores the phase. Both quantities are, however, accessible using WSI.

The Fourier transform of the intensity given in Equation 2.37, computed according to Equation 2.10, may be written as

$$\begin{aligned}\tilde{I}(\hat{f}) = &\left(A_r^2 + \sum_{j=1}^{N_s} A_j^2\right)\tilde{W}(\hat{f}) + A_r \sum_{j=1}^{N_s} A_j \exp\left(\pm i\phi_j\right)\tilde{W}(\hat{f} \pm \hat{f}_{\Lambda,j}) \\ &+ \sum_{j=1}^{N_s}\sum_{k=j+1}^{N_s} A_j A_k \exp\left[\pm i(\phi_j - \phi_k)\right]\tilde{W}\left[\hat{f} \pm (\hat{f}_{\Lambda,j} - \hat{f}_{\Lambda,k})\right].\end{aligned} \tag{2.42}$$

Unlike the case of the single scattering surface (Equation 2.11), all the terms above the Nyquist limit have been dropped for simplicity. The lower plot in Figure 2.10 schematically shows the Fourier transform $\tilde{I}(\hat{f})$ for the volume scattering case after conversion of the $\hat{f}$ variable to Λ according to Equations 2.7 and 2.8. The Fourier coefficients $\tilde{I}(\hat{f}_{\Lambda,j})$ evaluated from the upper half of the transform have a magnitude proportional to the amplitudes of the scattered waves from each of the layers, and a phase that encodes displacement in exactly the same way as for the single scattering surface (Equation 2.14). Also shown in the lower plot is the window function $\tilde{W}(\hat{f})$ centered on the three scattering layers, which shows the range of Λ values that can contribute to the measured amplitude of a given scattering surface. In the more realistic case of a continuum of scatterers – as opposed to the set of discrete layers modeled here – the reconstructed amplitude distribution is the convolution of $\tilde{W}(\hat{f})$ with the true amplitude distribution.

2.3.2.1 **Proof-of-Principle Experiments: Two Scattering Layers**

Benchmarking experiments on a model geometry consisting of just two scattering layers have been described in Ref. [32]. The experiments included some of the main features of the model in the previous section, that is, a smooth reference wavefront coming from a fixed surface, R, interfering with two speckle patterns coming from two semitransparent and independently tiltable scattering surfaces, S_1 and S_2, located one behind the other (see Figure 2.11). The light source used was a solid-state tunable laser (TL), whose wavenumber was varied approximately linearly in time by signal generator (SG) and laser controller (LC). Images of the superposed object surfaces S_1 and S_2 were recorded by high-speed camera, C. P_1 and P_2 were polarizers, used to optimize the ratio of object and reference wave intensities, and B was a beam block that removed the specular component from the light reflected from S_1.

Figure 2.12a shows the normalized intensity $I(m, n, t)$ (minus the DC term) at one pixel in a region of interest (ROI) of 200 × 200 pixels during a scan of 405 frames over a total wavelength range of 0.116 nm. The positive side of the frequency spectrum of $I(m, n, t)$ is shown in Figure 2.12b, after averaging over a single column of the ROI to reduce speckle noise. The peaks labeled RS_1, RS_2, and S_1S_2 correspond to the interference signal for surfaces R and S_1, R and S_2, and S_1 and S_2, respectively. Prior to the Fourier transform, the mean value of the intensity signal was subtracted and the signal multiplied by a Hanning window.

Figure 2.13a and b shows the measured wrapped phase difference maps resulting from small independent tilts applied to surfaces S_1 and S_2, respectively. Each fringe represents an out-of-plane displacement $u_z = \lambda_c/2 \sim 317$ nm. The gauge volume was $\sim 50\ \mu m \times 50\ \mu m \times 7$ mm and the system depth range was $z_M \approx 350$ mm. In

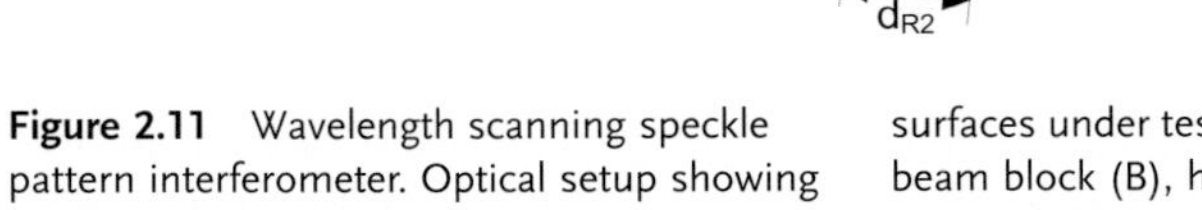

Figure 2.11 Wavelength scanning speckle pattern interferometer. Optical setup showing tunable laser (TL), laser controller (LC), signal generator (SG), lenses (L_1, L_2, L_3), steering mirror (M), reference surface (R), scattering surfaces under test (S_1, S_2), polarizers (P_1, P_2), beam block (B), high-speed camera (C), and personal computer (PC). (Reprinted with permission from Ref. [32].)

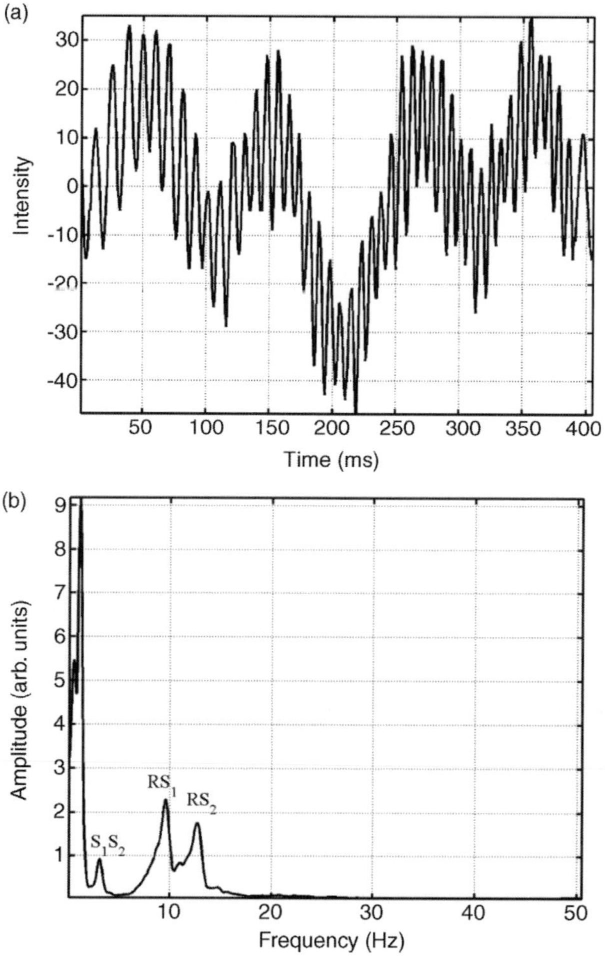

Figure 2.12 (a) Intensity signal (minus the DC term) recorded at a single pixel during a wavelength scan of the sample shown in Figure 2.11. (b) Frequency spectrum of the intensity signal after averaging over 200 pixels. Peaks RS_1, RS_2, and S_1S_2 correspond to the interference signal for surfaces R and S_1, R and S_2, and S_1 and S_2, respectively. (Reprinted with permission from Ref. [32].)

Figure 2.13c and d, standard ESPI subtraction fringes at a fixed wavelength $\lambda_c = 635.05$ nm representing the tilt of surfaces S_1 and S_2, respectively, are shown to illustrate the close agreement with the phase distributions measured using WSI. It is worth emphasizing that while the depth-resolving capabilities of WSI allowed the simultaneous measurement of the displacement field on the two scattering surfaces in Figure 2.13a and b, the ESPI results in Figure 2.13c and d required a sequential measurement process in which one of the surfaces had to be removed for the measurement of the other.

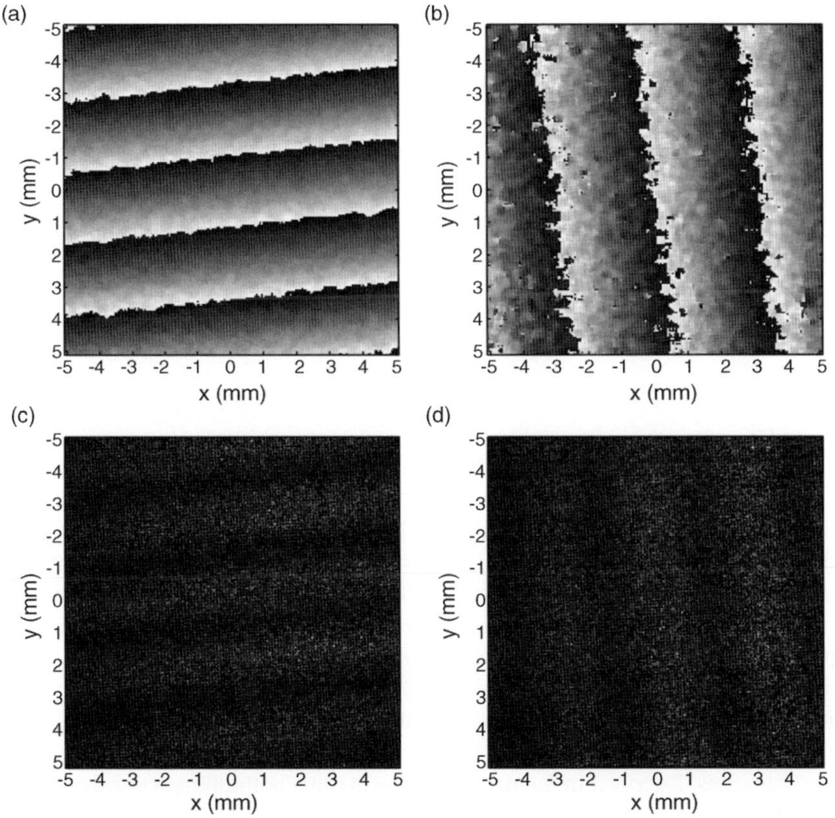

Figure 2.13 Wrapped phase difference map of surfaces S_1 (a) and S_2 (b) measured simultaneously using wavelength scanning interferometry. Black represents $-\pi$ radians and white $+\pi$ radians. Correlation fringes of surfaces S_1 (c) and S_2 (d) measured sequentially using standard digital speckle pattern interferometry. (Reprinted with permission from Ref. [32].)

2.3.3 Comparison of WSI with LCSI

Some of the advantages and disadvantages of WSI over LCSI for depth-resolved displacement field mapping are discussed in this section. One of the advantages is that the depth range of the displacement field is not limited by the mechanical scan range of the reference arm of the interferometer as in LCSI, but by the number of frames that can be recorded while scanning both the whole tuning range of the source and the tuning precision. Depth ranges that would be regarded as unfeasible for LCSI (of order 1 m or more) are in principle measurable using WSI.

The second advantage is that the limited dynamic range (typically 256–4096 gray levels) of whole-field image sensors based on CCD or CMOS technology limits the performance of WSI to a much lesser extent than that of LCSI. If δz is the slice thickness and Z the overall thickness of a sample containing uniformly distributed scatterers, then only the fraction $\delta z/Z$ of scattered photons contribute to signal modulation in the case of LCSI, while the rest merely produce a DC offset to the intensity image. For example, in the case $\delta z/Z = 1/100$, with an 8-bit camera, the signal would be no more than 2–3 gray levels deep and therefore rather susceptible to noise. WSI, on the other hand, ensures that the full dynamic range of the camera is utilized.

Finally, in systems with broadband light sources, dispersion may be a significant cause of fringe contrast reduction. In WSI, the fringes are produced at high visibility at all times by a single wavelength, ensuring that the full dynamic range of the camera is utilized, and therefore the reduction in signal-to-noise ratio due to dispersion does not arise.

WSI has two disadvantages, however, compared to LCSI. First, all the images in a wavelength scan need to be acquired before even a single slice can be selected. Drift within the sample or optical system during the scan time will introduce artifacts that are absent in an LCSI system with spatial phase shifting since the latter provides the displacement data on a single slice from just one pair of images. Second, it is technically easier to manufacture a stable low-coherence light source than a laser that is able to repeatably tune over the same bandwidth. Commercially available superluminescent LEDs, for example, have spectral ranges of up to 200 nm, whereas external cavity diode lasers have much smaller tuning ranges (typically from $\sim$1 to 10 nm). Ti:sapphire lasers can provide tunability over a 100–200 nm range, but are complex and not very repeatable. Low-cost tunable sources in the infrared (1.3–1.6 μm range) are more readily available, but require specialized and expensive cameras. The approaches described in the next section overcome both of these disadvantages of the WSI technique.

2.4
Spectral Optical Coherence Tomography

The two main drawbacks of wavelength scanning interferometry identified in the previous section, that is, long scan times and complexity of the tunable light source, are avoided by the use of a method known in the OCT literature as spectral optical coherence tomography (SOCT) or Fourier-domain OCT. In its simplest form, spectral OCT uses a broadband source focused to a narrow beam that is scanned in an x–y raster fashion across the sample. The backscattered light is separated into its constituent wavelength bands by a diffraction grating and imaged onto a linear photodetector array. At any given beam position, the 1D intensity distribution measured by the photodetector array is therefore the one given by Equation 2.37 with $N_m = N_n = 1$ and where t now represents the index of the pixel in the photodetector array instead of time. Fourier transformation of this distribution according to Equation 2.10 thus reconstructs the 1D scattered amplitude distribution along the bisector of the illumination and observation directions [44]. Benefits of this approach over the corresponding time-domain OCT technique include "single-shot" collection of the data and improved signal-to-noise ratio [45]. However, the x–y scan mechanism is needed in order to build up 3D volumes.

An extension to this approach, which removes the need for one of the two scan axes, involves the use of a 2D, rather than 1D, photodetector array. The sample is now illuminated with a sheet of light, rather than a narrow beam, which extends into the sample from the surface. The intersection of the sheet with the sample surface forms a line, which is once again imaged onto the sensor through a diffraction grating. The resulting images have one spatial axis and one wavenumber axis, the latter being converted to a second spatial (z) coordinate by Fourier transformation. This approach has been used for surface profilometry [46], for measurement of internal structure [47], and, more recently, using a phase contrast (PC-SOCT) version, for depth-resolved displacement field measurement [48–53].

In this section, three optical arrangements are described: first a 2D PC-SOCT system for single-shot measurement of out-of-plane displacement fields (Section 2.4.1), followed by an adaptation to allow measurement of both in-plane and out-of-plane displacement fields (Section 2.4.2). Finally, in Section 2.4.3, a technique called hyperspectral interferometry is described that allows 3D profiles to be measured in a single shot. Despite the differences in optical setup, the numerical analysis for the extraction of optical path length and phase change distributions is identical to that described in Section 2.3.1 for WSI, provided that t is reinterpreted as a spatial rather than a temporal wavenumber index.

2.4.1 Phase Contrast SOCT for 2D Out-of-Plane Displacement Field Measurement

Figure 2.14 shows the phase contrast SOCT (PC-SOCT) system described in Ref. [48]. In addition to the main broadband light source SLD ($\lambda_c = 840$ nm, $\Delta\lambda = 50$ nm, and optical power of 15 mW), a CW laser is included to verify the accuracy of the measured displacement fields. A cube beam splitter (CBS) brings the collimated beams from

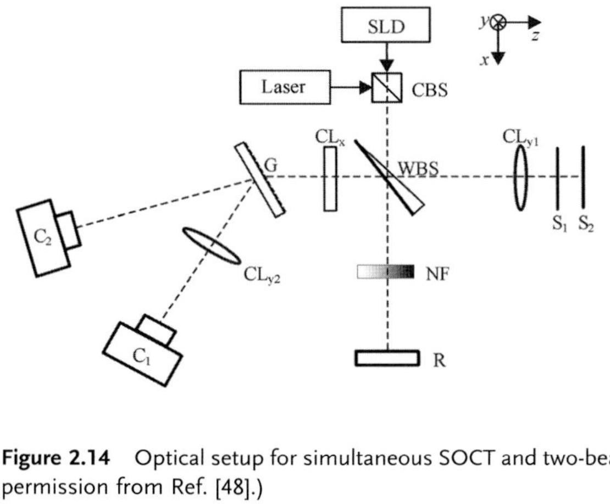

Figure 2.14 Optical setup for simultaneous SOCT and two-beam interferometry. (Reprinted with permission from Ref. [48].)

the two sources into alignment. Object and reference beams are produced by means of a wedge beam splitter (WBS). Test objects S_1 and S_2 (in this case, microscope cover slips) are oriented normal to the z-axis within the depth range of the system. The reference beam consists of a plane wave reflected from the reference mirror R. Upon recombination through WBS, the reference and the object beams are diffracted by a transmission grating G with 1200 lines/mm and optimized efficiency at 840 nm. NF is a neutral density filter used to optimize the speckle contrast upon interference of the reference and the object beams. Cameras C_1 and C_2 record the interference patterns from the SLD and from the laser, respectively. Three cylindrical lenses CL_x, CL_{y1}, and CL_{y2} are used to produce the necessary line illumination parallel to the y-axis on the sample and to image this line onto a corresponding column of photodetectors on each of the two cameras. The bandwidth of the illumination falling onto C_1 is 24 nm. From Equations 2.17 and 2.19, the resulting depth range and depth resolution (using a Hanning filter and 640 pixels to sample the λ-axis) are $\sim$4.7 mm and $\sim$58.8 μm, respectively.

The first sample, which consists of two thin glass sheets, each with a thickness of $\sim$300 μm, provides a simple illustration of the technique. By painting the rear cover slip black on the reverse side, reflections are suppressed from that interface. Figure 2.15a shows a raw fringe pattern from C_1 where the λ-axis gives the range of detected wavelengths for the sensor. Each image of this type was resampled along the λ-axis so as to give uniformly sampled points along the k-axis. Figure 2.15b is the spectrum of the pattern along one line parallel to the k-axis, where frequency has been converted to optical depth ($\Lambda/2$ for this illumination/observation geometry) using Equation 2.39. Peaks RS_1 and RS_2 are due to the interference between beams reflected at the reference surface R and surfaces S_1 and S_2, respectively. RS_1' is the interference peak coming from the reference and the back surface of the first cover slip, S_1'. The autocorrelation term representing interference between S_1 and S_1' generates the first peak in Figure 2.15b, located at an apparent depth of 440 μm, the same as the distance between RS_1 and RS_1'. These values nearly match the optical thickness of the cover slip.

Applying Equation 2.14 to the spectra of the type shown in Figure 2.15a provides wrapped phase data on the illuminated sheet extending into the medium. By repeating the measurement after small independent rotations have been made to S_1 and S_2, wrapped phase difference maps can be obtained as shown in Figure 2.16a. For clarity, regions containing no useful phase information have been masked out by applying a binary mask derived from the magnitude image. RS_1 and RS_1' have the same phase difference, as expected since they both come from the same cover slip undergoing a rigid body rotation around an axis parallel to its surface. The phase difference measured at the autocorrelation peak (situated at 440 μm) is also very close to zero for the same reason. Figure 2.16b shows the unwrapped phase map where the tilt is clearly observed for each cover slip interface. RS_2 has a smaller tilt than RS_1 and RS_1' due to the limited range of the PZT driven tilting stage. The unwrapped phase map is then converted to optical path change at each interface using Equation 2.35 with $n_1 = 1$. Figure 2.16c shows a comparison of the displacement profiles measured at the two surfaces by phase contrast SOCT and two-beam

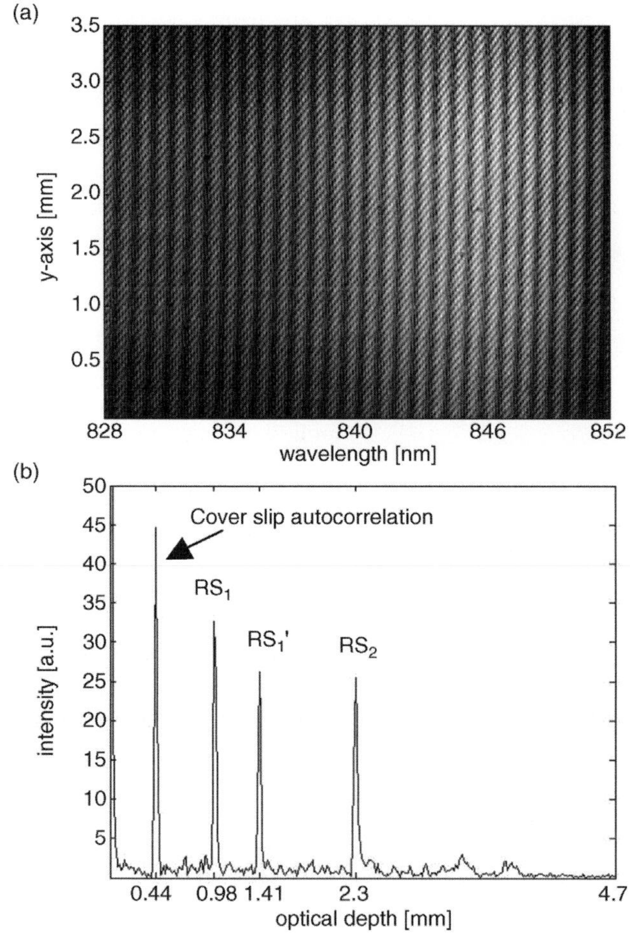

Figure 2.15 (a) Interference pattern with both cover slips present, (b) spectrum of intensity along line $y = 1.9$ mm with frequency converted to optical depth (linear scale). (Reprinted with permission from Ref. [48].)

interferometry. The estimated tilt angle from the two techniques agrees to better than 0.5%.

Application of the technique to a scattering volume, rather than to a set of specularly reflecting surfaces, is illustrated in Figure 2.17. Figure 2.17a shows the magnitude and Figure 2.17b the phase difference images within a porcine cornea subjected to a small pressure change. In this case, the theoretical depth range and depth resolution were 3.6 mm and 28 μm, respectively, with 1024 pixels sampling a total bandwidth of 50 nm. In practice, dispersion effects reduce the depth resolution as one gets deeper into the tissue. Dispersion compensation could be implemented by using plate compensators in the reference arm to account both for the extra optical elements in the sample arm of the interferometer and for corneal dispersion.

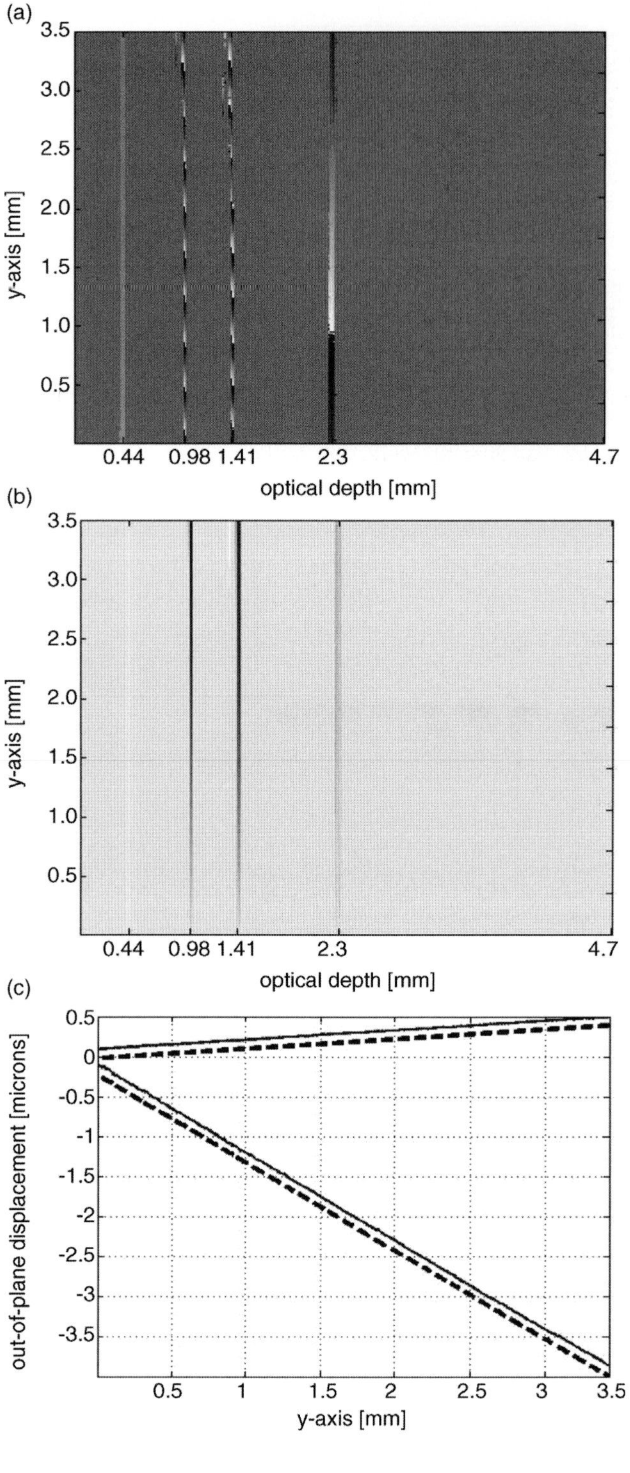
(a)
3.5
3.0
2.5
2.0
1.5
1.0
0.5
y-axis [mm]
0.44
0.98
1.41
2.3
4.7
optical depth [mm]
(b)
3.5
3.0
2.5
2.0
1.5
1.0
0.5
y-axis [mm]
0.44
0.98
1.41
2.3
4.7
optical depth [mm]
(c)
0.5
0
-0.5
-1
-1.5
-2
-2.5
-3
-3.5
out-of-plane displacement [microns]
0.5
1
1.5
2
2.5
3
3.5
y-axis [mm]

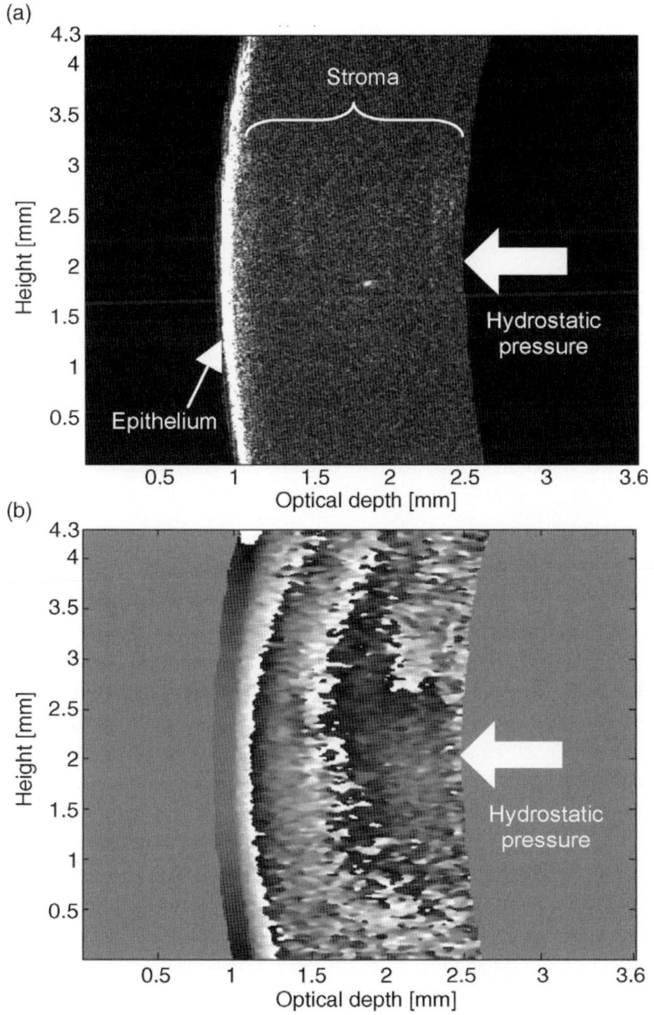

Figure 2.17 (a) Magnitude and (b) phase difference map from slice through porcine cornea. (Reprinted with permission from Ref. [48].)

This approach, however, does not correct the depth-dependent dispersion in the cornea. A better solution and particularly easy to implement in SOCT (and WSI) is numerical dispersion compensation, which allows depth-dependent compensation and higher orders of dispersion to be simultaneously corrected [54]. As dispersion

Figure 2.16 (a) Wrapped and (b) unwrapped phase difference map representing out-of-plane displacement field at the interfaces shown in Figure 2.15b as RS_1, RS_1', and RS_2; (c) comparison of displacement field profiles measured using phase contrast SOCT (solid lines) and standard two-beam interferometry (dashed lines). An arbitrary offset of 0.1 μm has been added for clarity. (Reprinted with permission from Ref. [48].)

involves the introduction of an extra wavelength-dependent phase term in Equation 2.6, as long as that term does not change when a load is applied, it will disappear when the phase difference is evaluated between the loaded and the reference states. In this case, dispersion will reduce only the axial resolution and the measured phase will correspond to an average over a bigger gauge area.

2.4.2
PC-SOCT for 2D In-Plane and Out-of-Plane Displacement Field Measurement

An adaptation of the system from the previous section allows the measurement of depth-resolved in-plane and out-of-plane displacement fields [50]. A mechanism – in this case a PZT actuator – is introduced to switch the illumination angle symmetrically about the observation direction. The rotation of the illuminating sheet, through an angle of $\pm\theta_0$, takes place within the plane of the sheet, that is, about the normal to the y–z plane. A total of four interferograms are recorded: two before the deformation and two after, where each pair is produced by a $+\theta_0$ and $-\theta_0$ illumination direction.

Figure 2.18 shows the geometry for simple shear experiments to illustrate the technique. The sample S consists of a transparent water-based polymer rubber layer with refractive index $n_2 = 1.48$ cured between two glass plates G_1 and G_2 with refractive index $n_1 = 1.51$. G_2 is fixed to a rigid support and G_1 is displaced a known distance within its plane by a micropositioning stage. For small deformations, the displacement field in the polymer has a constant in-plane gradient with virtually no out-of-plane component.

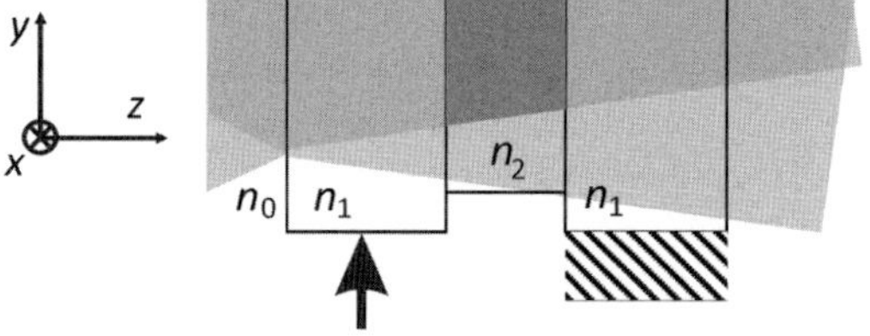

Figure 2.18 Oblique illumination of polymer layer S constrained between two glass plates G_1 and G_2 for combined in-plane and out-of-plane displacement measurements by PC-SOCT (Reprinted with permission from Ref. [50].)

For oblique illumination in the y–z plane at an angle θ_0 to the z-axis, the phase difference due to object deformation can be calculated from Equation 2.33 by substituting the values $\mathbf{e}_1 = (0, \sin\theta_1, \cos\theta_1)^T$ and $\mathbf{e}_2 = (0, 0, -1)^T$ to give

$$\Delta\phi(x, y, z) = n_1 k[u_y(x, y, z)\sin\theta_1 + u_z(x, y, z)(1 + \cos\theta_1)], \tag{2.43}$$

where θ_1 is the angle of incidence of the refracted beam within G_1. No additional correction terms to reflect changes to the n_0–n_1 interface geometry are required here since the motion of G_1 is purely in-plane.

By choosing equal and opposite illumination angles, θ_0 and $-\theta_0$, two phase differences can be evaluated, the sum of which is sensitive to the out-of-plane component and hence is denoted here $\Delta\phi_z$, and the difference of which is sensitive to the in-plane component (denoted $\Delta\phi_y$). The in-plane and out-of-plane displacements are thus obtained as

$$u_y(x, y, z) = \frac{\lambda\Delta\phi_y(x, y, z)}{4\pi n_1 \sin\theta_1}, \tag{2.44}$$

$$u_z(x, y, z) = \frac{\lambda\Delta\phi_z(x, y, z)}{4\pi n_1 (1 + \cos\theta_1)}. \tag{2.45}$$

According to Equations 2.44 and 2.45, with $\theta_0 = 5.7°$ and $\theta_1 = 3.7°$, the resulting out-of-plane and in-plane sensitivities are 0.14 and 4.2 μm per fringe, respectively. Refraction at the glass/elastomer interface changes the out-of-plane sensitivity by only 2% due to the slight difference between refractive indexes n_1 and n_2.

The interferograms are processed in the same way as for the out-of-plane configuration discussed in the previous section. One effect of the oblique illumination, however, is a tilt of the reconstructed cross section of the object in the y–z plane. This tilt is also observed with the tilt-scanning interferometry technique described in Section 2.5 and is removed here by reregistering the scattering potential corresponding to each illumination direction in the same way as for TSI. The out-of-plane and in-plane phase components are then obtained by adding and subtracting, respectively, the reregistered phase differences.

Figure 2.19a shows the in-plane wrapped phase map for the simple shear test. As expected, a linear phase profile is observed in the region between G_1 and G_2, with virtually no phase change for G_2, indicating that it remained static. Figure 2.19b shows the wrapped phase corresponding to the out-of-plane sensitivity. Figure 2.20 shows the corresponding displacement profiles obtained by averaging the displacement fields obtained from Figure 2.19a and b and the corresponding sensitivities for each phase component along the position axis between $y = 1.7$ and 3.7 mm for optical depth values between $z_o = 2.3$ and 2.9 mm. The expected displacement field and its uncertainty introduced by the translation stage, 1.6 ± 0.1 μm, are shown as two solid lines in Figure 2.20. For the in-plane displacements, an RMS error of ~6% was found between the optical measurements and the expected average displacement profile.

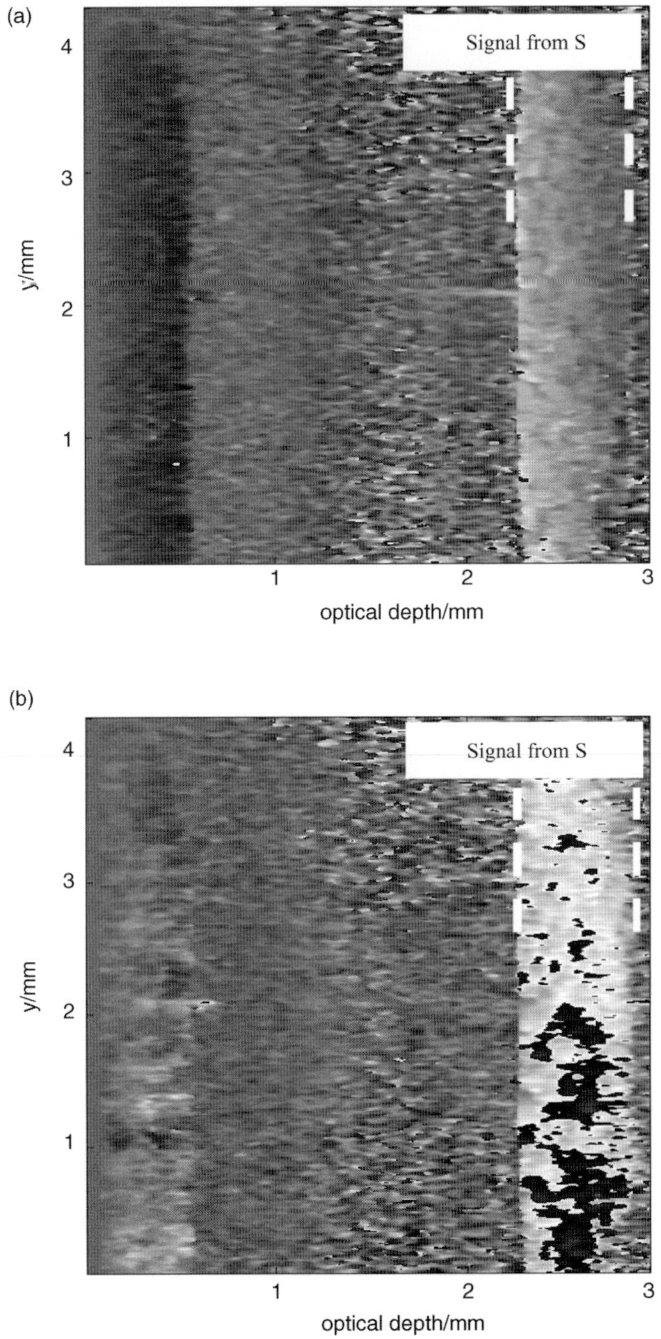

Figure 2.19 In-plane (a) and out-of-plane (b) wrapped phase (black and white represent $-\pi$ and π, respectively) due to simple shear of sample S from Figure 2.18. The regions between 0 and 1 mm and between ~2.2 and 3 mm correspond to the cross-correlation terms and to the polymer layer, respectively. (Reprinted with permission from Ref. [50].)

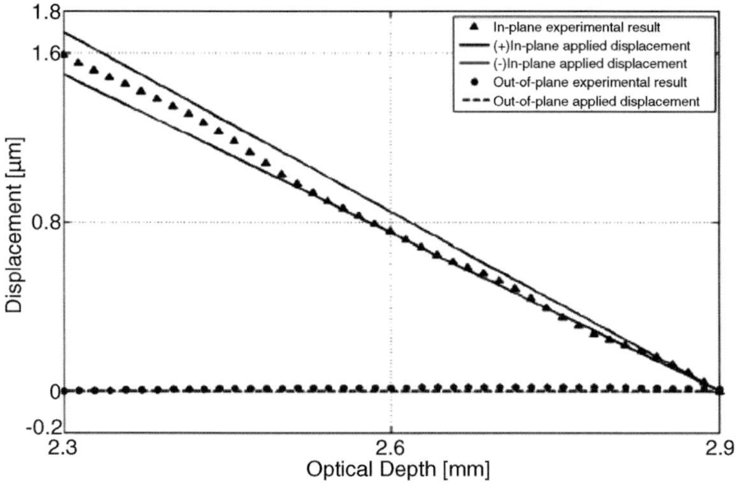

Figure 2.20 In-plane and out-of-plane displacement profiles through the thickness of a polymer layer under pure shear. (Reprinted with permission from Ref. [50].)

2.4.3 Hyperspectral Interferometry for 3D Surface Profilometry

Hyperspectral interferometry is a recently proposed method for the measurement of surface profiles, which can be thought of as a fully parallel version of the SOCT technique [36]. As with SOCT, a broadband light source is used; however, instead of illuminating along a line or sheet, an entire 3D region of interest is illuminated at the same time and observed by a conventional interference microscope. The resulting 2D white light interferogram is then separated by a hyperspectral imaging system into a set of narrowband interferograms, which are simultaneously recorded in different spatial locations on a high-resolution image sensor. These are then registered spatially and assembled to form a three-dimensional intensity distribution $I(m, n, t)$, as shown schematically in Figure 2.6. Provided the individual narrowband interferograms are separated by a constant wavenumber, the $I(m, n, t)$ distribution is entirely equivalent to that recorded by a WSI system, but where t now refers to the narrowband interferogram index, rather than to the frame number in a wavelength scan. As with WSI, Fourier transformation along the wavenumber axis, therefore, allows an absolute optical path difference to be obtained for each pixel independent of the other pixels in the field of view. The big advantage over WSI is that all the data are obtained simultaneously so that motion artifacts can be effectively eliminated by sufficiently reducing the exposure time. The main drawback is that the number of available pixels in the resulting profiles is reduced by a factor of at least N_t compared to a WSI system with the same image sensor.

An optical setup to demonstrate the principle of the HSI technique is shown in Figure 2.21. Light from two sources, denoted LS_1 and LS_2, is combined by a fiber coupler (FC). LS_1 is a broadband superluminescent LED (SLED) light source

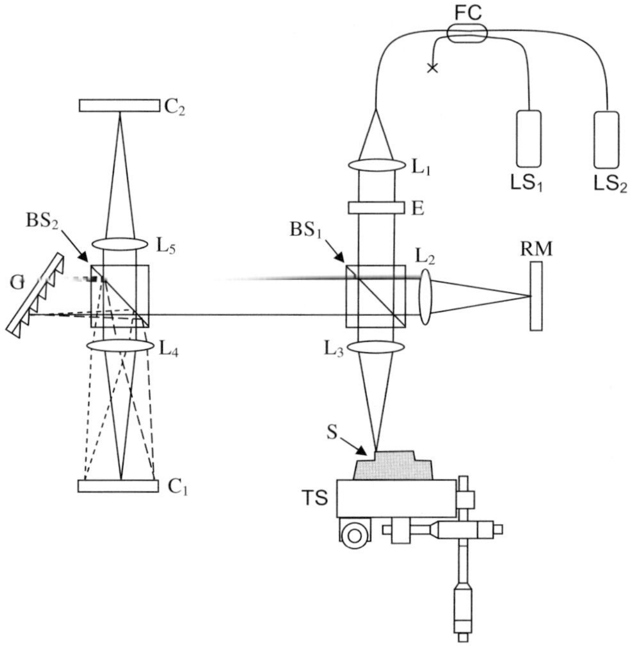

Figure 2.21 Hyperspectral interferometer for single-shot 3D shape measurement. (Reprinted with permission from Ref. [36].)

(Superlum Diodes Ltd., 840 HP1) with center wavelength 840 nm, and full width half maximum 50 nm. LS_2 is a narrowband source (He–Ne laser; wavelength 632.8 nm) that is useful for alignment purposes, but otherwise contributes nothing to the HSI. The output from the optical fiber passes first through an etalon, E, is next collimated by lens L_1 and then enters a Linnik interferometer head comprising beam splitter BS_1, lenses L_2 and L_3, and reference mirror, RM. The sample S is mounted on a translation stage, TS, to allow the region of interest to be adjusted.

Light from a single-point P on S produces a bundle of parallel rays that enter the hyperspectral imaging system that comprises beam splitter BS_2, diffraction grating, G, and digital camera, C_1. The blazed grating is arranged in the Littrow configuration to maximize diffraction efficiency. The parallel rays are brought to a focus in the plane of the photodetector array of C_1. Without E in place, the effect of the grating is to smear the image of P along a line on C_1. The etalon, which has a free spectral range of 0.5 nm at 840 nm and a finesse >15, modifies the spectral content of the illuminating beam to produce a broadband comb with uniform δk spacing, and thus produces instead a set of N_t discrete spots. Points close to P on the sample are, in turn, imaged onto a set of N_t neighboring points on C_1. Thus, a set of N_t images of the sample are replicated across the sensor array of camera C_1, with a wavenumber shift of δk from one image to the next. The second camera, C_2, and imaging lens L_5 are used to produce a single high-resolution broadband image of the object to assist in alignment of the optical system. Once again, C_2 and L_5 are not an essential part of the HSI system.

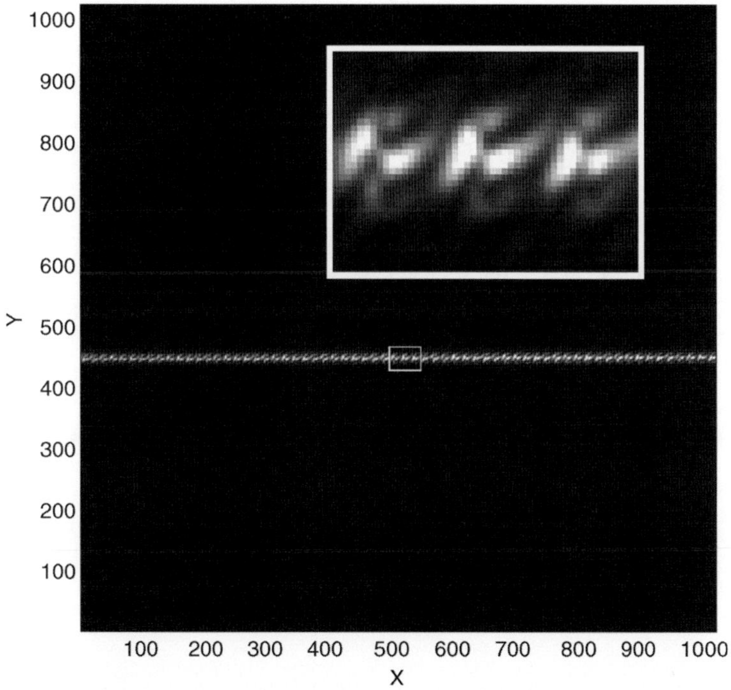

Figure 2.22 Set of $N_t = 62$ hyperspectral images of a stepped surface spread across the horizontal axis of camera C_1. *Inset*: Three of the hyperspectral images within the central white box enlarged by a factor 10×. (Reprinted with permission from Ref. [36].)

An example from a proof-of-principle experiment, in which a set of 62 interferograms (3 of them enlarged 10 ×) with a 0.5 nm separation were acquired in a single shot, is shown in Figure 2.22, together with the resultant reconstructed surface profile of a 120 μm stepped object in Figure 2.23. Although the technique has so far been used only for single scattering surface applications, the formal equivalence between the data from HSI and WSI means that, in principle, HSI can also be used to measure, in a single shot, depth-resolved displacement fields within a 3D scattering volume.

2.5 Tilt Scanning Interferometry

As seen in the previous sections, while WSI, PC-SOCT, and HSI have different optical implementations, they are from a mathematical viewpoint essentially the same technique. In this section, a different approach to measuring depth-resolved displacements within optically scattering materials is described, which is based on tilting a monochromatic illuminating beam during the acquisition of the image sequences [55]. This provides the necessary depth-dependent phase shifts that allow

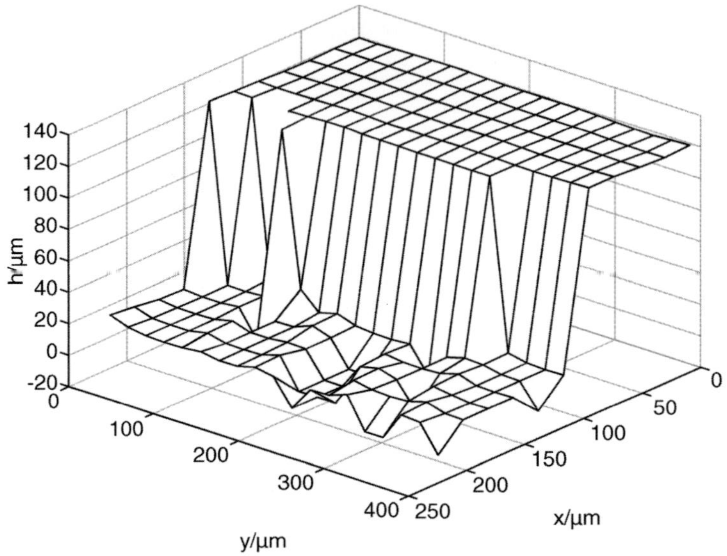

Figure 2.23 Surface profile of a stepped sample on an 11 × 19 pixel grid measured using the single-shot hyperspectral interferometry system. (Reprinted with permission from Ref. [36].)

the reconstruction of the object structure and its internal displacements. The method is distinct from the other techniques in that only a single wavelength is present throughout the entire data recording process, thereby reducing the expense and complexity of the light source. The depth-encoding frequency shift can be regarded as coming from a position-dependent Doppler shift of the photons in the object illumination beam path. As with the WSI version of OCT, the displacement sensitivity of TSI is decoupled from the intrinsic depth resolution of the technique and is a few tens of nanometers at optical wavelengths.

Extraction of depth and displacement information of nontransparent object surfaces has been previously reported using speckle contouring techniques based on source displacement [56]. The concept of angular spectrum scanning using a spatial light modulator (SLM) for profilometry applications has also been proposed recently [57]. A related method for high-resolution 3D imaging was described in Ref. [58], in which a sample is successively illuminated by a series of plane waves with different directions and the scattered amplitude distribution measured by phase shifting interferometery. TSI extends the capabilities of these approaches to allow the study of depth-resolved displacement fields inside semitransparent materials. This section provides an overview of the main aspects of the TSI technique as described in Ref. [55].

2.5.1 Depth-Dependent Phase Shift Introduced by a Tilting Wavefront

Figure 2.24a shows a semitransparent scattering material of refractive index n_1 immersed in a medium of refractive index n_0, illuminated by a collimated beam

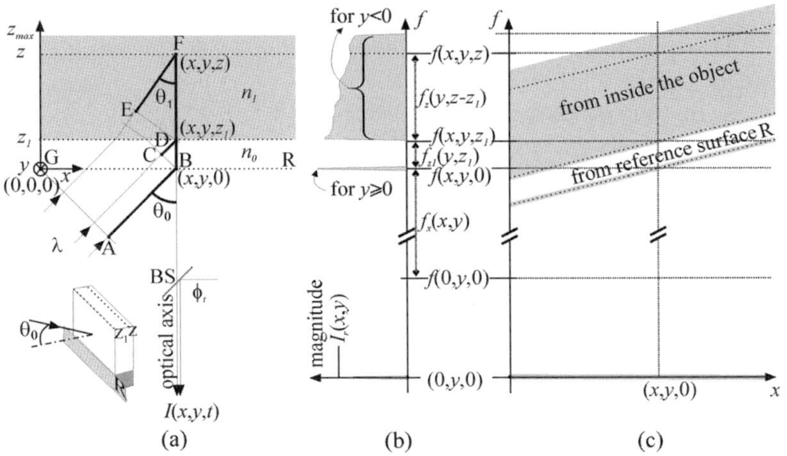

Figure 2.24 (a) Ray diagram for tilt scanning interferometry. A variation in the illumination angle modulates the interference signal with a depth-dependent Doppler shift. (b) Spectrum of interference signal associated with points along BF. (c) Shift of the spectrum for different positions in the horizontal direction. (Reprinted with permission from Ref. [55].)

of wavelength λ at an angle θ_0 to the optical axis of the system. As with the theory of WSI (Section 2.3.1), the interface is assumed to be planar, though this restriction can be relaxed through a suitable extension of the analysis. It is experimentally convenient to place a flat but microscopically rough opaque surface over the region $y \geq 0$ of the plane $z = 0$, where the coordinate system (x, y, z) is as defined in Figure 2.24a. This reference surface serves two purposes: first allowing correction of the nonlinearity of the tilting device and second (since it does not strain during the loading of the sample) enabling registration of the before- and after-load scans of the sample. It should be emphasized, however, that it does not provide the reference wave for the interferometer, which rather is introduced by means of a separate beam splitter, BS. For simplicity, we assume the backscattered light of interest follows a path parallel to the z-axis, that is, the imaging system is telecentric on the object side. As a result, there is a simple linear z-independent relationship between the sample coordinates (x, y) and the pixel coordinates (m, n), which will be interchangeably used for the remainder of this section.

The illumination beam is refracted at the object surface $z_1(x, y)$ and reaches point F with coordinates (x, y, z) at an angle $\theta_1 = \sin^{-1}(n_0 \sin\theta_0 / n_1)$ to the optical axis. Some of the light scattered at F travels vertically downward, is recombined by BS with a reference wave derived from the same laser light source as the object beam, and is imaged onto a pixel lying within a two-dimensional photodetector array.

The phase difference between light scattered at F and a reference wavefront can be expressed relative to the phase difference at point G, which lies on a rough opaque reference surface R at the origin (0, 0, 0), as

$$\phi(x, y, z) = \phi(0,0,0) + k\left[n_0 x \sin\theta_0 + n_0 z_1 (1 + \cos\theta_0) + n_1 (z - z_1)(1 + \cos\theta_1)\right], \tag{2.46}$$

where we assumed that $z \geq z_1 \geq 0$. The phase differences due to the first, second, and third terms between square brackets in Equation 2.46 account for the optical paths $\overline{AB}$, $\overline{CD} + \overline{DB}$, and $\overline{EF} + \overline{FD}$, respectively. The random distribution of scattering centers within the material gives rise to a speckle phase distribution along the x-axis in this two-dimensional representation.

Suppose the illumination angle θ_0 changes linearly with nondimensional time t about the center angle θ_c, that is,

$$\theta_0(t) = \theta_c + \dot{\theta}_0 t, \tag{2.47}$$

where $\dot{\theta}_0$ is the angular velocity $\delta\theta_0/\delta t$ and $\delta\theta_0$ is the tilt angle between successive frames ($\delta t = 1$). The phase ϕ (x, y, z) will vary as

$$\frac{\partial\phi(x,y,z,t)}{\partial t} = \frac{\partial\phi(0,0,0,t)}{\partial t} + kn_0\dot{\theta}_0\,[x\cos\theta_0 - z_1\sin\theta_0 - (z-z_1)\xi] \tag{2.48}$$

with t ranging from $-\Delta\theta_0/2\delta\theta_0$ to $+\Delta\theta_0/2\delta\theta_0$, where $\Delta\theta_0$ is the total tilt angle. In the last term between the square brackets, the following relationship, derived from Snell's law of refraction, is used:

$$\frac{\partial\theta_1}{\partial\theta_0} = \frac{\chi\cos\theta_0}{\sqrt{1-\chi^2\sin^2\theta_0}}, \tag{2.49}$$

with $\chi = n_0/n_1$. For convenience, a parameter ξ may be defined as

$$\xi = \frac{\chi\cos\theta_0\sin\theta_0}{\sqrt{1-\chi^2\sin^2\theta_0}}. \tag{2.50}$$

Figure 2.25 shows the variation of $|\xi|$ with θ_0 for a range of typical $1/\chi$ values. Equation 2.48 can be expressed in terms of nondimensional temporal frequencies (units of cycles per frame) as

$$f(x,y,z) = f(0,0,0) + \frac{kn_0}{2\pi}\dot{\theta}_0\,[x\cos\theta_0 - z_1\sin\theta_0 - (z-z_1)\xi], \tag{2.51}$$

$$f(x,y,z) = f(0,0,0) + f_x(x,y) + f_{z_1}(y,z_1) + f_z(y,z-z_1). \tag{2.52}$$

$f(0, 0, 0)$ is carrier frequency due to the rigid body translation or piston term of wavefront GA as it tilts around an axis perpendicular to the plane of Figure 2.24a and is zero if that axis passes through point G. The frequency of the second term in Equation 2.52 varies linearly with x, whereas $f_{z_1}(y, z_1)$ accounts for the distance from the object surface to the reference surface, z_1, and $f_z(y, z - z_1)$ to the depth $z - z_1$ of scattering points within the sample. The last two terms can be interpreted as depth-encoding heterodyne frequency shifts due to the Doppler shift of the tilting beam.

At $z = z_1$, $f_z(y, z - z_1) = 0$ and $f(x,y,z_1) = f(0,0,0) + f_x(x,y) + f_{z_1}(y,z_1)$ is the frequency associated with point D at position (x, y, z_1) on the object surface in

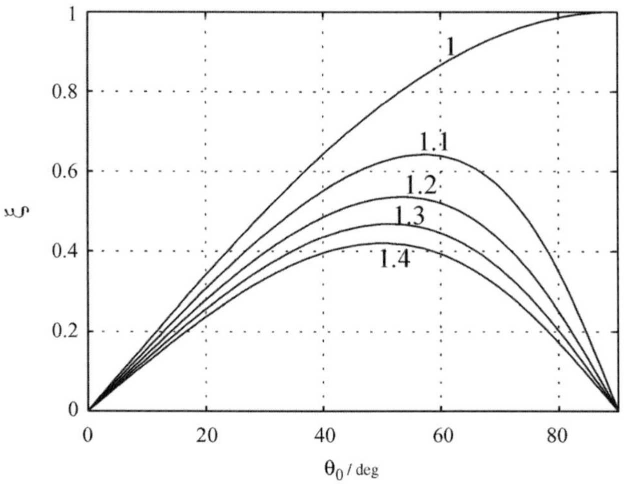

Figure 2.25 Variation in refraction parameter ξ with the illumination angle θ_0 for five different values of $1/\chi = n_1/n_0 = 1$, 1.1, 1.2, 1.3, and 1.4. (Reprinted with permission from Ref. [55].)

Figure 2.24a. If we put $z = z_1 = 0, f_z(y, z-z_1) = f_{z_1}(y, z_1) = 0$ and $f(x, y, 0) = f(0, 0, 0) + f_x(x, y)$ is the frequency associated with point B lying at $(x, 0, 0)$ on the reference surface R in Figure 2.24a.

2.5.2
Extraction of the Scattered Amplitude Distribution

The amplitude distribution of the scattered waves from within the sample can be extracted by mapping depth from frequency. It will be assumed, as for standard OCT, that the contributions from multiple scattering within the material can be neglected. The analysis for WSI (Section 2.3.2) can be repeated to show that the intensity signal due to the interference between light coming from all the scattering points along $\overline{DF}$ in Figure 2.24a and the reference wavefront is modulated with multiple frequencies as follows:

$$I(m,n,t) = A_r^2(x,y) + 2A_r(x,y)\int_{z_1}^{z_{max}} A_o(x,y,z)\cos[2\pi f(x,y,z)t]dz + 2\int_{z_1}^{z_{max}}\int_{z_1}^{z_{max}} A_o(x,y,z)A_o(x,y,z')\cos\{2\pi[f(x,y,z)-f(x,y,z')]t\}dz\,dz'. \quad (2.53)$$

$A_o(x,y,z)$ represents the amplitude of the object wave originating from a small volume element centered on (x,y,z) and which is imaged, in turn, onto pixel (m, n). Equation 2.53 is the TSI continuum equivalent to the WSI "discrete scattering layers" Equation 2.37. While the first term on the right-hand side of Equation 2.53

represents the DC component of the reference beam, the second term corresponds to the modulation due to interference between the reference beam and the light scattered within the material. The integration limit z_{max} represents either the object back surface or the maximum penetration depth. As is the case with WSI, the double integral in the third term is due to cross interference between light coming from within the object and contributes to the DC component and low-frequency components in the interference signal. The frequency of each term on the right-hand side of Equations 2.51 and 2.52 in general depends on θ_0 and therefore changes during the course of the scan. If the total scan range $\Delta\theta_0$ is small enough, the resulting frequency shifts can be neglected, otherwise linearization of the signal may be necessary [55].

A one-dimensional Fourier transform of the intensity signal $I(m, n, t)$ along the time axis gives rise to a spectrum as schematically shown in Figure 2.24b. Any given pixel sees either the reference surface or the sample, but not both; this figure can therefore be interpreted as a top view of the spectrum over all y onto the (f, x) plane, that is, as a superposition of the spectra corresponding to the reference surface R and the interior of the sample. There is a DC peak at $f=0$ for all x, another peak at $f(x, 0)$ corresponding to the reference surface R, and a band associated with the object. The position of the peaks is therefore linked to the internal structure of the object and its position relative to the reference surface, whereas their amplitudes are related to the degree of scattering or reflection coefficient at each point within the object or at the reference surface. The position of the object surface relative to the reference surface R, z_1, is proportional to f_{z_1}. This frequency difference term is not present in a spectrum from a single pixel, but rather is calculated from two or more pixels imaging the reference and sample at the same x-value (i.e., from the same column of the photodetector array), so that all other terms on the right-hand side of Equation 2.52 are common to both. From the third term of Equation 2.52, we have

$$z_1 = -\frac{2\pi f_{z_1}}{kn_0\dot{\theta}_0 \sin\theta_0}. \tag{2.54}$$

For simplicity, the analysis in Ref. [55] is limited to the case where $z_1(x)$ is a constant. For objects with a surface of arbitrary shape, the refracted angle θ_1 at point F would depend on the coordinates (x, y, z) and the angle between the incident illumination beam and the normal of the object surface to the optical axis at the point where the ray that ends in F intersects the surface.

Once z_1 has been evaluated through Equation 2.54, the position z of a scattering point underneath the object surface can be obtained as

$$z(x, y) = z_1 - \frac{2\pi\,[f(x, y, z) - f(x, y, z_1)]}{\xi kn_0\dot{\theta}_0}. \tag{2.55}$$

The difference between the terms in square brackets is simply $f_z(y, z - z_1)$ (as seen in Figure 2.24b) and depends on the refractive indices of the material and surrounding medium and on the illumination angle through the parameter ξ.

The spectral bandwidth Δf associated with a thickness Δz within the object can be obtained using Equation 2.55 as

$$\Delta f = -\frac{kn_0|\xi|\dot{\theta}_0}{2\pi}\Delta z. \tag{2.56}$$

The amplitude distribution of the scattered light is obtained by mapping the modulation amplitude $|\tilde{I}(m,n,f)|$ from spectrum coordinates (m, n, f) into spatial coordinates (x, y, z) through the relationship between frequency and position shown in Equation 2.55.

2.5.3
Depth-Resolved Displacements

Changes in the value of ϕ at a particular voxel (x, y, z) in the specimen due to load-induced displacements can be calculated, in exactly the same way as for WSI, from the real and imaginary parts of the spectrum $\tilde{I}(m,n,f)$ at a frequency given by Equation 2.55. For the system shown in Figure 2.26, displacements in the y-direction

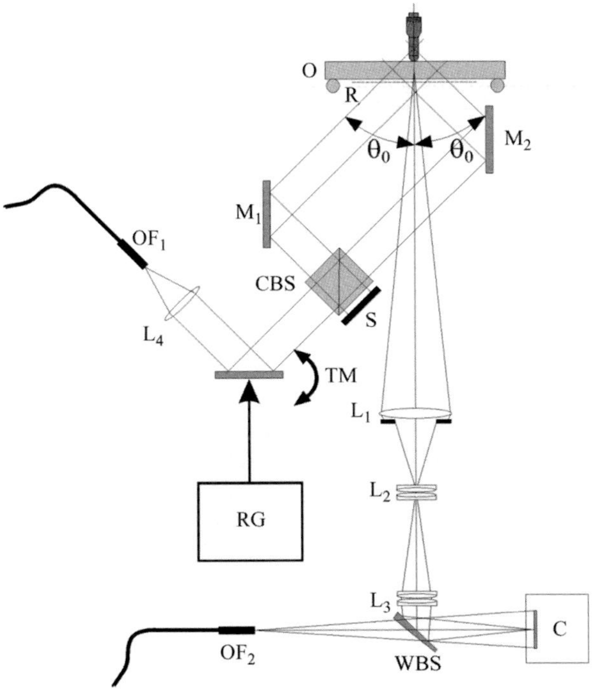

Figure 2.26 Optical setup of a tilt scanning interferometer showing optical fibers OF_1 and OF_2; lenses L_1–L_4; tilting mirror TM; cube beam splitter CBS; mirrors M_1, M_2; alignment screen S; wedge beam splitter WBS; camera C; ramp generator RG; and object O. (Reprinted with permission from Ref. [55].)

cause no phase change due to the illumination geometry. If point F at (x, y, z) moves to a new position F′ with coordinates $(x + u_x, y, z + u_z)$ after deformation, then from Equation 2.46 the phase difference after displacement can be written as

$$\begin{aligned}\phi(x+u_x, y, z+u_z) = \phi(0,0,0) + k\{&[x+u_x(x,y,z)]n_0 \sin\theta_0 \\ &+ [z_1+u_z(x,y,z_1)]n_0(1+\cos\theta_0) \\ &+ [z+u_z(x,y,z)-z_1-u_z(x,y,z_1)]n_1(1+\cos\theta_1)\}\end{aligned} \tag{2.57}$$

While a single illumination direction is sufficient to extract the scattering amplitude distribution and a single displacement component, at least two illumination directions are essential to determine the in-plane (x) and the out-of-plane (z) displacement components. For right and left lateral illuminations, where θ_0 takes the values $\theta_{0R} > 0$ and $\theta_{0L} < 0$, respectively, the phase difference $\Delta\phi\,(x, y, z) = \phi\,(x+u_x, y, z+u_z) - \phi\,(x, y, z)$ due to object deformation is

$$\begin{aligned}\Delta\phi_R(x,y,z) &= k\{u_x(x,y,z)n_0 \sin\theta_{0R} + u_z(x,y,z_1)n_0(1+\cos\theta_{0R}) \\ &\quad + [u_z(x,y,z)-u_z(x,y,z_1)]n_1(1+\cos\theta_{1R})\}, \\ \Delta\phi_L(x,y,z) &= k\{-u_x(x,y,z)n_0 \sin\theta_{0L} + u_z(x,y,z_1)n_0(1+\cos\theta_{0L}) \\ &\quad + [u_z(x,y,z)-u_z(x,y,z_1)]n_1(1+\cos\theta_{1L})\},\end{aligned} \tag{2.58}$$

where $u_z(x, y, z_1)$ is the out-of-plane displacement of point D at (x, y, z_1) to D′ at $(x, y, z_1 + u_z(z_1))$. By choosing $\theta_{0R} = -\theta_{0L} = \theta_0$, the in-plane (x) and out-of-plane (z) phase difference components $\Delta\phi_x(x,y,z) = \Delta\phi_R(x,y,z) - \Delta\phi_L(x,y,z)$ and $\Delta\phi_z(x,y,z) = \Delta\phi_R(x,y,z) + \Delta\phi_L(x,y,z)$ are obtained as

$$\Delta\phi_x(x,y,z) = 2kn_0u_x(x,y,z)\sin\theta_0, \tag{2.59}$$

$$\begin{aligned}\Delta\phi_z(x,y,z) = 2k\{&u_z(x,y,z_1)[n_0(1+\cos\theta_0) - n_1(1+\cos\theta_1)] \\ &+ u_z(x,y,z)n_1(1+\cos\theta_1)\}.\end{aligned} \tag{2.60}$$

The first term within the $\{\ldots\}$ in Equation 2.60 corresponds to the phase change due to a displacement of the object surface relative to the reference surface, while the second one is due to the displacement of point F to F′ along the optical axis. The in-plane and out-of-plane displacements can be extracted from Equations 2.59 and 2.60 as

$$u_x(x,y,z) = \frac{\lambda\Delta\phi_x(x,y,z)}{4\pi n_0 \sin\theta_0}, \tag{2.61}$$

$$u_z(x,y,z) = \frac{\lambda\Delta\phi_z(x,y,z)}{4\pi n_1(1+\cos\theta_1)} - \frac{\lambda\Delta\phi_z(x,y,z_1)}{4\pi}\left[\frac{1}{n_1(1+\cos\theta_1)} - \frac{1}{n_0(1+\cos\theta_0)}\right]. \tag{2.62}$$

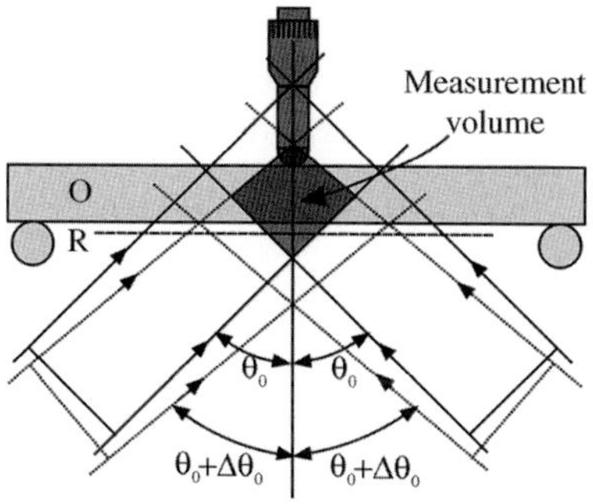

Figure 2.27 The measurement volume is defined by the intersection of the illumination beams during the tilt scan. (Reprinted with permission from Ref. [55].)

2.5.4 Gauge Volume, Depth Range, and Displacement Sensitivity

In a symmetric lateral illumination system setup, the measurement volume is limited by the intersection of the illumination beams during the whole tilt scan (see Figure 2.27). The depth resolution can be defined in the same way as for WSI, that is, the frequency difference Δf between two neighboring peaks has to be at least twice the distance from their centers to their first zero. A rectangular window of duration T results in a sinc function of width $\Delta f = 2/T$, while a Hanning window has a spectral width of $\Delta f = 4/T$. From Equation 2.56, the depth resolution is therefore

$$\delta z = \gamma \frac{\lambda}{n_0 |\xi| \Delta\theta_0}, \tag{2.63}$$

where $\gamma = 2$ and $\gamma = 4$ for rectangular and Hanning windows, respectively. The variation in $|\xi|$ with θ_0 plotted in Figure 2.25 shows that the optimum angle of incidence to minimize δz is $\theta_0 = 50$–$60°$ for typical values of the ratio n_1/n_0. The case $n_0 = n_1$, on the other hand, has an optimal depth resolution of $\theta_0 \rightarrow 90°$. In general, the depth resolution outside the object is better than that within it.

The number of frames N_t required to adequately sample the intensity signal may be calculated, using Equation 2.48, from the total phase change introduced in the wavefront from F as the illumination beam tilts through an angle $\Delta\theta_0$. In order to satisfy the Shannon sampling theorem with at least two samples per cycle, we must have

$$N_t \geq \frac{\Delta\phi(0,0,0)}{\pi} + \frac{k n_0 \Delta\theta_0}{\pi}[x\cos\theta_0 - z_1 \sin\theta_0 - (z - z_1)\xi]. \tag{2.64}$$

The use of off-axis illumination means that, unlike the case with on-axis WSI, the required N_t depends on both x and z. Thus for a given N_t, the allowed depth range depends on the range of x-coordinates of the illuminated region of the sample according to the right-hand side of inequality (2.64).

As with WSI, the displacement resolution σ_z (sometimes called the sensitivity of the technique) is decoupled from the depth resolution δz and depends only on the wavelength of the laser source and the degree of speckle decorrelation. For out-of-plane sensitivity, this is typically better than $\lambda_c/30$.

2.5.5
Experimental Implementation

Proof-of-principle experiments were presented in Ref. [55] in which the tilt scanning technique was used to measure displacements fields within a partially transparent scattering sample. The optical setup is shown in Figure 2.26. A collimated continuous wave (CW) beam is steered by mirror TM mounted on a tilting stage. Left and right illumination beams are obtained with the aid of a cube beam splitter and mirrors M_1 and M_2. TM is tilted by means of a piezoelectric (PZT) actuator controlled by a ramp generator (RG). The imaging system consists of an imaging lens L_1, field and relay lenses L_2 and L_3, respectively, wedged beam splitter WBS, and high-speed camera C. WBS serves to recombine reference and object beams onto the camera sensor. A smooth on-axis reference wave is used rather than a second speckled wave in order to maximize the signal-to-noise ratio.

Figure 2.27 shows a close up of the illumination beams in the region surrounding the object, before and after tilt of the beams by an angle $\Delta\theta_0$. It can be seen that the measurement volume is the intersection of the beams at the beginning and end of the scan range of the tilt angle. Scattering points within this volume give rise to a continuously modulated interference signal throughout the whole tilt scanning sequence.

Results are presented here from a test object consisting of a beam that underwent a three-point bending test as shown in Figure 2.27. A rough reference surface R was placed just in front of the object, so as to cover approximately 20% of the lower portion of the area imaged. As described earlier, this served to compensate for the shift of the peaks along the horizontal axis x and to allow correction of the nonlinear response of the tilting stage at TM. A total of $N_t = 480$ frames were acquired over the tilt angle scanning range $\Delta\theta_0$ of 0.0048 rad. The illumination angle θ_c was 45°; refractive indices n_0 and n_1 took the values 1.0 and 1.4, respectively; and the wavelength λ was 532 nm. From Equation 2.50, the refraction parameter is $\xi = 0.4138$, close to the maximum for this illumination angle and refractive index ratio. The resulting depth resolution was $\delta z \sim 1.1$ mm. The number of frames N_t was sufficient to guarantee a depth range bigger than the object depth $d = 7.8$ mm. Two intensity distributions from the predeformation state were recorded, one from the left and one from the right illumination direction, denoted here by $I_{L_1}(m, n, t)$ and $I_{R_1}(m, n, t)$, respectively. $I_{L_2}(m, n, t)$ and $I_{R_2}(m, n, t)$ represent the corresponding postdeformation data volumes.

Figure 2.28 shows the interference intensity signal from a pixel imaging part of the reference surface (bottom) and from another pixel imaging part of the epoxy resin beam (top). The former shows a single frequency and corresponds to signal coming from a single depth, while the latter shows a more complex wave train due to a mixing

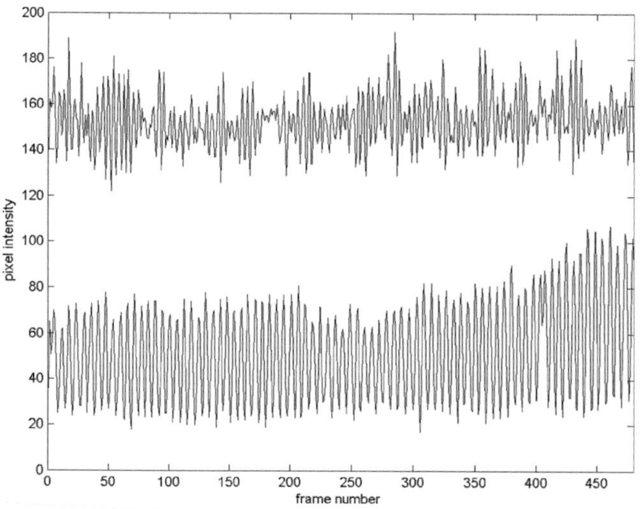

Figure 2.28 Interference intensity signal due to tilt scanning from a pixel on the epoxy resin beam (*top*) and from a pixel on the reference surface (*bottom*). (Reprinted with permission from Ref. [55].)

of frequencies coming from within the thickness of the sample. Figure 2.29 shows the magnitude spectrum representing the amplitude distribution of the scattered waves for left and right illumination directions, obtained along the horizontal axis, x, and averaged along the columns of $\tilde{I}_{L_1}(m, n, f)$ and $\tilde{I}_{R_1}(m, n, f)$ to reduce the noise content. These correspond closely to the scheme shown in Figure 2.24c. The peak due to the reference surface and the band corresponding to scattering points through the whole thickness d of the beam can be clearly seen. Figure 2.29 should be interpreted as a "top view" where both the reference surface and the beam cross section are visible. The noisy appearance on the right-hand side of Figure 2.29a is an artifact of the linearization routine that was used to correct for the nonlinearity of the tilting stage [55].

A 1D Fourier transform along the time axis of the four data volumes I_{L_1}, I_{R_1}, I_{L_2}, and I_{R_2} allows the calculation of the optical phase for each (m, n, f) coordinate in the conjugate spectrum volumes. This results in two 3D phase change volumes $\Delta\phi_L(m, n, f)$ and $\Delta\phi_R(m, n, f)$ corresponding to each illumination direction, the sum of which gives the out-of-plane-sensitive phase change volume and the difference between them gives the in-plane phase change volume. Figure 2.30a shows the wrapped in-plane phase change distribution for different slices within the epoxy resin beam starting at the object surface $z - z_1 = 0$ mm (left) in steps of 1.74 mm down to $z - z_1 = 5.22$ mm (right). The wrapped out-of-plane phase change distributions for the same depth slices are shown in Figure 2.30b. The corresponding in-plane and out-of-plane displacement fields produced by unwrapping the phase of each slice were compared with those predicted by finite element analysis in Ref. [55] and were found to be in good agreement.

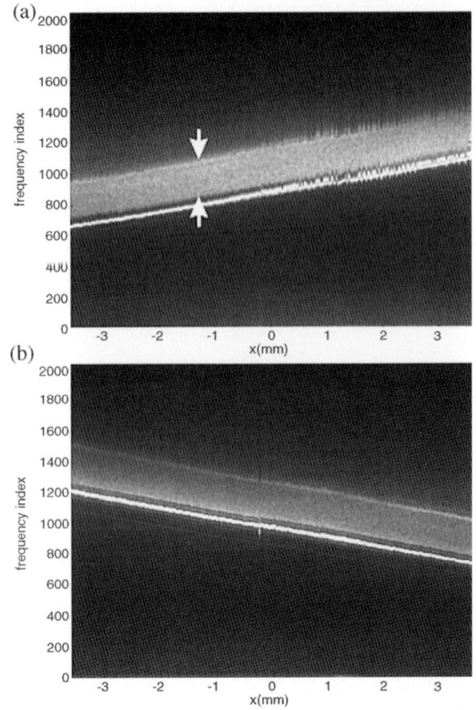

Figure 2.29 Magnitude spectrum, or scattering amplitude distribution, along the horizontal axis x averaged along the columns of (a) left illumination and (b) right illumination. (Reprinted with permission from Ref. [55].)

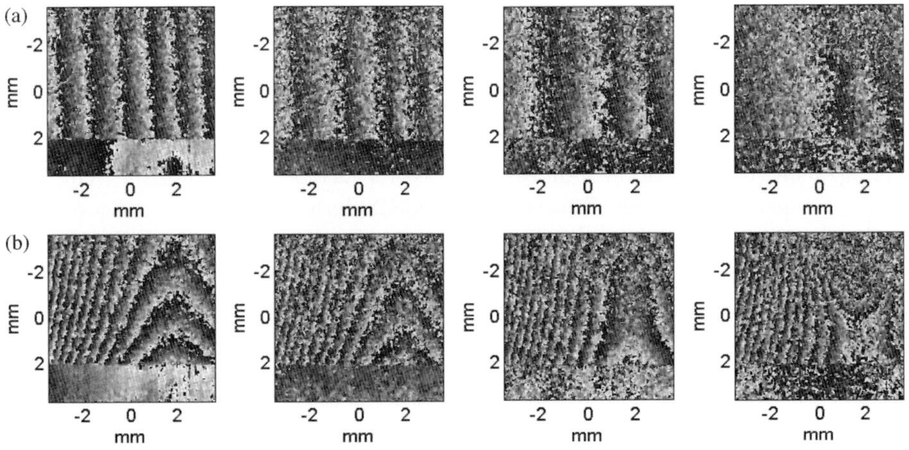

Figure 2.30 In-plane (a) and out-of-plane (b) wrapped phase change distribution for different slices within the beam. Black represents $-\pi$ and white $+\pi$. (Reprinted with permission from Ref. [55].)

2.6
Depth-Resolved Techniques Viewed as Linear Filtering Operations

In this section, we present an alternative model to the "discrete scattering layers" model introduced earlier. While the previous model was based on ray tracing in the spatial domain, the one considered here uses a 3D spatial frequency domain representation. Although some simplifying assumptions are required, the benefit of this model is that it allows one to describe all the techniques within a common unifying framework and to derive the key features such as depth resolution and displacement sensitivity from a simple geometric construction known as the Ewald sphere.

The measurement of 3D structures from interferometric measurement of the scattered field (i.e., holography) was first considered by Wolf [59] and expressed in a frequency space representation by a number of other authors [60, 61]. More recently, a mathematical framework was proposed that considered holography, tomography, and methods of 3D microscopy as linear filtering operations [62, 63] and this approach will be summarized here to compare some of the depth-resolved, displacement measuring systems that are the subject of this chapter.

Consider a weakly scattering, nondispersive material of complex refractive index $n_1'(\mathbf{r}) = n_1 + \Delta n_1(\mathbf{r})$, with small variations $\Delta n_1(\mathbf{r})$ around a uniform index n_1 and $\mathbf{r}$ a position vector in 3D space as shown in Figure 2.31a. A small volume V in this material, which we will refer to as the object, is illuminated by a plane wave $A_i \exp(i\bar{\mathbf{k}}_i \cdot \mathbf{r})$ with wave vector $\bar{\mathbf{k}}_i$ and constant amplitude A_i. The scattered field is measured at a great distance R away from V, along the observation wave vector $\bar{\mathbf{k}}_o$. The overbar symbol on both $\bar{\mathbf{k}}_i$ and $\bar{\mathbf{k}}_o$ is used to denote the fact that, unlike the wavenumbers used elsewhere in the chapter, these are measured in the material, that is,

$$\left|\bar{\mathbf{k}}_i\right| = \left|\bar{\mathbf{k}}_o\right| = \bar{k} = 2\pi/\bar{\lambda}, \tag{2.65}$$

with $\bar{\lambda} = \lambda/n_1$ and where λ is the vacuum wavelength.

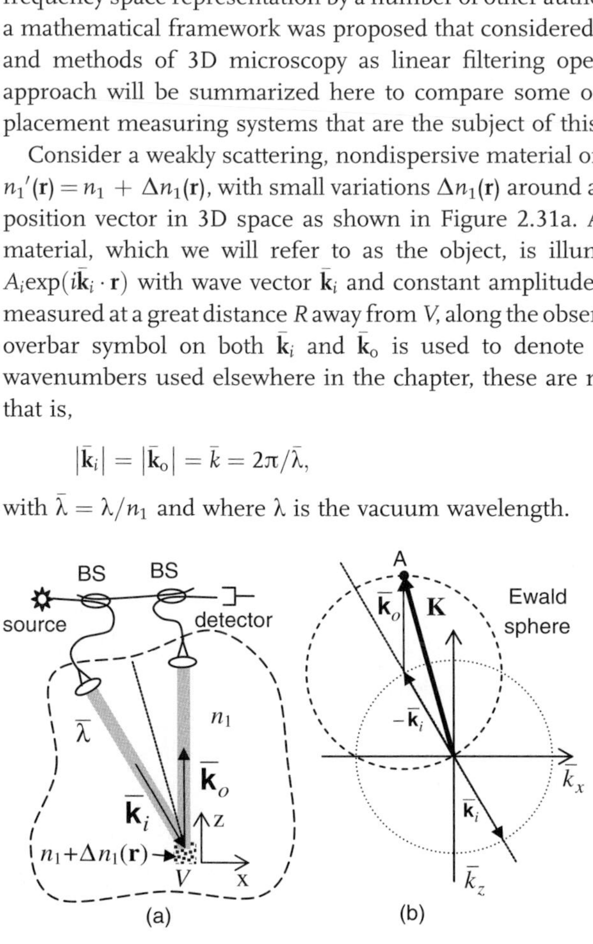

Figure 2.31 Ewald sphere representation of illumination and scattered fields for optical tomography: (a) collimated monochromatic illumination and detection; (b) corresponding scattering vector **K** and Ewald sphere.

Using scalar diffraction theory, and neglecting multiple scattering effects, it has been shown [44, 61, 63] that the measured field U observed at $\mathbf{r} \sim R\bar{\mathbf{k}}_o/|\bar{\mathbf{k}}_o|$ is given by

$$U(\mathbf{K}) = -\frac{A_i}{4\pi}\frac{\exp(i\bar{k}R)}{R}\int_V \Phi(\mathbf{r})\exp(-i\mathbf{K}\cdot\mathbf{r})\mathrm{d}^3\mathbf{r}, \tag{2.66}$$

where

$$\mathbf{K} = \bar{\mathbf{k}}_o - \bar{\mathbf{k}}_i \tag{2.67}$$

is the *scattering vector* defined in terms of the observation and illumination wave vectors, and $\Phi(\mathbf{r})$ is the *scattering potential* that represents the object microstructure within V:

$$\Phi(\mathbf{r}) = -(\bar{k}^2/n_1^2)\,[2n_1\Delta n_1(\mathbf{r}) + \Delta n_1^2(\mathbf{r})]. \tag{2.68}$$

Equation 2.66 is a simplified version of the well-known theorem of diffraction tomography [64]. The scattered wave is seen to be a spherical wave $\exp(i\bar{k}R)/R$ with a complex amplitude that is given by a volume integral of the scattering potential weighted by a phase factor $\exp(-i\mathbf{K}\cdot\mathbf{r})$. This phase fractor can be thought of as resulting from the variation of optical path length with position in the sample that has already been derived in Section 2.3.1.5.

The significance of Equation 2.66 is that the measured amplitude is proportional to one component of the 3D Fourier transform of the scattering potential. Physically, the microstructure can be regarded as a superposition of many sets of 3D parallel sinusoidal fringes known in holography and crystallography as Bragg planes. U measures the amplitude of the particular set of Bragg planes that have their normal along $\mathbf{K}$ (i.e., along the bisector of the illumination and observation directions) and have a pitch of $2\pi/|\mathbf{K}|$. In what follows, we will use the term "**K** space" to indicate the 3D Fourier transform of the scattering potential for a given sample microstructure. A scattering vector or set of scattering vectors, defined by Equation 2.67, can be drawn on a **K**-space diagram to indicate which Fourier components are measurable by the interferometer.

2.6.1
Methods Viewed as Linear Filtering Operations

In principle, if one were to measure $U(\mathbf{K})$ on a sufficiently fine and extended 3D mesh of $\mathbf{K}$ values, Equation 2.66 shows that the required spatial variation in scattering potential could be computed as

$$\Phi(\mathbf{r}) \propto \Im^{-1}[U(\mathbf{K})], \tag{2.69}$$

where $\Im^{-1}$ is the 3D inverse Fourier transform operator. In practice, however, one can access only a limited subset of $U(\mathbf{K})$. For example, if the illumination consists of a single plane wave of wave vector $\bar{\mathbf{k}}_i$, all Fourier components of the scattering potential accessible by scattering at different observation directions $\bar{\mathbf{k}}_o$ are limited to those located on the surface of a sphere described by the arrowhead of the scattering vector **K**. This is known as the Ewald sphere for the specific wavelength $\bar{\lambda}$; it has radius $\bar{k}$ and is centered at $\bar{\mathbf{k}} = -\bar{\mathbf{k}}_i$ (see Figure 2.31b). If a single illumination and observation direction is used, for example, in speckle interferometry with collimated illumination and observation, then a single point in **K** space is obtained, labeled A in Figure 2.31b, which represents a single spatial frequency of the scattering potential. Such an isolated Fourier component provides no depth-resolving capability.

The region of **K** space that is accessible by a given measurement technique can be specified by a window function $W(\mathbf{K})$ that is nonzero only wherever a valid measurement can be made. The reconstructed scattering potential is then given by

$$\Phi'(\mathbf{r}) \propto \Im^{-1}[U(\mathbf{K}) \cdot W(\mathbf{K})]. \tag{2.70}$$

The relationship between $\Phi'(\mathbf{r})$ and the true scattering potential $\Phi(\mathbf{r})$ follows from the convolution theorem as

$$\Phi'(\mathbf{r}) \propto \Phi(\mathbf{r}) \otimes H(\mathbf{r}), \tag{2.71}$$

where $\otimes$ denotes convolution and where $H(\mathbf{r})$ is the 3D impulse response or point spread function of the measurement apparatus that is related to $W(\mathbf{K})$ as follows:

$$H(\mathbf{r}) \propto \Im^{-1}[W(\mathbf{K})]. \tag{2.72}$$

Equations 2.70–2.72 express the reconstruction problem as a linear, shift invariant, filtering operation [62, 63]. $H(\mathbf{r})$ is a key function for all the depth-resolving techniques since it determines both their spatial resolution and their sensitivity. These two aspects are now considered individually in Sections 2.6.2 and 2.6.3, before moving on to consider how to evaluate $H(\mathbf{r})$ for two specific examples (WSI and TSI) in Sections 2.6.4–2.6.6.

2.6.2
Relationship Between W(K) and Spatial Resolution

The relationship between $W(\mathbf{K})$ and $H(\mathbf{r})$ given by Equation 2.72 is illustrated in Figure 2.32. In this example, $W(\mathbf{K})$ is nonzero only near to the **K**-space origin, which physically corresponds to measurements made close to the forward scattering

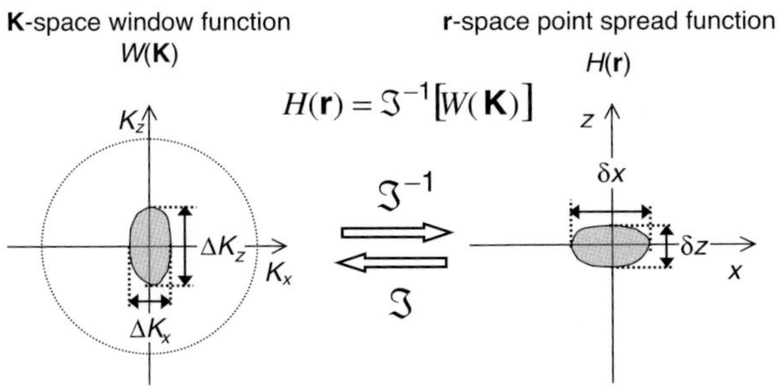

Figure 2.32 Relationship between the **K**-space window function $W(\mathbf{K})$ and real-space point spread function $H(\mathbf{r})$.

direction. If we denote the characteristic dimensions of W along the K_x, K_y, and K_z directions by ΔK_x, ΔK_y, and ΔK_z, respectively, the spatial extent of the point spread function along the x-, y-, and z-axes, that is, the resolution of the tomographic imaging system, is given to a first approximation by

$$\delta x = \gamma \frac{2\pi}{\Delta K_x}, \quad \delta y = \gamma \frac{2\pi}{\Delta K_y}, \quad \delta z = \gamma \frac{2\pi}{\Delta K_z}. \tag{2.73}$$

The constant γ reflects the influence of the shape of the window function. For example, if $W(\mathbf{K})$ is a rectangular cuboid taking the value 1 inside the cuboid and 0 outside, then $H(\mathbf{r})$ is a product of sinc functions along each of the three orthogonal spatial axes. Equation 2.73 with $\gamma = 2$ then defines the full width of the point spread function along each axis as measured between the zero crossing points. Although more precise values for the axial and lateral resolutions can be obtained by direct application of Equation 2.72, the inherent simplicity of Equation 2.73 is attractive when comparing different depth-resolving techniques, or when optimizing their performance.

2.6.3
Relationship Between W(K) and Displacement Sensitivity

The point spread function $H(\mathbf{r})$ is in general a complex function of position. Figure 2.33 illustrates this point with the window function $W = W_1(\mathbf{K})$ from Figure 2.32, in which the real and imaginary parts of $H = H_1(\mathbf{r})$ are displayed along

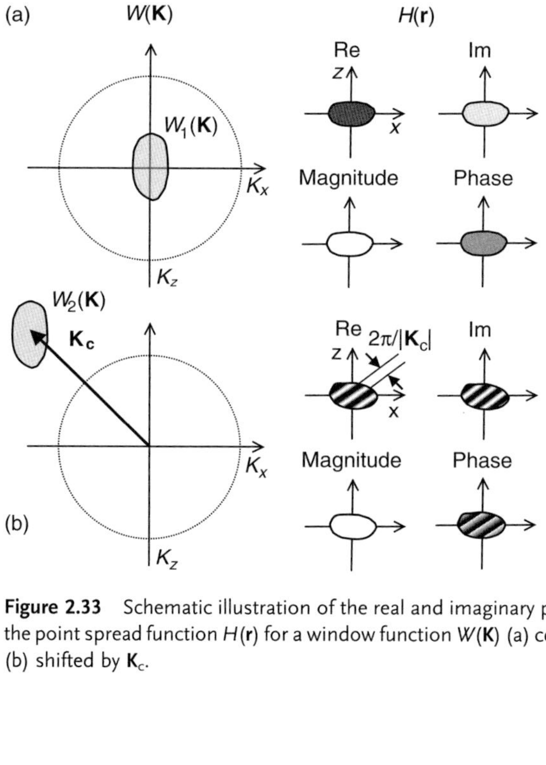

Figure 2.33 Schematic illustration of the real and imaginary parts, and magnitude and phase, of the point spread function $H(\mathbf{r})$ for a window function $W(\mathbf{K})$ (a) centered on the origin of $\mathbf{K}$ space and (b) shifted by $\mathbf{K}_c$.

with the equivalent representation in terms of magnitude and phase. Of direct interest for depth-resolved displacement measurement is the phase variation within the point spread function since this dictates the measured phase change at a given point in the tomographic reconstruction as a scattering point moves. The phase gradient in Figure 2.33a is, however, very low because all the measured frequency components are clustered around the origin of **K** space. As a result, measurements made with any interferometric depth-resolving technique close to the forward-scattering direction will have poor displacement sensitivity.

Suppose the interferometer is modified to allow it to measure a different region of **K** space, defined by a second window function, $W = W_2(\mathbf{K})$, that is identical to W_1 except for a translation by the vector $\mathbf{K}_c$ as shown in Figure 2.33b, that is,

$$W_2(\mathbf{K}) = W_1(\mathbf{K} - \mathbf{K}_c). \tag{2.74}$$

By the Fourier shift theorem, the resulting point spread function $H = H_2(\mathbf{r})$ is related to H_1 as follows:

$$H_2(\mathbf{r}) = \exp(i\mathbf{K}_c \cdot \mathbf{r}) H_1(\mathbf{r}). \tag{2.75}$$

The dimensions of H_2 are identical to H_1, and therefore the spatial resolutions of the original and modified interferometers are identical to one another. As a result of the $\exp(i\mathbf{K}_c \cdot \mathbf{r})$ term in Equation 2.75, however, the real and imaginary parts of H_2 are now modulated by parallel sinusoidal fringes orientated normal to $\mathbf{K}_c$ and with spacing $2\pi/|\mathbf{K}_c|$, as shown in Figure 2.33b. The corresponding phase distribution within H_2 has a gradient of 2π per cycle of these fringes. The tomographic reconstruction of a scattering point will therefore show a phase change of 2π for a movement of $2\pi/|\mathbf{K}_c|$ along the $-\mathbf{K}_c$ direction. This should come as no surprise once it is realized that the *scattering vector* **K** defined by Equation 2.67 and the *sensitivity vector* defined by Equation 2.34 are essentially the same quantity, related through

$$\mathbf{K} = -\mathbf{S}. \tag{2.76}$$

The beauty of the Ewald sphere construction for interpreting depth-resolved displacement field measurements is that it shows in a simple pictorial way

- the spatial frequencies of the sample microstructure that may be imaged;
- the spatial resolution along all three axes (from the window function shape);
- the displacement component (from the direction of $\mathbf{K}_c$); and
- the displacement sensitivity (from the magnitude of $\mathbf{K}_c$).

In the next two sections, we show the Ewald sphere construction for two of the techniques presented earlier in the chapter, namely, wavelength scanning and tilt scanning interferometry.

2.6.4
Ewald Sphere for a Wavelength Scanning Interferometer

A wavelength scanning interferometer and its Ewald spheres are shown schematically in Figure 2.34a and b. If the illuminating beam is collimated vertically

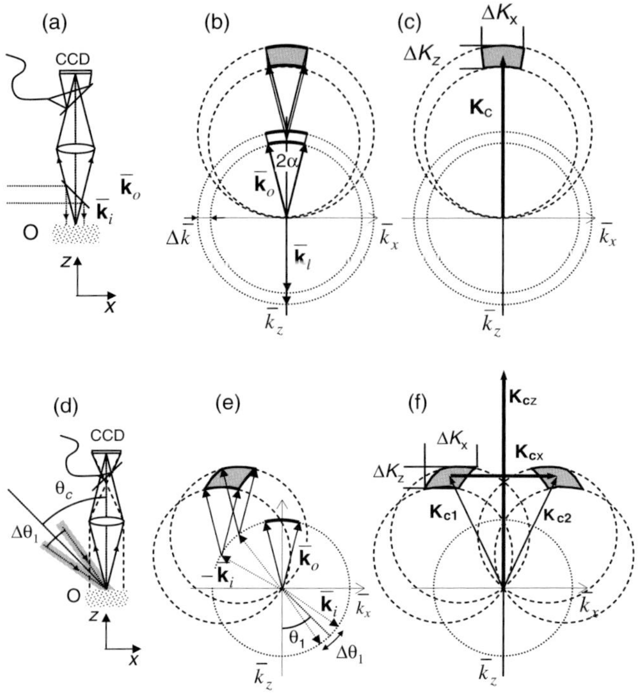

Figure 2.34 (a) Optical setup, (b) Ewald sphere, (c) window function and sensitivity vector for wavelength scanning interferometry, (d) optical setup, (e) Ewald sphere, and (f) window function and dual (in-plane and out-of-plane) sensitivity vectors for tilt scanning interferometry.

downward, then $\bar{\mathbf{k}}_i$ is aligned along the $-z$-direction. In common with all far-field optical instruments, only a fraction of the scattered field is collected by the entrance pupil of the interferometer, defined by the object-space numerical aperture of the system $N_A = n_1 \sin(\alpha)$, where α is the half angle subtended by the cone of rays accepted by the aperture from a point in the object. The measurable $\bar{\mathbf{k}}_o$ for this interferometer at the start of a wavelength scan, therefore, covers a range that is represented by the bold curve on the inner continuous circle of Figure 2.34b. The allowable $\mathbf{K}$ vectors are produced by adding this range of $\bar{\mathbf{k}}_o$ to the constant $-\bar{\mathbf{k}}_i$ resulting in the bold curve on the inner dashed circle. In three dimensions, the allowable $\mathbf{K}$ vectors lie on a cap at the top of the Ewald sphere.

As the wavelength scan proceeds, the bold curve sweeps out the region of $\mathbf{K}$ space shown in gray in Figure 2.34b, which is therefore the window function $W(\mathbf{K})$. The scattering vector $\mathbf{K}_c$ is shown in Figure 2.34c. From the orientation of $\mathbf{K}_c$, this interferometer measures the out-of-plane displacement component u_z. The magnitude of $\mathbf{K}_c$ ($|\mathbf{K}_c| = 4\pi n_1/\lambda_c$) shows that the displacement sensitivity is $\lambda_c/2n_1$ per 2π phase change, in agreement with Equation 2.35.

Figure 2.34c also shows the window function dimensions ΔK_x and ΔK_z. ΔK_z is $2\Delta\bar{k} = 2n_1\Delta k$. Using the relations in Equation 2.73, the axial resolution is therefore

$\delta z = \gamma\pi/n_1\Delta k \approx \gamma\lambda_c{}^2/2n_1\Delta\lambda$, in agreement with the value derived previously and given in Equation 2.32. ΔK_x and ΔK_y both take the value $2\bar{k}\sin\alpha = 2n_1 k \sin\alpha$, and therefore the lateral resolution is $\delta x = \delta y = \gamma\pi/n_1 k_c \sin\alpha = \gamma\lambda_c/2n_1\sin\alpha$ that is, within a numerical factor, the usual Rayleigh criterion for the diffraction-limited lateral resolution of an imaging system [65].

2.6.5
Ewald Sphere for a Tilt Scanning Interferometer

An optical layout and corresponding Ewald sphere for the tilt scanning interferometry technique are shown schematically in Figure 2.34d and e. This uses the same normal observation direction as used for the WSI system, but has an off-axis illuminating beam that changes direction during the scan. There is now a range of $\bar{\mathbf{k}}_i$ directions, but the length of the $\bar{\mathbf{k}}_i$ vector remains unchanged because of the fixed wavelength. The Ewald sphere of constant radius therefore pivots around the origin of $\mathbf{K}$ space as shown in Figure 2.34e. As with WSI, the allowable $\mathbf{K}$ vectors lie on a cap at the top of the Ewald sphere, and trace out the distorted diamond-shaped window function that is shaded gray as the sphere rotates.

The sensitivity vector for this illumination direction, $\mathbf{K}_{c1}$, is shown in Figure 2.34f. The second sensitivity vector, $\mathbf{K}_{c2}$, is produced with a tilt scan again centered on an angle θ_c, but this time to the right of the observation direction. The sum and difference of the resulting phase maps gives rise to displacement fields with sensitivity vectors $\mathbf{K}_{cz}$ and $\mathbf{K}_{cx}$, respectively.

2.6.6
Comparison of Spatial Resolution for WSI and TSI

It is instructive to compare the effect on the spatial resolution of the two techniques produced by varying the center illumination direction. Simple analytical expressions for the dimensions of the window function's bounding box were given in Ref. [63] and are plotted in Figure 2.35 as a function of θ_c after normalization by the center wavenumber k_c. A numerical aperture of 0.08 was used for both WSI and TSI, with $n_1 = 1$. A total fractional wavenumber change of 0.1 was assumed for the WSI case (Figure 2.35a), and an angular tilt range of 0.2 rad for the TSI results (Figure 2.35b).

In both cases, the continuous line representing ΔK_z approaches zero for $\theta_c = 180°$. The equivalent spatial domain point spread function has $\delta z \to \infty$ by Equation 2.73, showing that neither approach can resolve axially in the forward-scattering direction. The situation is different in the backscattering direction ($\theta_c = 0°$), however, where the TSI resolution also tends to infinity, but where the WSI resolution is at its optimum. The optimal depth resolution for TSI, by contrast, is at $\theta_c = 90°$. This observation is consistent with the results from the spatial domain model (Equation 2.63 and Figure 2.25) that shows that depth resolution is minimized at $\theta_c = 90°$ for the case where $n_1/n_0 = 1$. The Ewald sphere construction for the case $\theta_c = 90°$ shows that $\Delta K_z \approx \bar{k}\Delta\theta_0$ and hence the predicted depth resolution is, using Equation 2.73,

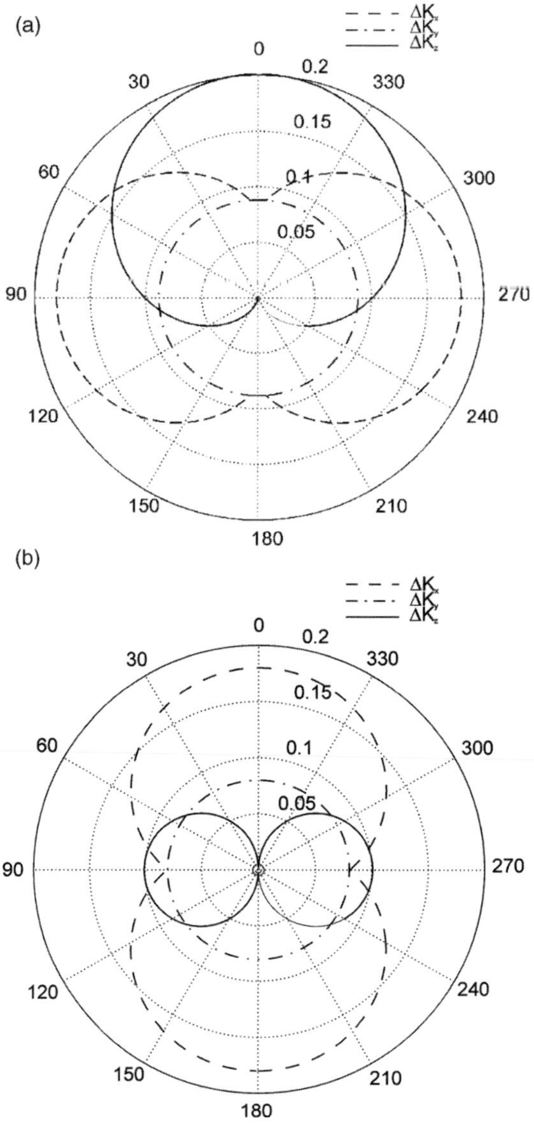

Figure 2.35 Dimensions of the window function normalized by $\bar{k}$ of (a) a WSI system and (b) a TSI system as a function of the illumination angle θ_1 in the xz plane. Numerical aperture $N_A = 0.08$ for both cases; $n_1 = 1$; $\Delta\bar{k}/\bar{k}_c = 0.1$ (a) and $\Delta\theta_1 = 0.2$ rad (b).

$\delta z = \gamma\lambda/n_1\Delta\theta_1$. As with WSI, this is again in agreement with the prediction of the spatial domain model given by Equation 2.63. Substitution of typical values $\gamma = 2$, $n_1 = 1.4$, $\lambda = 0.7$ μm, and a potential scan range of $\Delta\theta_1 = 0.2$ rad gives a theoretical depth resolution of 5 μm.

2.7 Phase Unwrapping in Three Dimensions

The techniques described in the previous sections of this chapter provide phase maps in two or three dimensions whose values are wrapped onto the range $-\pi$ to π. These require unwrapping, that is restoration of the unknown integral multiple of 2π to each pixel, in order to obtain the corresponding component of the 2D or 3D displacement field. Lower dimensional phase unwrapping algorithms are now relatively well established in the literature. The overview in Ref. [6], for example, covers 1D and 2D unwrapping and the textbook by Ghiglia and Pritt [66] provides a detailed description of some solutions to the 2D phase unwrapping problem. Three-dimensional unwrapping is, however, a more recent development.

It turns out that phase unwrapping in 3D can be simpler than in 2D and more robust than in either 1D or 2D. In this section, we explain why this is the case using the concept of "phase singularity loops (PSL)" that was introduced in Ref. [67] and developed further in Refs [68–72]. A few alternative approaches to the 3D problem have also been proposed over the past few years. For example, an *N*-dimensional phase unwrapping method is based on minimizing a cost function that measures the difference in phase values at the interface of adjacent regions containing no phase jumps [73]. Extension to three dimensions of the classical two-dimensional minimum L^p-norm framework has also been attempted and implemented using an iterative Monte Carlo algorithm [74].

2.7.1 Phase Singularities in Two-Dimensional Phase Data

Phase unwrapping is a trivial process on data that are free of noise and have been sampled in accordance with the Shannon sampling theorem; however, practical data sets almost always contain regions that are noisy or subsampled leading to "singular" points in a 2D phase field. The effect of such singular points on the unwrapping process is illustrated in the wrapped two-dimensional phase map of Figure 2.36, which corresponds to the case of four horizontal fringes corrupted by noise near the center of the field. The number of 2π phase jumps required to unwrap point Q, given the phase at P, is path-dependent: path A crosses three phase jumps whereas path B crosses two. The presence of the two phase singularities, points 1 and 2, caused by local undersampling of the phase map, can therefore result in the global propagation of large (2π) phase errors.

While from a rigorous mathematical viewpoint, it is impossible to recover the unwrapped phase unambiguously in the presence of singularities, additional *ad hoc* assumptions can often be made to come to a solution that is acceptably close to the true unwrapped phase in many situations. A large variety of different two-dimensional unwrapping algorithms have been developed to reduce the influence of the singular points [66]. They can be broadly classified into path-following methods, which guide the unwrapping path in order to prevent error propagation, and minimum norm methods, which seek to minimize a cost function that measures

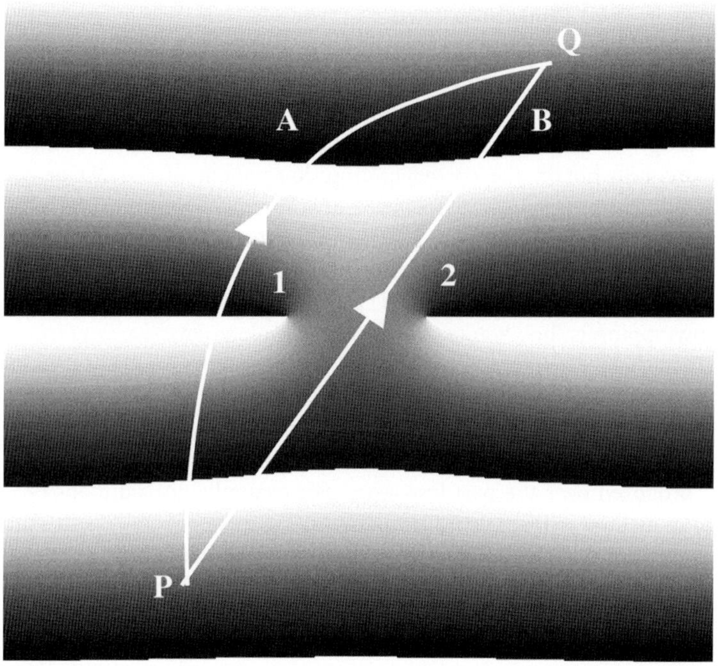

Figure 2.36 Wrapped phase map (black and white representing, respectively, $-\pi$ and $+\pi$) containing two singular points 1 and 2 results in path-dependent unwrapped phase at point Q. (Reprinted with permission from Ref. [79].)

the difference between the gradients of the original wrapped phase and the estimated unwrapped phase. The minimum norm methods have some drawbacks such as requiring the use of iterative solution methods and in certain cases introducing systematic errors to the reconstructed phase field [66].

Among the path-following methods, one of the most direct is to place branch cuts within the phase map between pairs of singular points of opposing signs to make the unwrapped results independent of the unwrapping path [66]. The singular points are identified by counting the number of 2π phase jumps around all possible 2×2 pixel squares in the image (see Figure 2.37a). For a given square, the phase values at the corner pixels, ordered in an anticlockwise sense, are denoted ϕ_1, ϕ_2, ϕ_3, and ϕ_4. The number of phase jumps is then

$$s = \Delta_{21} + \Delta_{32} + \Delta_{43} + \Delta_{14}, \tag{2.77}$$

where

$$\Delta_{kl} = \text{NINT}\left(\frac{\phi_k - \phi_l}{2\pi}\right), \tag{2.78}$$

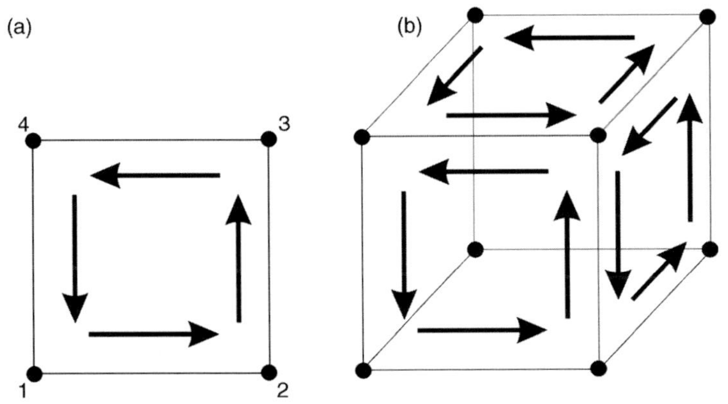

Figure 2.37 Unwrapping paths for calculation of (a) the number of singularities within a 2×2 pixel square (2D unwrapping) and (b) the singularity distribution on three adjacent faces of a $2 \times 2 \times 2$ voxel cube (3D unwrapping).

where NINT denotes rounding to the nearest integer. A positive or $+1$ singularity is present in the square if $s = +1$, and a negative or -1 singularity is present if the result is $s = -1$. The only other possible resulting value is 0, in which case no singularity is present. The term singularity (as opposed to the terms "pole" and "residue" that are also used in the literature) is used here for consistency with [67]. In the example shown in Figure 2.36, point 1 would be joined to point 2 by the cut and path B would not be allowed. All paths not crossing the branch cut then agree on the integral multiple of 2π to be added to the phase at Q.

2.7.2 Phase Singularity Loops in Three-Dimensional Phase Data

In three dimensions, the equivalent of the 2D elemental square of 2×2 pixels is an elemental cube of $2 \times 2 \times 2$ voxels (see Figure 2.37b). The presence of a singularity s_i on a given face i of the cube is detected by summing the phase jumps that are crossed as one follows a closed path along the four edges of the cube face using Equations 2.77 and 2.78.

Now consider the sum of the singularities over the six faces of the cube,

$$S = \sum_{i=1}^{6} s_i. \tag{2.79}$$

For every contribution Δ_{kl} to S from face i, there is a contribution Δ_{lk} to S from the face that shares the common edge between voxels k and l. In Figure 2.37b, this is illustrated for three faces by the opposing directions of the unwrapping arrows on their adjacent edges. However, by definition (Equation 2.78), $\Delta_{kl} = -\Delta_{lk}$ and therefore S must be identically equal to zero.

The fact that S is zero means that a singularity s_i on a given face i of the cube must be balanced by a singularity of opposite sign $s_j = -s_i$ on one of the other five faces of the cube, denoted j. Face j is shared with a neighboring cube of $2 \times 2 \times 2$ voxels, and as calculated on this neighbor's face (for which the outward normal is in the opposite direction to face j on the first cube), the singularity therefore takes the value s_i from face i of the first cube.

The argument can then be repeated as above for the second cube, then for the third, and so on, the process terminating only when we reach either face i of the first cube again or the boundary of the phase volume. A list of these cubes can be regarded as defining a container for a line in space that we call a phase singularity loop. Each cube in the list contains a face with a -1 singularity and the second face with a $+1$ singularity, both measured in the outward normal direction. The -1 singularity corresponds to a $+1$ singularity measured in the inward normal direction and can be thought of as the face through which the loop "enters" the cube, whereas the $+1$ singularity occurs on the face through which the loop "leaves" the cube.

This simple picture leads to two interesting conclusions. First, a 2D cross section through a phase volume with PSLs will automatically generate dipole pairs of $+1/-1$ singular points. Figure 2.38a shows an example: the arrows on the PSL indicate the direction that the thumb of the right hand would point in as it follows the loop through the volume. Any plane intersecting this loop will contain two singular points, one with the thumb pointing into the plane, the other with the thumb pointing out. The singular points therefore form a $+1/-1$ pair. One of the arguments sometimes made against the branch cut method in two dimensions is that the pairing of the singular points, when placing the cuts, is somewhat arbitrary [75]. There is no way of knowing which singular point should be paired with which, thereby requiring the use of statistical approaches. In one example, the maximum-likelihood pairing was shown to involve a global minimization of the sums of the squares of the cut lengths [76]. With the above interpretation, however, it is clear that each singular point is, in fact, linked to another singular point by a line that happens to be invisible to an observer confined to the 2D surface. The computational effort involved in the

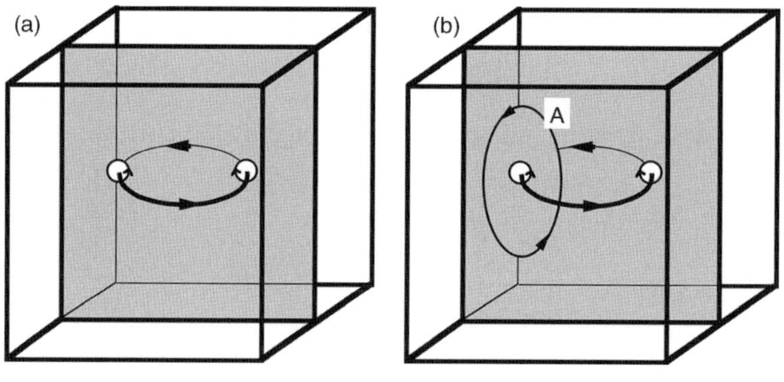

Figure 2.38 (a) Example of the formation of a pair of singular points within a 2D phase map through the intersection of the plane (shaded) with a phase singularity loop. (b) Closed path A linking a PSL encircles a single phase singularity when represented on a surface. (Reprinted with permission from Ref. [67].)

global minimization, which takes most of the time in a 2D phase unwrapping, is therefore eliminated in the 3D case.

The second conclusion that follows from this viewpoint is that a closed unwrapping path that is linked with a PSL will automatically have a nonzero total number of 2π phase jumps. "Linked" means that, like two links in a chain, the closed path and the PSL cannot be separated without the two curves passing through one another. This result can be seen by constructing a surface that contains the complete closed unwrapping path (see, for example, Figure 2.38b). Such a surface will intersect the PSL, generating a dipole pair of singular points, only one of which will be enclosed within the loop. This conclusion is important since it shows that in order to create a path-independent phase unwrapping algorithm in three dimensions, one must place branch surfaces across all PSLs so as to prevent potential unwrapping paths from passing through them.

2.7.3
3D Phase Unwrapping Algorithm

A path-independent algorithm applicable to three-dimensional phase distributions follows directly from the concept of PSLs outlined in the previous section. The following description is a summary of [67] that has since been extended in Refs [68–70]. Three basic steps are required: first, the identification of the loops; second, the placing of branch cut surfaces to prevent unwrapping paths passing through the loops; and, finally, the unwrapping itself.

The first step involves calculating the singular point distributions along each of the planes that are normal to each of the three orthogonal axes. The list of singular points is then sorted into loops using local neighborhood rules, based on the fact that a phase singularity entering one face of a cube element must leave through another face of the same cube.

Step 2 involves creating arrays of flags to specify whether a potential link in a valid unwrapping path exists between a given voxel and its six nearest neighbors. Initially, all potential links are allowed, but the links within a loop are removed by placing an imaginary loop that shadows the PSL, and then shrinking the shadow loop down to zero size. As the loop shrinks, it passes through successive links. Each time it passes through a link, the corresponding flag is set to 1, denoting removal of the link from the list of valid unwrapping paths. A useful physical analogy is that of a highly elastic rubber band under tension (representing the shadow loop) snaking through a cubic lattice of weak matchsticks (representing the initial set of links). As the band shrinks under the tension, it cuts through successive sticks, which can then no longer act as part of a valid unwrapping path. Any conceivable closed unwrapping path linking a PSL will be broken by the shrinking shadow loop, which, as stated earlier, is required in order to guarantee path independence of the unwrapped phase map.

The third step is implemented by a flood-fill type algorithm, just as for the two-dimensional branch cut algorithms [66]. A queue is used to store the coordinates of voxels that have been unwrapped and that have nearest neighbors that have not yet been unwrapped. Each entry in the queue contains the three coordinates of the unwrapped voxel, and the three coordinates of a nearest neighbor that is connected to the unwrapped voxel by a valid link. Each time an entry is removed from the queue, the phase of the neighbor is unwrapped by bringing the phase difference between the two voxels into the range $(-\pi, \pi)$, and its nearest neighbors are examined in turn. Those that are joined by valid links to the first nearest neighbor, and have also not yet been unwrapped, are added to the queue. This process is repeated until the queue is empty.

2.7.4 Remaining Ambiguities

Although the concept of phase singularity loops removes one of the main ambiguities encountered in two-dimensional unwrapping – that is, how to pair the $+1/-1$ singular points – there are still two remaining questions that cannot be answered unambiguously from the data alone. First, the line defined by the PSL does not define uniquely the position of the branch cut surface. It is often possible to find (at least) two different surfaces for a given PSL that have the same area, so that a "minimum area" criterion does not help distinguish between them. Although such surfaces produce the same global unwrapped phase distribution, the results differ by 2π at some of the voxels in the neighborhood of the PSL. Additional knowledge of the physics of the problem being analyzed (see, for example, Ref. [70]) can be used to reduce these local unwrapping errors.

The second ambiguity arises from the possibility that two (or even three) PSLs pass through a single elemental cube of $2 \times 2 \times 2$ voxels, forming a so-called knot point. Knot points lead to ambiguities because when a PSL enters the cube, it is not clear which of the two exiting PSLs it should be joined to. One approach to solving this problem was developed in Ref. [68] in which a recursive tree structure was implemented that builds all possible loop constellations and chooses the set with the shortest loops. A related problem can occur when two or more loops leave the surface of the phase volume because it is not clear whether any such loop reenters the volume elsewhere. Optimization techniques can be used to choose the pairing that minimizes a cost function based on the total branch cut surface area [74] or the total distance between the ends of the loops [77].

2.7.5 Example: Dynamic Deformation of Carbon-Fiber Composite Panel

An example of the behavior of spatial and temporal unwrapping algorithms in a demanding situation is shown in Figure 2.39. Data from a speckle interferometer have been deliberately undersampled along both the temporal and the spatial axes to illustrate how these two approaches start to fail as the relevant Nyquist limit is approached. The wrapped phase (top row) encodes the out-of-plane displacement component from the front surface of a carbon–fiber composite panel that contains a subsurface delamination crack. Four successive frames show the increasing deformation as air pressure is progressively removed from the top surface of the panel.

(a)
(b)
(c)
(d)
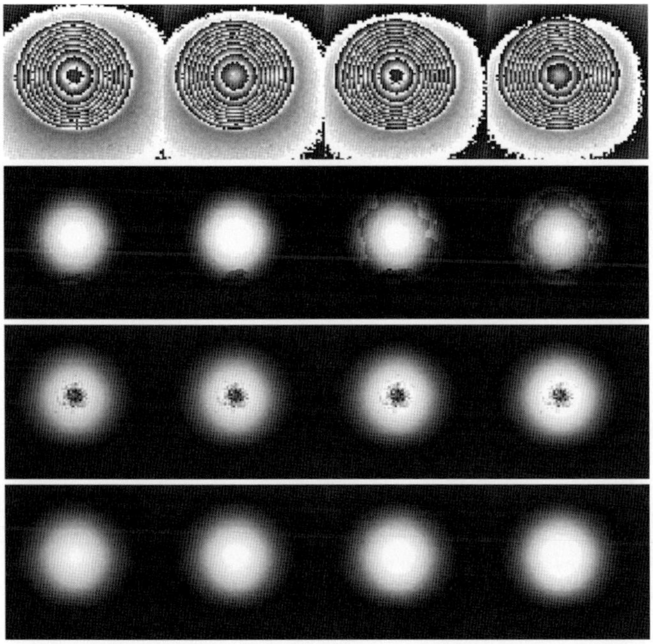

Figure 2.39 Four successive phase maps, measured using speckle interferometry, from a carbon–fiber sample containing a subsurface delamination crack. Wrapped phase (a), result of unwrapping using a branch cut-based 2D spatial unwrapping algorithm (b), temporal unwrapping algorithm (c), and full 3D unwrapping algorithm (d).

Figure 2.39b–d compare the unwrapped phase distributions produced by (i) a two-dimensional branch cut algorithm based on minimizing the sums of the squares of the cuts [76], (ii) a one-dimensional temporal unwrapping, and (iii) a full 3D unwrapping using the branch cut surface concept outlined above. The increased dimensionality of the data analysis procedure has resulted in a clear improvement in the quality of the recovered phase distribution in this particular application.

The reason for the partial failure of the other two algorithms can be seen by examining the distribution of phase singularity loops (see Figure 2.40). The mushroom-shaped cloud of PSLs consists of a central "stalk" where the loop normals are oriented primarily parallel to the t-axis. This loop orientation is indicative of temporal undersampling and so occurs most frequently where the sample is moving the fastest. The temporally unwrapped results therefore show errors mostly near the center of the field. The "head" of the mushroom, on the other hand, contains loops whose normals lie predominantly in the horizontal (m, n) plane. Such loops indicate the presence of spatial undersampling, which is most likely to occur in the regions where the steepest gradient in the deformed surface occurs. The spatially unwrapped data therefore show errors approximately midway between the regions of maximum and the regions of minimum deflections.

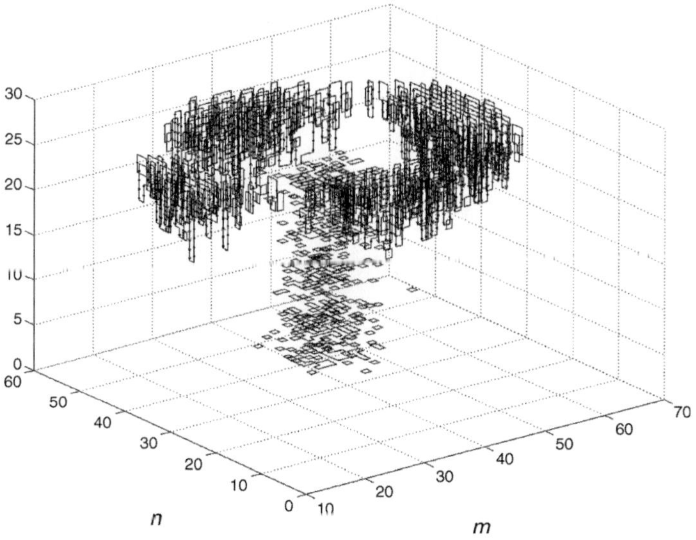

Figure 2.40 Phase singularity loops for the full time-varying phase distribution from which the subset of four frames shown in Figure 2.39 was taken. The vertical axis is frame number.

2.8 Concluding Remarks

In this chapter, we have summarized some of the main interferometric methods for measuring depth-resolved displacement fields within weakly scattering media. As with traditional speckle interferometry, these methods use the phase information in the scattered light, which is measured using standard phase shifting techniques, to encode the displacement component along the sensitivity vector. The additional feature of the techniques that provides the depth-localizing capability is the use of measurements at multiple wavevectors, which is achieved either by changing the magnitude of the wavevector (WSI, PC-SOCT) or by changing its direction (TSI). Although the spatial resolution and displacement sensitivity of these two classes of techniques can be individually analyzed by geometrical ray tracing, a more general framework is provided in the spatial frequency domain through the use of the Ewald sphere construction.

The addition of depth-sensing capability to other traditional experimental mechanics techniques seems likely to be one direction of future research in this field. Depth-resolved photoelasticity, for example, has recently been proposed using polarization-sensitive OCT [78]. Another direction is likely to be how to address the many complicating factors that have been largely neglected in the models discussed so far. These include issues such as dispersion, internal refraction due to gradients in refractive index, multiple scattering, speckle decorrelation, and phase changes due to load-induced changes in sample geometry, to name a few. The development of solutions to these problems, and of improved hardware to implement the wavenumber scans, is likely to be important in encouraging the future adoption of these techniques as practical engineering tools.

Acknowledgments

The authors thank J. M. Coupland for the data shown in Figure 2.3, and A. Kraft, G. Gülker, K. D. Hinsch, and A. Eljarad for permission to reproduce Figures 2.4 and 2.5. Past and present support from the Engineering and Physical Sciences Research Council, the Royal Society, and the Leverhulme Trust is also gratefully acknowledged.

References

1 McRobbie, D.W., Moore, E.A., Graves, M.J., and Prince, M.R. (2006) *MRI from Picture to Proton*, 2nd edn, Cambridge University Press, Cambridge.

2 Forsberg, F., Sjödahl, M., Mooser, R., Hack, E., and Wyss, P. (2010) Full three-dimensional strain measurements on wood exposed to three-point bending: analysis by use of digital volume correlation applied to synchrotron radiation micro-computed tomography image data. *Strain*, **46** (1), 47–60.

3 Lenoir, N., Bornert, M., Desrues, J., Besuelle, P., and Viggiani, G. (2007) Volumetric digital image correlation applied to X-ray microtomography images from triaxial compression tests on argillaceous rock. *Strain*, **43** (3), 193–205.

4 Avril, S., Huntley, J.M., Pierron, F., and Steele, D.D. (2008) 3D heterogeneous stiffness reconstruction using MRI and the virtual fields method. *Exp. Mech.*, **48** (4), 479–494.

5 Maranon, A., Ruiz, P.D., Nurse, A.D., Huntley, J.M., Rivera, L., and Zhou, G. (2007) Identification of subsurface delaminations in composite laminates. *Compos. Sci. Technol.*, **67**, 2817–2826.

6 Huntley, J.M. (2001) Automated analysis of speckle interferograms, in *Digital Speckle Pattern Interferometry and Related Techniques* (ed. P. Rastogi), John Wiley & Sons, Inc., pp. 59–139.

7 Balasubramanian, N. (1982) Optical system for surface topography measurement. United States Patent No. 4340306.

8 Deck, L. and de Groot, P. (1994) High-speed noncontact profiler based on scanning white-light interferometry. *Appl. Optics.*, **33** (31), 7334–7338.

9 Dresel, T., Häusler, G., and Venzke, H. (1992) 3-dimensional sensing of rough surfaces by coherence radar. *Appl. Optics.*, **31** (7), 919–925.

10 Lee, B.S. and Strand, T.C. (1990) Profilometry with a coherence scanning microscope. *Appl. Optics.*, **29** (26), 3784–3788.

11 Huang, D., Swanson, E.A., Lin, C.P., Schuman, J.S., Stinson, W.G., Chang, W., Hee, M.R., Flotte, T., Gregory, K., Puliafito, C.A., and Fujimoto, J.G. (1991) Optical coherence tomography. *Science*, **254** (5035), 1178–1181.

12 Hitzenberger, C.K. (1991) Optical measurement of the axial eye length by laser Doppler interferometry. *Invest. Ophthalmol. Vis. Sci.*, **32** (3), 616–624.

13 Fercher, A.F., Hitzenberger, C.K., Drexler, W., Kamp, G., and Sattmann, H. (1993) *In-vivo* optical coherence tomography. *Am. J. Ophthalmol.*, **116** (1), 113–115.

14 Swanson, E.A., Huang, D., Hee, M.R., Fujimoto, J.G., Lin, C.P., and Puliafito, C.A. (1992) High-speed optical coherence domain reflectometry. *Opt. Lett.*, **17** (2), 151–153.

15 Fercher, A.F., Drexler, W., Hitzenberger, C.K., and Lasser, T. (2003) Optical

coherence tomography: principles and applications. *Rep. Prog. Phys.*, **66** (2), 239–303.

16 Stifter, D. (2007) Beyond biomedicine: a review of alternative applications and developments for optical coherence tomography. *Appl. Phys. B*, **88** (3), 337–357.

17 Drexler, W. and Fujimoto, J.G. (eds) (2008) *Optical Coherence Tomography: Technology and Applications*, Springer-Verlag, Berlin.

18 Schmitt, J.M. (1998) OCT elastography: imaging microscopic deformation and strain of tissue. *Opt. Express*, **3** (6), 199–211.

19 Kraft, A., Gulker, G., Hinsch, K.D., and Eljarad, A. (2000) Low-coherence videoholography for sub-surface deformation measurements in layered objects. *Proc. SPIE*, **4101**, 89–96.

20 Gülker, G., Hinsch, K.D., and Kraft, A. (2003) Low-coherence ESPI in the investigation of ancient terracotta warriors. *Proc. SPIE*, **4933**, 53–58.

21 Gastinger, K., Winther, S., and Hinsch, K.D. (2003) Low-coherence speckle interferometry (LCSI) for characterisation of adhesion in adhesive bonded joints. *Proc. SPIE*, **4933**, 59–65.

22 Gastinger, K. (2008) Low coherence speckle interferometry (LCSI): when speckle interferometry goes sub-surface. *Proc. SPIE*, 7008111–70081110.

23 Macdonald, R.I. (1981) Frequency-domain optical reflectometer. *Appl. Optics.*, **20** (10), 1840–1844.

24 Eickhoff, W. and Ulrich, R. (1981) Optical frequency-domain reflectometry in single-mode fiber. *Appl. Phys. Lett.*, **39** (9), 693–695.

25 Fercher, A.F., Hitzenberger, C.K., Kamp, G., and Elzaiat, S.Y. (1995) Measurement of intraocular distances by backscattering spectral interferometry. *Opt. Commun.*, **117** (1–2), 43–48.

26 Lexer, F., Hitzenberger, C.K., Fercher, A.F., and Kulhavy, M. (1997) Wavelength-tuning interferometry of intraocular distances. *Appl. Optics.*, **36** (25), 6548–6553.

27 Coe, P.A., Howell, D.F., and Nickerson, R.B. (2004) Frequency scanning interferometry in ATLAS: remote, multiple, simultaneous and precise distance measurements in a hostile environment. *Meas. Sci. Technol.*, **15** (11), 2175–2187.

28 Takeda, M. and Yamamoto, H. (1994) Fourier-transform speckle profilometry: 3-dimensional shape measurements of diffuse objects with large height steps and/or spatially isolated surfaces. *Appl. Optics.*, **33** (34), 7829–7837.

29 Kuwamura, S. and Yamaguchi, I. (1997) Wavelength scanning profilometry for real-time surface shape measurement. *Appl. Optics.*, **36** (19), 4473–4482.

30 de Groot, P. (2000) Measurement of transparent plates with wavelength-tuned phase-shifting interferometry. *Appl. Optics.*, **39** (16), 2658–2663.

31 Ruiz, P.D., Zhou, Y.Z., Huntley, J.M., and Wildman, R.D. (2004) Depth-resolved whole-field displacement measurement using wavelength scanning interferometry. *J. Opt. A: Pure Appl. Opt.*, **6** (7), 679–683.

32 Ruiz, P.D., Huntley, J.M., and Wildman, R.D. (2005) Depth-resolved whole-field displacement measurement by wavelength-scanning electronic speckle pattern interferometry. *Appl. Optics.*, **44** (19), 3945–3953.

33 de Groot, P. (1995) Derivation of algorithms for phase-shifting interferometry using the concept of a data-sampling window. *Appl. Optics.*, **34** (22), 4723–4730.

34 Surrel, Y. (1997) Design of phase-detection algorithms insensitive to bias modulation. *Appl. Optics.*, **36** (4), 805–807.

35 Surrel, Y. (1996) Design of algorithms for phase measurements by the use of phase stepping. *Appl. Optics.*, **35** (1), 51–60.

36 Huntley, J.M., Widjanarko, T., and Ruiz, P.D. (2010) Hyperspectral interferometry for single-shot absolute measurement of two-dimensional optical path distributions. *Meas. Sci. Technol.*, **21**, 075304, pp 9.

37 Bruning, J.H., Herriott, D.R., Gallagher, J.E., Rosenfeld, D.P., White, A.D., and Brangaccio, D.J. (1974) Digital wavefront measuring interferometer for testing

optical surfaces and lenses. *Appl. Optics.*, **13** (11), 2693–2703.

38 Zhou, Y.Z., Wildman, R.D., and Huntley, J.M. (2010) Measurement of the mechanical properties of granular packs by wavelength-scanning interferometry. *Proc. R. Soc. A: Math. Phys. Eng. Sci.*, **466** (2115), 789–808.

39 Kaufmann, G.H., Ennos, A.E., Gale, B., and Pugh, D.J. (1980) An electro-optical readout system for analysis of speckle photographs. *J. Phys. E: Sci. Instrum.*, **13** (5), 579–584.

40 Heikkilä, J. and Silvén, O. (1997) A four-step camera calibration procedure with implicit image correction, in *1997 IEEE Computer Society Conference on Computer Vision and Pattern Recognition (CVPR 97)*, IEEE Computer Society Press, Los Alamitos.

41 Kidger, M.J. (2002) *Fundamental Optical Design*, vol. 92 SPIE Press.

42 Jones, R. and Wykes, C. (1989) *Holographic and Speckle Interferometry*, Cambridge University Press, Cambridge.

43 Wojtkowski, M., Leitgeb, R., Kowalczyk, A., Bajraszewski, T., and Fercher, A.F. (2002) *In vivo* human retinal imaging by Fourier domain optical coherence tomography. *J. Biomed. Opt.*, **7** (3), 457–463.

44 Fercher, A.F., Hitzenberger, C.K., Kamp, G., and El-Zaiat, S.Y. (1995) Measurement of intraocular distances by backscattering spectral interferometry. *Opt. Comm.*, **117**, 43–48.

45 Leitgeb, R., Hitzenberger, C.K., and Fercher, A.F. (2003) Performance of Fourier domain vs. time domain optical coherence tomography. *Opt. Express*, **11** (8), 889–894.

46 Schwider, J. and Zhou, L. (1994) Dispersive interferometric profilometer. *Opt. Lett.*, **19** (13), 995–997.

47 Zuluaga, A.F. and Richards-Kortum, R. (1999) Spatially resolved spectral interferometry for determination of subsurface structure. *Opt. Lett.*, **24** (8), 519–521.

48 De la Torre-Ibarra, M.H., Ruiz, P.D., and Huntley, J.M. (2006) Double-shot depth-resolved displacement field measurement using phase-contrast spectral optical coherence tomography. *Opt. Express*, **14** (21), 9643–9656.

49 Ruiz, P.D., de la Torre-Ibarra, M., and Huntley, J.M. (2006) Single-shot depth-resolved displacement field measurement using phase-contrast polychromatic speckle interferometry. *Proc. SPIE*, **6341**, U194–U200.

50 De la Torre Ibarra, M.H., Ruiz, P.D., and Huntley, J.M. (2009) Simultaneous measurement of in-plane and out-of-plane displacement fields in scattering media using phase-contrast spectral optical coherence tomography. *Opt. Lett.*, **34** (6), 806–808.

51 Kirkpatrick, S.J., Wang, R.K., and Duncan, D.D. (2006) OCT-based elastography for large and small deformations. *Opt. Express*, **14** (24), 11585–11597.

52 Wang, R.K.K., Kirkpatrick, S., and Hinds, M. (2007) Phase-sensitive optical coherence elastography for mapping tissue microstrains in real time. *Appl. Phys. Lett.*, **90** (16), 164105.

53 Wang, R.K.K., Ma, Z.H., and Kirkpatrick, S.J. (2006) Tissue Doppler optical coherence elastography for real time strain rate and strain mapping of soft tissue. *Appl. Phys. Lett.*, **89** (14), 144103.

54 Wojtkowski, M., Srinivasan, V.J., Ko, T.H., Fujimoto, J.G., Kowalczyk, A., and Duker, J.S. (2004) Ultrahigh-resolution, high-speed, Fourier domain optical coherence tomography and methods for dispersion compensation. *Opt. Express*, **12** (11), 2404–2422.

55 Ruiz, P.D., Huntley, J.M., and Maranon, A. (2006) Tilt scanning interferometry: a novel technique for mapping structure and three-dimensional displacement fields within optically scattering media. *Proc. R. Soc. A: Math. Phys. Eng. Sci.*, **462** (2072) 2481–2502.

56 Rodríguez-Vera, R., Kerr, D., and Mendoza-Santoyo, F. (1992) Electronic speckle contouring. *J. Opt. Soc. Am. A*, **9** (11), 2000–2008.

57 Duan, Z.H., Miyamoto, Y., and Takeda, M. (2006) Dispersion-free absolute

interferometry based on angular spectrum scanning. *Opt. Express*, **14** (2), 655–663.

58 Lauer, V. (2002) New approach to optical diffraction tomography yielding a vector equation of diffraction tomography and a novel tomographic microscope. *J. Microsc.*, **205**, 165–176.

59 Wolf, E. (1969) Three-dimensional structure determination of semi-transparent objects from holographic data. *Opt. Commun.*, **1** (4), 153–156.

60 Dändliker, R. and Weiss, K. (1970) Reconstruction of the three-dimensional refractive index from scattered waves. *Opt. Commun.*, **1** (7), 323–328.

61 Fercher, A.F., Bartelt, H., Becker, H., and Wiltschko, E. (1979) Image-formation by inversion of scattered field data: experiments and computational simulation. *Appl. Optics.*, **18** (14), 2427–2439.

62 Coupland, J.M. and Lobera, J. (2008) Holography, tomography and 3D microscopy as linear filtering operations. *Meas. Sci. Technol.*, **19** (7), 070101, pp 12.

63 Ruiz, P.D., Huntley, J.M., and Coupland, J.M. (2010) Depth-resolved imaging and displacement measurement techniques viewed as linear filtering operations. *Exp. Mech.* doi 10.1007/s11340-010-9393-8

64 Born, M. and Wolf, E. (1999) *Principles of Optics: Electromagnetic Theory of Propagation, Interference and Diffraction of Light*, 7th edn, Cambridge University Press, Cambridge.

65 Spencer, M. (1982) *Fundamentals of Light Microscopy*, Cambridge University Press, Cambridge.

66 Ghiglia, D.C. and Pritt, M.D. (1998) *Two-Dimensional Phase Unwrapping*, John Wiley & Sons, Inc., New York.

67 Huntley, J.M. (2001) Three-dimensional noise-immune phase unwrapping algorithm. *Appl. Optics.*, **40** (23), 3901–3908.

68 Marklund, O., Huntley, J.M., and Cusack, R. (2007) Robust unwrapping algorithm for three-dimensional phase volumes of arbitrary shape containing knotted phase singularity loops. *Opt. Eng.*, **46**, 085601-1–0855601-13.

69 Salfity, M.F., Huntley, J.M., Graves, M.J., Marklund, O., Cusack, R., and Beauregard, D.A. (2006) Extending the dynamic range of phase contrast magnetic resonance velocity imaging using advanced higher-dimensional phase unwrapping algorithms. *J. R. Soc. Interface*, **3** (8), 415–427.

70 Salfity, M.F., Ruiz, P.D., Huntley, J.M., Graves, M.J., Cusack, R., and Beauregard, D.A. (2006) Branch cut surface placement for unwrapping of undersampled three-dimensional phase data: application to magnetic resonance imaging arterial flow mapping. *Appl. Optics*, **45** (12), 2711–2722.

71 Arevalillo-Herraez, M., Gdeisat, M.A., and Burton, D.R. (2009) Hybrid robust and fast algorithm for three-dimensional phase unwrapping. *Appl. Optics.*, **48** (32), 6313–6323.

72 Abdul-Rahman, H., Arevalillo-Herraez, M., Gdeisat, M., Burton, D., Lalor, M., Lilley, F., Moore, C., Sheltraw, D., and Qudeisat, M. (2009) Robust three-dimensional best-path phase-unwrapping algorithm that avoids singularity loops. *Appl. Optics.*, **48** (23), 4582–4596.

73 Jenkinson, M. (2003) Fast, automated, *N*-dimensional phase-unwrapping algorithm. *Magnet. Reson. Med.*, **49** (1), 193–197.

74 Hooper, A. and Zebker, H.A. (2007) Phase unwrapping in three dimensions with application to InSAR time series. *J. Opt. Soc. Am. A*, **24** (9), 2737–2747.

75 Ghiglia, D.C. and Romero, L.A. (1996) Minimum L^p-norm two-dimensional phase unwrapping. *J. Opt. Soc. Am. A*, **13** (10), 1999–2013.

76 Buckland, J.R., Huntley, J.M., and Turner, S.R.E. (1995) Unwrapping noisy phase maps by use of a minimum-cost-matching algorithm. *Appl. Optics.*, **34** (23), 5100–5108.

77 Gdeisat, M., Arevalillo-Herraez, M., Burton, D., and Lilley, F. (2009) Three-dimensional phase unwrapping using the Hungarian algorithm. *Opt. Lett.*, **34** (19), 2994–2996.

78 Heise, B., Wiesauer, K., Gotzinger, E., Pircher, M., Hitzenberger, C.K., Engelke, R., Ahrens, G., Grutzner, G., and Stifter, D. (2010) Spatially resolved stress measurements in materials with polarisation-sensitive optical coherence tomography: image acquisition and processing aspects. *Strain*, **46** (1), 61–68.

79 Huntley, J.M. and Saldner, H. (1993) Temporal phase-unwrapping algorithm for automated interferogram analysis. *Appl. Optics.*, **32** (17), 3047–3052.

3
Single-Image Interferogram Demodulation

Manuel Servin, Julio Estrada, and Antonio Quiroga

3.1
Introduction

In this chapter, we are going to analyze several techniques that apply when only a single interferogram image is available for measuring a physical quantity through an interferometric experiment. Analyzing a single interferogram image is probably the oldest way to extract the phase information from interferometric data. In the early days of optical metrology, this was done using a digitizing table where one would label by hand some points along the maxima and/or minima of the interferometric fringes and assign them a "fringe order." After collecting enough fringe-order data and its corresponding spatial location, one would fit, using least square method, a low-order polynomial to obtain a graph of the spatial variation in the physical quantity being analyzed.

However, the good old days of the digitizing tables have gone for good and new automatic and more accurate techniques have emerged to phase demodulate a single interferogram. Single-image interferogram demodulation can be broadly divided into two large and clearly distinct categories. One is when a spatial carrier may be introduced in the interferometric data by, for example, introducing a large tilt in the reference wavefront of the interferometer. The other category is the one where the introduction of a spatial carrier is not possible and one has to deal with the more difficult task of automatic demodulation of a single interferogram containing closed fringes. These two approaches to single interferogram demodulation are dealt with in this chapter

3.1.1
Spatial Carrier Frequency Techniques

Historically, spatial carrier frequency techniques were the first real and sound approaches to successfully analyze a single interferogram image. These spatial carrier techniques normally come in two flavors: one is the Fourier transformed techniques originally initiated by Takeda *et al.* in 1982 and the other one are the

Advances in Speckle Metrology and Related Techniques. Edited by Guillermo H. Kaufmann

ISBN: 978-3-527-40957-0

so-called spatial phase shifting techniques. The Fourier technique, as its name implies, uses the Fourier transform to demodulate the interferometric data. This technique works by manipulating the interferometric data as a whole and is probably the most widely used technique and is the theme treated in Section 3.2. On the other hand, the spatial phase shifting technique uses a small quadrature convolution filter to obtain the demodulated phase from the interferogram. These small convolution quadrature filters are the same as the ones used in temporal phase shifting techniques, but instead of using the temporal dimension, they use one spatial direction. This is the theme of Section 3.3.

3.1.2 Spatial Demodulation Without Carrier

Automatic demodulation of a single interferogram containing closed fringes is a much more complicated achievement. That is why this theme is one of the most recent research themes in optical metrology. The first fully automatic technique was proposed some 10 years ago by Servin *et al.* using what is being called a regularized phase tracker (RPT). This attempt was immediately followed by a technique named adaptive robust quadrature filters by Marroquin *et al.* More recently, another perspective into this problem was given by Larkin *et al.* using a two-dimensional generalization of the Hilbert transform, and generalized to any dimension in the Euclidean space by Servin *et al.* In this last view, the demodulation problem of a single interferogram image containing closed fringes may be seen as composed of two different albeit related problems. One of these is the orientation of the fringes and the other one is a two or higher dimension Hilbert filter. All these themes will be treated in the following sections.

3.2 The Fourier Spatial Demodulating Method

Optical metrology techniques have a very extensive application fields; among these, the most common use of optical metrology is for testing surfaces, fluid dynamics, and temperatures gradients [1]. A very attractive characteristic of these techniques is that they are noninvasive methods. Their basic principle is the wave properties of light and its interference phenomena. These interferometers are designed in such a way that the information of the event under test is phase modulated by the interference pattern. Therefore, after a series of interferograms are taken, it is necessary to apply angle demodulation methods to recover the phase.

The *Fourier transform method* (FTM) is one of the most widely used tools for recovering the phase of a single interferogram with spatial carrier frequency [2–4]. Actually, this is the *de facto* tool in interferometry techniques such as *digital TV holographic* [5]. As we said before, the FTM works with interferograms that have a spatial frequency carrier. An example of this kind of interferograms is shown in Figure 3.1. In Figure 3.1a, we see an interferogram image without spatial frequency

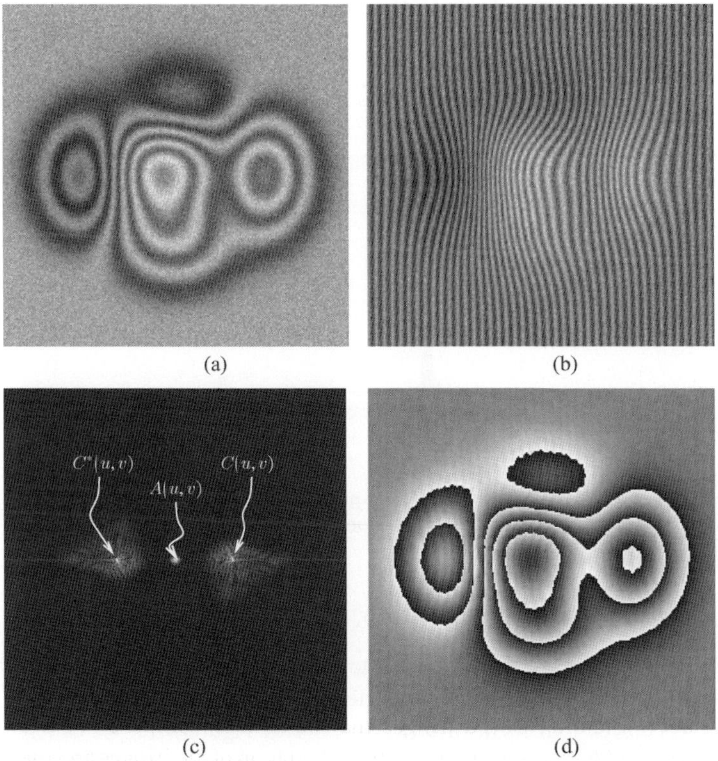

Figure 3.1 (a) An interferogram without frequency carrier (closed fringes). (b) The same interferogram with spatial frequency carrier (open fringes). (c) The Fourier transform of (b). (d) The recovered phase using the FTM.

carrier and in Figure 3.1b we show the same interferogram with the spatial frequency carrier added.

Let us explain how we can use the FTM using the interferogram in Figure 3.1b. In this example, the frequency carrier is only in the x-direction of the image. Its mathematical description can be given as follows:

$$I(x,y) = a(x,y) + b(x,y)\cos[\phi(x,y) + \omega_x x], \tag{3.1}$$

where ω_x is the spatial carrier frequency along x-direction, $a(x,y)$ the background illumination, $b(x,y)$ the contrast or modulation term, and $\phi(x,y)$ the unknown modulating phase that we want to recover. For convenience, we rewrite this in the following way:

$$I(x,y) = a(x,y) + c(x,y)\exp(i\omega_0 x) + c^*(x,y)\exp(-i\omega_0 x), \tag{3.2}$$

with $c(x,y) = (b(x,y)/2)\exp[i\phi(x,y)]$, where * denotes complex conjugate. Its Fourier transform can be described in the following way:

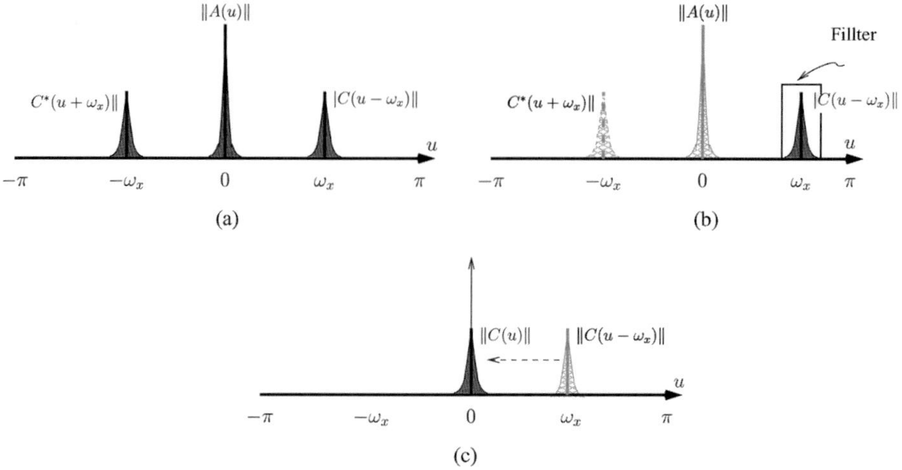

Figure 3.2 Graphical representations of the Fourier transform method. (a) The frequency distribution (in one dimension) of (3.3), (b) the application of a square band-pass filter, and (c) the translation to the origin of the filtered distribution.

$$\hat{I}(u, v) = A(u, v) + C(u - \omega_x, v) + C^*(u + \omega_x, v), \tag{3.3}$$

where space (u, v) is the transformed frequency domain, $A(u, v)$ is the Fourier transform of $a(x, y)$, $C(u - \omega_x, v)$ is the Fourier transform of $c(x, y)e^{i\omega_x x}$, and $C^*(u + \omega_x)$ is the Fourier transform of $c^*(x, y)e^{-i\omega_x x}$. In Figure 3.1c, we can see the Fourier transform of the image shown in Figure 3.1b. However, for illustration purposes, let us use the one-dimensional graphic description of Equation 3.3 shown in Figure 3.2a.

The idea of recovering the phase $\phi(x, y)$ of the carrier interferogram is to preserve only the complex term $C(u - \omega_x, v)$ and take its angle, which is the searched phase. This is done by filtering out the frequency distributions given by $A(u, v)$ and $C^*(u + \omega_x, v)$. Thus, assuming that we have used a pass-band square filter like the one shown in Figure 3.2b, we keep only the component $C(u - \omega_x, v)$. Translating this remaining distribution to the origin of the frequency domain, we obtain the distribution as shown in Figure 3.2c. This last operation removes the carrier frequency from the modulating phase. After all these manipulations, we end up having a filtered and translated frequency distribution that corresponds to the term $C(u, v)$. Taking into account the inverse Fourier transform, we have the following:

$$\mathsf{F}^{-1}\{C(u, v)\} = c(x, y) = \frac{b(x, y)}{2} \exp i[\phi(x, y)], \tag{3.4}$$

where $\mathsf{F}^{-1}\{C(u, v)\}$ denotes the inverse Fourier transform. Finally, the modulating phase is obtained (modulus 2π) as the angle of this complex signal in the following manner:

$$\phi(x, y) = \arctan\left[\frac{\mathrm{Im}\{c(x, y)\}}{\mathrm{Re}\{c(x, y)\}}\right]. \tag{3.5}$$

With this in mind, the FTM process to recover the modulating phase given an interferogram with carrier frequency may be stated as follows:

1) Take the Fourier transform of the interferogram.
2) Looking the image of its Fourier transform, locate the lobes corresponding to the frequency response of component $c(x,y)e^{i\omega_x x}$ and its complex conjugate (see Equation 3.3).
3) Filter out one of the two lobes (the right one in this case) along with the lobe of the center (DC) and translate the remaining distribution to the origin of the Fourier domain to remove the spatial carrier.
4) Take the inverse Fourier transform of this filtered spectrum.
5) Compute the phase as shown in Equation 3.5. Note that the phase will be obtained modulus 2π.

The difference of filtering either the component $C(u-\omega_x, v)$ or $C^*(u+\omega_x, v)$ is a global change of the sign in the recovered phase. In Figure 3.1d, we see the phase map obtained from Figure 3.1b using the FTM process as described above.

In general, the carrier frequency of the interferogram can be given as a linear combination of carriers along x- and y-directions. When this is the case, the interferogram is given in the following way:

$$I(x,y) = a(x,y) + b(x,y)\cos[\phi(x,y) + \omega_x x + \omega_y y]. \tag{3.6}$$

Having this, the procedure of recovering the phase using the Fourier transform method is the same as the shown above. However, in this case the component that we must filter is located at frequency coordinates (ω_x, ω_y) of the Fourier domain, and not in one axis (as shown in Figure 3.1a).

3.3 Linear Spatial Phase Shifting

Spectral analysis in PSI algorithm theory has been systematically developed since 1990, and it is based on the work of Freishlad and Koliopoulos (F&K) [6]. Application of the F&K spectral analysis to PSI algorithm synthesis may be seen in Ref. [7]. In 1996, Surrel [8] developed an algebraic approach to analyze PSI algorithms based on what he called the characteristic polynomial associated with the quadrature filter. Even though the properties of any PSI algorithm may be deduced from the roots of this polynomial, no spectral plot (similar to the one proposed in Ref. [6]) was proposed, where one may gauge in a glance the frequency response of PSI algorithms. Over the years, the F&K spectral plot emerged as the standard way of gauging the frequency response of PSI algorithms [1]. However, the F&K spectral analysis has a major drawback, namely, the spectrum varies when the PSI algorithm's reference signal is time shifted. It is well known that the estimated phase of a reference time-shifted PSI algorithm remains unchanged, except for an irrelevant piston ([1], p. 239). In Ref. [9], we have recently proposed a new way to analyze the spectra of PSI algorithms based on the frequency transfer function (FTF). This new way of spectral analysis of PSI algorithms is, however, not new in engineering. In particular in electrical engineering, the use of FTF has been the standard way of analyzing the spectra of digital filters for decades now [9, 10].

Considering PSI algorithms as linear quadrature filters, we can apply this theory to demodulate spatial frequency carrier interferograms. Although originally phase shifting interferometry methods assume that we have a time sequence of interferograms with a constant phase variation (temporal carrier) among them, we can apply the temporal phase shifting algorithms to demodulate interferograms having spatial carrier, just by changing the temporal axis to a spatial dimension. First, the basic statement of PSI algorithms assumes that we have a temporal interferogram sequence modeled as

$$I_n(x,y) = a(x,y) + b(x,y)\cos[\phi(x,y) + \alpha n], \tag{3.7}$$

where subindex I_n represents the nth interferogram, taken in an instant of time $t = n$, and α is its phase shift. Here, we may call α, the temporal carrier, instead of the phase shift because it is used in temporal PSI. A quality of PSI quadrature filters is that they are designed to use a smallest possible number of samples. Hence, it is very common to find algorithms that take only 3 or around 11 samples (or interferograms) [11–16]. For example, let us show the temporal three-step phase shifting algorithm. This algorithm needs three interferograms given in sequence as $\{I_0, I_1, I_2\}$, and its *formula* to recover the phase is the following:

$$\phi(x,y) = \arctan\left[\frac{I_0(x,y) - I_2(x,y)}{2I_0(x,y) - I_1(x,y) - I_2(x,y)}\tan(\alpha/2)\right]. \tag{3.8}$$

We say that this is a tunable phase shifting algorithm since the carrier frequency can be given as a parameter of the three-step phase shifting filter. Another very common PSI algorithm is the five-step phase shifting algorithm [15, 16]. The principal characteristic of this algorithm is its tolerance to small carrier detuning (frequency carrier mismatch). Its formula is written as

$$\phi(x,y) = \arctan\left[\frac{2I_{-1}(x,y) - 2I_1(x,y)}{2I_0(x,y) - I_{-2}(x,y) - I_2(x,y)}\right]. \tag{3.9}$$

In this case, the interferogram sequence is given as $\{I_{-2}, I_{-1}, I_0, I_1, I_2\}$ having a phase shift $\alpha = \pi/2$. However, this algorithm is not a tunable algorithm since it works only for a temporal frequency carrier of $\alpha = \pi/2$. In the work of Estrada *et al.*, it is shown that we may systematically design PSI filters or algorithms directly in the Fourier domain by using two basic building blocks [17]. These building blocks represent first- and second-order digital filters. In this paper, the tunable version of the Schwider–Hariharan algorithm is given; this is expressed in the following formula:

$$\phi(x,y) = \arctan\left[\frac{2I_{-1}(x,y) - 2I_1(x,y) - [I_{-2}(x,y) - I_2(x,y)]\cos(\alpha)}{2I_0(x,y) - [I_{-2}(x,y) + I_2(x,y)]\sin(\alpha)}\right]. \tag{3.10}$$

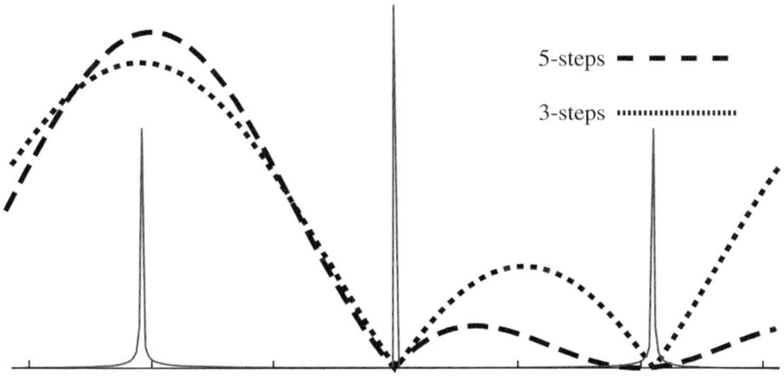

Figure 3.3 Frequency response of the tunable three-step (in dark gray) and five-step (in gray) algorithms (see Equation 3.8 and 3.10). The horizontal axis ω, measured in radians per pixel, runs from $-\pi$ to π. The light gray graph is the Fourier transform of the spatial (or temporal) interferometry signal. The peak at the center corresponds to the DC term and the other two peaks to the interference terms of the signal. The vertical axes are shown for illustration purposes and do not represent any scale.

In Figure 3.3, we can see the magnitude of the frequency response of the three-step and five-step algorithms [9, 17]. This spectrum is obtained directly by taking the Fourier transform of the quadrature filter that represents each algorithm [9]. As we can see in the graphics, the frequency response of the five-step phase shifting algorithm touches zero tangentially at the right peak of the signal response making this algorithm very tolerant to detuning. This fact contrasts with the response of the three-step algorithm that crosses the right peak of the signal. This makes a notable difference between these algorithms, and we say that the five-step phase shifting algorithm is more tolerant to signal carrier detuning than the three-step phase shifting algorithm.

As we said, phase shifting algorithms can be used as spatial algorithms applying them to neighborhood pixels of an interferogram. Like in the Fourier transform method, this approach needs an interferogram with frequency carrier. Assuming that the frequency carrier is along x-direction, the interferogram with spatial carrier is given exactly as in Equation 3.1 in the following way:

$$I(x, y) = a(x, y) + b(x, y)\cos[\phi(x, y) + \omega_x x]. \tag{3.11}$$

To apply the phase shifting algorithms given above, we can directly translate these temporal algorithms to demodulate spatial carrier frequency interferograms by simply changing the temporal coordinate to the discrete pixel coordinate. For example, the three-step phase shifting algorithm takes the following form:

$$\hat{\phi}(x, y) + \omega_x x = \arctan\left[\frac{I(x-1, y) - I(x+1, y)}{2I(x, y) - I(x-1, y) - I(x+1, y)}\tan(\omega_x/2)\right]. \tag{3.12}$$

The spatial phase that we obtain using a PSI algorithm is obtained along with the frequency carrier. To remove the frequency carrier to obtain $\hat{\phi}(x, y)$, we must know the amount of the carrier ω_x in radians. Knowing the frequency carrier, the procedure

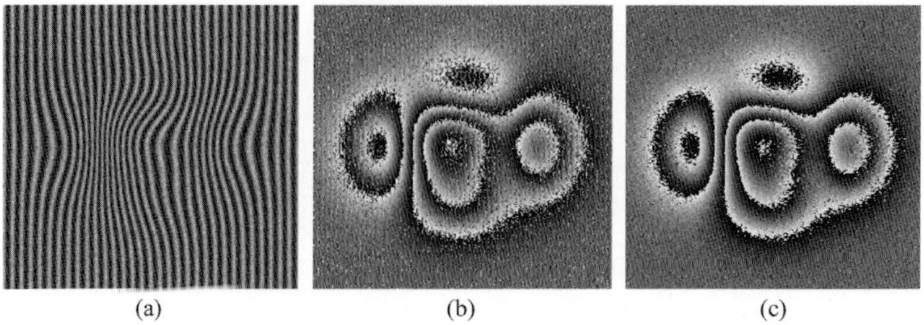

Figure 3.4 Testing the three-step and five-step PSI algorithms for recovering spatial modulating phase. In (a) we see the input interferogram. This interferogram has normal phase noise having zero mean and variance of 0.2 rad. In (b) we show the demodulated phase map obtained with the three-step PSI algorithm and in (c) the estimated phase map obtained using the five-step PSI algorithm.

to remove this is (i) construct a complex image with the obtained phase $\hat{\phi}(x, y)$ as follows:

$$g(x, y) = \cos[\hat{\phi}(x, y) + \omega_x] + i \sin[\hat{\phi}(x, y) + \omega_x], \tag{3.13}$$

and (ii) given this, the modulating phase without spatial carrier is obtained as follows:

$$\phi(x, y) = \arctan\left[\frac{\mathrm{Im}\{g(x, y)e^{-i\omega_x x}\}}{\mathrm{Re}\{g(x, y)e^{-i\omega_x x}\}}\right]. \tag{3.14}$$

To show the behavior of the three- and five-step spatial algorithms, let us compare them with a simulated interferogram. In Figure 3.4a, we can see a simulated interferogram along with its recovered phase using the tunable three- and five-step spatial algorithms shown above (Equations 3.9 and 3.10).

Figure 3.4a shows a simulated noisy interferogram, in Figure 3.4b we can see the obtained phase using the three-step algorithm, and Figure 3.4c shows the phase obtained with the five-step algorithm. We can see that the demodulated phase shown in Figure 3.4c has less noise than the phase shown in Figure 3.4c. This is because the area under the frequency transfer function associated with the five-step algorithm is smaller than the three-step one for the same output power signal [18]. The phase maps that we see in Figure 3.4b and c are shown with the frequency carrier removed, following Equations 3.13 and 3.14. These results also show that the five-step phase shifting algorithm is more robust to carrier detuning in the fringe pattern than the three-step algorithm.

On the other hand, not knowing the interferograms' spatial carrier, the resulting phase maps would have a remaining phase inclination, in addition to the phase errors caused by detuning. An example of this is shown in Figure 3.5. We show in Figure 3.5a an interferogram with spatial frequency carrier and in Figure 3.5b the recovered phase using a wrong frequency carrier. There we see that the recovered phase has a remaining phase inclination due to an erroneous frequency carrier used in Equation 3.14.

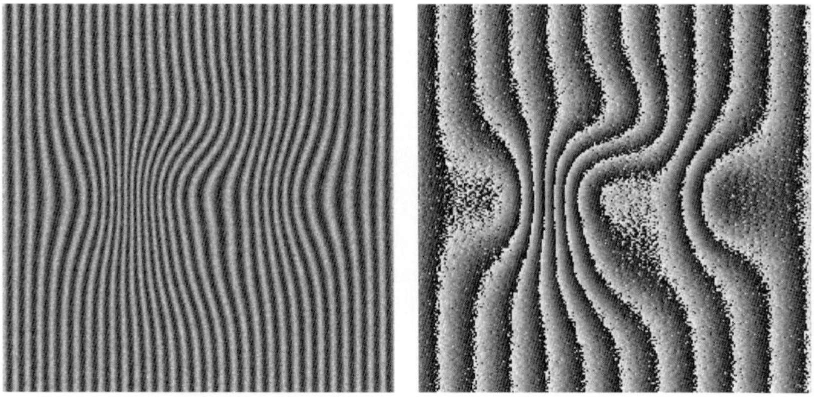

(a) (b)

Figure 3.5 (a) Simulated interferogram with a frequency carrier of $\omega_x = \pi/2$ rad. (b) Recovered phase using the three-step phase shifting algorithm of Equation 3.12, the phase obtained has a remaining phase inclination.

Therefore, it is far more important to know or previously to obtain a very good estimate of the spatial frequency carrier in the case of spatial phase shifting than in the Fourier technique.

3.4 Nonlinear Spatial Phase Shifting

In the previous section, we have shown how we can use the phase shifting algorithms to demodulate interferograms with spatial carrier. We saw that to obtain the phase it is necessary to know the carrier frequency. When we do not know the spatial frequency carrier, we obtain unexpected results like the one given in Figure 3.4c. In this section, we show another class of spatial phase shifting algorithms that do not require to know the carrier frequency. These algorithms are known in the literature as asynchronous or nonlinear phase shifting algorithms [11, 19–21].

To our knowledge, the first asynchronous phase shifting algorithm was proposed by Carré in 1966 [11]. After this, Ransom and Kokal [19] and independently Servin and Cuevas [20] presented a three-step nonlinear algorithm to demodulate interferograms with an unknown amount of spatial carrier. The formula presented by Servin and Cuevas [20] is as follows:

$$\phi(x,y) + \omega_x x = \arctan\left[\frac{I'(x-1,y) - I'(x+1,y)}{\mathrm{sign}[I'(x,y)]\sqrt{4I'(x,y)^2 - (I(x-1,y) + I(x+1,y))^2}}\right], \tag{3.15}$$

where the function $\mathrm{sign}[I'(x,y)]$ returns -1 if $I'(x,y) < 0$, otherwise it returns 1. However, this algorithm does not remove the background illumination of the interferogram; hence, the interferogram has to be filtered with a high-pass filter before using this formula. The filtered interferogram $I'(x,y)$ can be obtained using the following first-order difference:

$$I'(x,y) = \frac{I(x+1,y)-I(x-1,y)}{2}. \tag{3.16}$$

Another spatial asynchronous algorithm is the one presented by Gomez-Pedrero *et al.* [21]. Unlike the three-step algorithm shown in Equation 3.15, this algorithm implicitly removes the background illumination of interferogram. In this case, the algorithm uses five subsequent points to estimate the phase in a site. The formula of this algorithm can be written as

$$\phi(x,y) + \omega_x x = \arctan\left[\frac{O_i\{I(x,y)\}}{O_r\{I(x,y)\}}\right], \tag{3.17}$$

where $O_i\{I\}$ is given by

$$O_i\{I\} = sign[I(x-1,y)-I(x+1,y)] \times \sqrt{4(I(x-1,y)-I(x+1,y))^2-(I(x-2,y)-I(x+2,y))^2} \tag{3.18}$$

and $O_r\{I\}$ is given by

$$O_r\{I\} = 2I(x,y)-I(x-2)-I(x+2). \tag{3.19}$$

Let us illustrate the performance of these two asynchronous algorithms. In Figure 3.6, we show a comparison between the asynchronous algorithms of Equations 3.15 and 3.17. These tests were conducted with a simulated interferogram (shown on the left-hand side of this figure) with and without noise. In Figure 3.6a, we show the demodulated phase using the nonlinear three-step phase shifting algorithm of Equation 3.15, and in Figure 3.6b is shown the resulting phase of the five-step phase shifting algorithm of Equation 3.17. Phase maps in Figure 3.6a and b correspond to the interferogram without noise. Similarly, the noisier phase maps in Figure 3.6c and d show the resulting phases of the three and five steps, respectively, using the noisy interferogram. The noisy interferogram was simulated using normal phase noise with zero mean and variance equal to 0.5 rad. Unexpectedly, the result obtained with the five-step phase shifting algorithm has about the same demodulated phase noise than the three-step phase shifting algorithm. This result may be due to sensitivity to the noise of the nonlinear function sign(.) that in both cases is used.

In conclusion, the algorithms developed for temporal phase shifting interferometry (PSI) can be applied to spatially demodulate interferograms with carrier frequency as spatial phase shifting algorithms. If we use linear phase shifting algorithms, we must know the carrier frequency of the interferogram. On the other hand, when using asynchronous (nonlinear) algorithms, it is not necessary to know the carrier frequency. However, if we know the frequency carrier, it is recommended to use the linear algorithms instead of the asynchronous ones because high-order linear algorithms have better signal-to-noise response than the nonlinear algorithms.

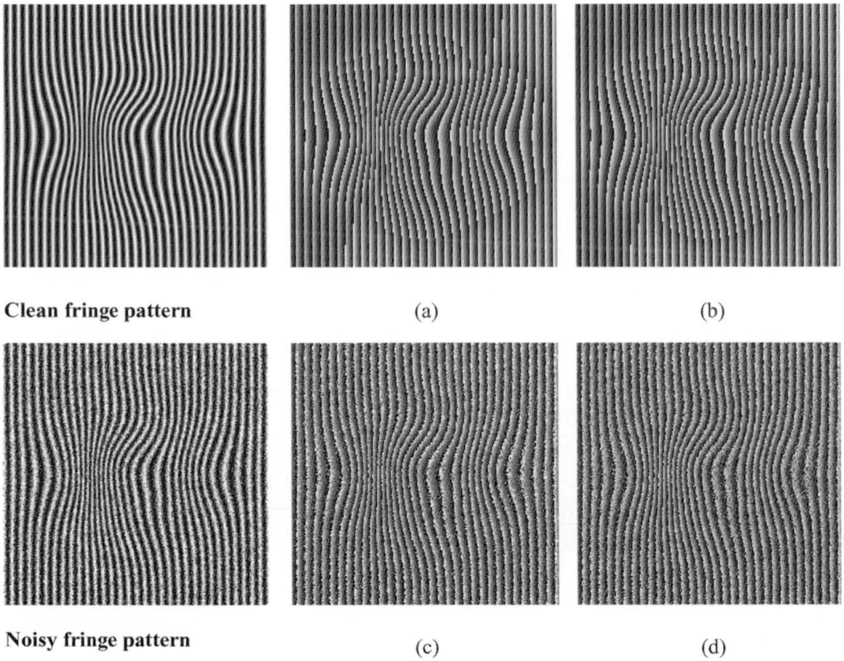

Figure 3.6 Two demodulation tests using asynchronous algorithms. The images in the first row (a) and (b) are noise free and the images in the second row (c) and (d) are noisy. In both cases, the unknown carrier was not removed.

3.5 Regularized Phase Tracking

Both in moiré interferometry and in other areas of optical metrology, one can have situations where it is necessary to deal with a single fringe pattern without carrier. In other words, we are forced to deal with interferograms with closed fringes. As the information of interest is phase modulated by these closed fringe images, it is necessary to apply a demodulation method that is able to recover the modulating phase of these closed fringe interferograms.

As we have seen, when we have an interferogram with spatial carrier frequency, it is easy to recover its modulating phase using spatial methods such as the Fourier transform method (PSI) or spatial phase shifting interferometry (SPSI) algorithms. However, when the experiment or the object under test does not allow us to introduce a spatial or temporal carrier, it is necessary to deal with a single-image interferogram with closed fringes. For the rest of this chapter, we are going to show several techniques that help us to recover the modulating phase from a single-image interferogram without carrier (closed fringes).

The first method that we analyze is the well-known *regularized phase tracker* [22, 23]. In this case, the interferogram is modeled as follows:

$$I(x, y) = a(x, y) + b(x, y)\cos[\phi(x, y)], \tag{3.20}$$

where $a(x, y)$ is the background illumination, $b(x, y)$ is the contrast, and $\phi(x, y)$ is the modulating phase that we want to estimate. For starters, the RPT method assumes that we have a normalized fringe pattern, that is, the interferogram is preprocessed to obtain the following normalized fringe pattern:

$$I'(x, y) = \cos[\phi(x, y)]. \tag{3.21}$$

To this end, we must use normalization techniques [24]. Basically, one can obtain a normalized fringe pattern by filtering the fringes to remove the background by applying a broadband one-sided quadrature filter. Then, we take the phase of the filtered complex image, and finally the normalized fringe pattern is obtained by taking the cosine of this phase. However, if our interferogram has a more complicated fringe structure, using more sophisticated normalizing methods such as the one published in Ref. [24] may be necessary.

The RPT method is a sequential demodulation technique that supposes that the interferogram has a locally spatial phase that may be modeled by a plane. This assumption implies that in a little neighborhood Γ of a given site (x, y), the interferogram can be approximated as

$$I'(\eta, \xi) = \cos[p(\eta, \xi)], \tag{3.22}$$

where $p(x, y)$ is the local phase plane approximation. This first-order approximation is given as

$$p(\eta, \xi) = \phi_0(x, y) + \omega_x(x-\eta) + \omega_y(y-\xi), \quad \forall(\eta, \xi) \in \Gamma, \tag{3.23}$$

where $\phi_0(x, y)$ is the phase being estimated at site (x, y) and ω_x and ω_y are the local spatial frequencies along the x-and y-directions, respectively. Then, a cost function to estimate the parameters of the local phase plane approximation is proposed. This cost function is given in the following way:

$$\begin{aligned} U(\phi_0, \omega_x, \omega_y) = \sum_{(\eta,\xi)\in\Gamma} & [\cos[p(\eta, \xi)] - I'(\eta, \xi)]^2 \\ & + [I'_x(\eta, \xi) + \omega_x \sin(p)]^2 + [I'_y(\eta, \xi) - \omega_y \sin(p)]^2 \\ & + \kappa[\hat{\phi}(\eta, \xi) - p(\eta, \xi)]^2 m(\eta, \xi), \end{aligned} \tag{3.24}$$

where Γ is the square neighborhood around the site (x, y). In practice, the size of neighborhood Γ usually ranges from 5×5 to 11×11 pixels. The function ϕ_0 is the estimated phase at site (x, y). The data term (the first term) makes consistent the approximation we are seeking with the interferometric data. The other three terms add further restrictions to the approximating phase. For example, the second and third terms constrain the quadrature of the solution with the quadrature of the data. To obtain this, the data quadrature is obtained as

$$I'_x(x, y) = \frac{I(x-1, y) - I(x+1, y)}{2}, \tag{3.25}$$

$$I'_y(x,y) = \frac{I(x,y-1) - I(x,y+1)}{2}. \quad (3.26)$$

The fourth term is a regularization term that looks for smooth solutions in the neighborhood Γ and κ is the regularization parameter that controls it. For example, in the examples shown below we use $\kappa = 5$. The regularization term uses an indicator function $m(\eta, \xi)$ of the previously demodulated pixels.

Then, to obtain the phase map, we minimize (3.24) for each site (x, y) of the interferogram image. To minimize (3.24), we compute its gradient and equal it to zero. Doing this, we obtain a nonlinear equation system. Thus, to solve this nonlinear equation system, we use the *steeped descent* method [25]. For the RPT, the iterations of the steeped descent method are given by

$$\hat{\phi}_0^{k+1} = \hat{\phi}_0^k - \mu \left.\frac{\partial U}{\partial \phi_0}\right|_{\hat{\phi}_0^k}, \quad (3.27)$$

$$\hat{\omega}_x^{k+1} = \hat{\omega}_x^k - \mu \frac{\partial U}{\partial \omega_x}\left\{\hat{\omega}_x^k\right\}, \quad (3.28)$$

$$\hat{\omega}_y^{k+1} = \hat{\omega}_y^k - \mu \frac{\partial U}{\partial \omega_y}\left\{\hat{\omega}_y^k\right\}, \quad (3.29)$$

where $\hat{\phi}_0^k$, $\hat{\omega}_x^k$, and $\hat{\omega}_y^k$ represent the value of the phase and spatial frequencies at the kth iteration. Value μ is the convergence speed of the steeped descent, and it must be small enough to ensure convergence. The steeped descent method, like all iterative methods, needs an initial starting point or seed. Here, for the first site being demodulated, we can take $(\hat{\phi}_0, \hat{\omega}_x, \hat{\omega}_y) = (0, 0, 0)$ as initial seed. For subsequent sites being demodulated, we use the values of the previous estimated site.

To obtain more robust solutions with the RPT, it is necessary to scan the interferogram image following isophase contours. To follow the isophase contours of the interferogram, we must obtain an isophase path. This isophase path can be obtained by following the path of the fringes as shown in Figure 3.7.

In Figure 3.6, we show an example of how the RPT method works. This example was taken from the work published by Servin *et al.* [23]. In Figure 3.6a, we see the interferogram image and the isophase path followed by the demodulating algorithm. In Figure 3.6b, we show how this path is followed while its phase is being estimated for each visited site. Finally, in Figure 3.6c we show the obtained quadrature of the interferogram and its recovered phase. One attractive characteristic of the RPT method is that it obtains the unwrapped phase already unwrapped. However, for illustration proposes to compare the demodulated phase with the original interferogram, the recovered phase is shown wrapped in Figure 3.6c.

This estimation process is effective only if the phase variation may be locally approximated as a plane, that is, if the fringes are (locally) open. So, to estimate the phase of closed fringes, as shown in Figure 3.6, the RPT method scans the image in such a way that it always sees open fringes that is why the scanning strategy

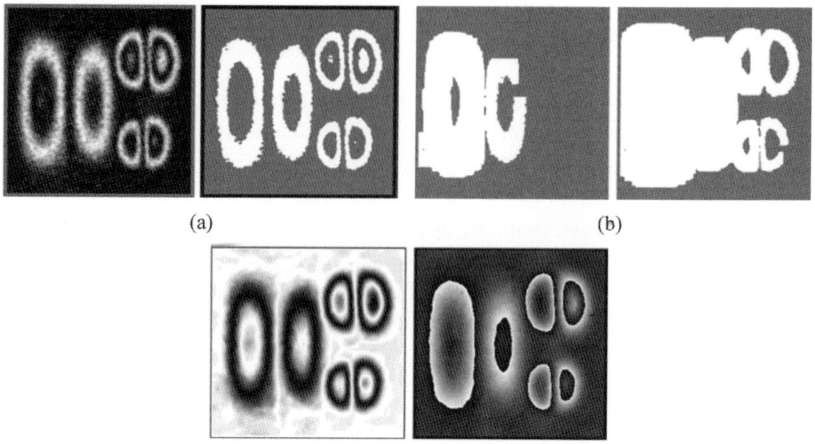

(a) (b)

(c)

Figure 3.7 (a) The normalized interferogram and the isophase path followed. (b) Two instants of the pixels demodulating process following isophase paths. (c) The quadrature image of the interferogram given in (a) and the recovered phase. We show the recovered phase wrapped for comparative purposes.

recommended follows the fringes. The robustness of the RPT method is improved if the size of the neighborhood Γ is increased, but its maximum size is limited by the smoothness of the underlying phase surface, that is, the size of the larger window neighborhood where a linear approximation to the phase remains valid.

3.6 Local Adaptive Robust Quadrature Filters

The RPT method is a very useful technique to demodulate a single fringe pattern with closed fringe. However, to correctly use it we must apply a normalization process before using the RPT [24]. In this section, we will see some useful techniques to demodulate a single interferogram with closed fringes that does not require previous normalization. These techniques use linear quadrature filters to demodulate the phase, unlike the RPT methods that are nonlinear systems. In particular, we are going to show a technique that we call *local adaptable robust quadrature filters* (LARQF).

In 1997, Marroquin *et al.* developed what we call robust quadrature filters (RQFs) [26]. As proposed by Marroquin, these RQF filters are full-field filters such as the convolution filters. The RQF filters have the advantage of being more robust (less sensitive) at the image borders, that is, these filters reduce the border effect that convolution filters introduce. Thus, the RQFs were proposed as an alternative to the convolution quadrature filters in order to reduce this border effect. As convolution filters, the RQFs can be used in the FTM technique and obtain better results at the image borders. However, this implies that the interferogram must have a carrier frequency (it must have open fringes). Therefore, the RQF cannot be applied to demodulate single interferograms with closed fringes. Here, we show a strategy for applying the RQF filters locally in such a way that we can demodulate a single-image interferogram with closed fringes. To do this successfully, we apply the RQFs in a small neighborhood Γ of each pixel of the interferogram, and each pixel is demodulated sequentially, in a close resemblance with the RPT. The local frequency on Γ is used as carrier frequency for these local RQFs. As the local carrier of the RQF is updated depending on the current site, the RQF gets adapted to the local frequency. That is why we call these filters local adaptive robust quadrature filters.

To apply LARQFs, we propose to minimize the following cost function:

$$U(f) = \sum_{(\eta,\xi)\in\Gamma} \|f(\eta,\xi) - I(\eta,\xi)\|^2 + \lambda\left[\sum_{C_a} \|f(\eta,\xi) - f(a,\xi)e^{-i\omega_x(\eta-a)}\|^2 \right.$$
$$\left. + \sum_{C_b} \|f(\eta,\xi) - f(\eta,b)e^{-i\omega_y(\xi-b)}\|^2\right], \tag{3.30}$$

where $f(\eta,\xi) = f_r(\eta,\xi) + i f_i(\eta,\xi)$ is the complex field that we obtain after minimization. The first term is the fidelity metric between the complex field we are estimating and the real valued observed data (the interferogram). The second and third terms constrain f into a smooth continuous complex field by first-order difference on f. The first-order difference on f is tuned to the local frequency (ω_x, ω_y). As we can see, these sums are over the sets called C_a and C_b. The set C_a is composed of all cliques in x-direction belonging to each site $(\eta,\xi) \in \Gamma$ and the set C_b is composed of all cliques in y-direction belonging to each site $(\eta,\xi) \in \Gamma$. In Figure 3.8, we illustrate the cliques in the x- and the y-directions belonging to the neighborhood Γ. As shown in this figure, a clique in x-direction is composed by

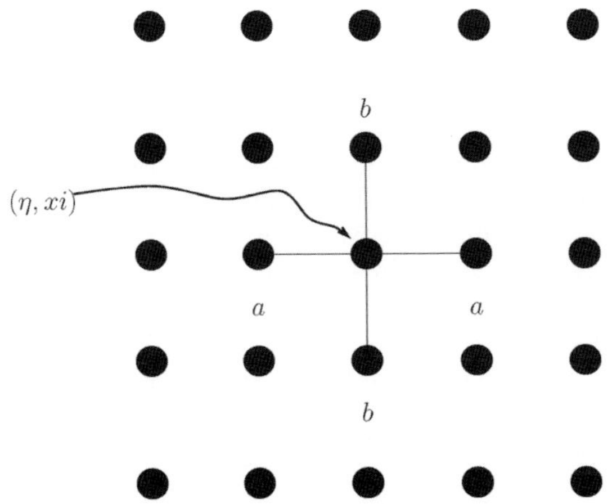

Figure 3.8 Square lattice of neighborhood Γ. Here, we show the clique set around site (η,ξ) into the neighborhood Γ. Each a forms a clique instance with site (η,ξ) along x-direction and each b forms a clique instance with site (η,ξ) along y-direction.

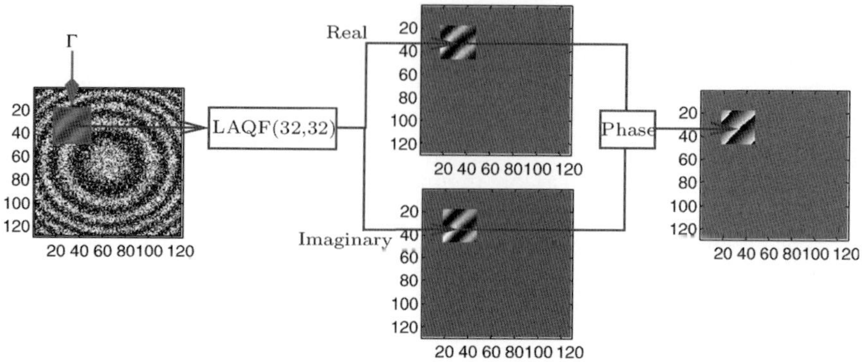

Figure 3.9 Here, we show how to apply the robust quadrature filter obtained by minimizing Equation 3.30 in a local neighborhood Γ. Once the complex field $\hat{f} \in \Gamma$ is obtained, the phase is computed using Equation 3.31.

the pixels located at (η, ξ), and $(\eta-1, \xi)$, when $a = \eta-1$. The other clique in x-direction is defined by the pixels located at (η, ξ), and $(\eta+1, \xi)$, when $a = \eta+1$.

If the interferogram $I(\eta, \xi)$ at (η, ξ) has a local carrier (ω_x, ω_y) and we take this frequency carrier for minimizing the LARQF (Equation 3.30), the complex field $\hat{f}$ that minimizes (3.30) will be the filtered complex field of the observed data. The filtered real interferogram is the real part, and the imaginary part is its filtered quadrature. Thus, the phase at the neighborhood Γ, around (x, y), is obtained as

$$\hat{\phi}(\eta, \xi) = \arctan\left[\frac{\mathrm{Im}\{f(\eta, \xi)\}}{\mathrm{Re}\{f(\eta, \xi)\}}\right]. \tag{3.31}$$

In Figure 3.9, it is shown how this process is applied in a neighborhood Γ around a site (x, y) of a fringe pattern with closed fringes. To minimize Equation 3.30, we take its gradient and equal it to zero. After this, we will have a linear equation system that can be solved using a far more efficient method than the steeped descent used for RPT. In particular, we have used the *Gauss–Seidel* numerical method to obtain $\hat{f}$ [25].

Like the RPT method, this method obtains the phase sequentially by following the isophase contours of the interferogram (the fringe path). Then, we use the same fringe following scanning strategy used for the RPT. However, unlike the RPT method, here we do not require previous fringe pattern normalization. In Figure 3.10, we show some experimental results where we obtain the modulating phase using the LARQF method.

In this test, we show two experimentally obtained interferograms. These interferograms were obtained from a defective plate and from a loaded elastic surface, respectively. Before applying the LARQF demodulator, we filter the interferograms using a conventional band-pass filter first to remove the background, as shown in Figure 3.10.

In addition, in Figure 3.11 we show a comparison between the RPT method and the LARQF method. There we show in a graph table the results obtained from simulated interferograms under different levels of noise. The first column shows the variance of

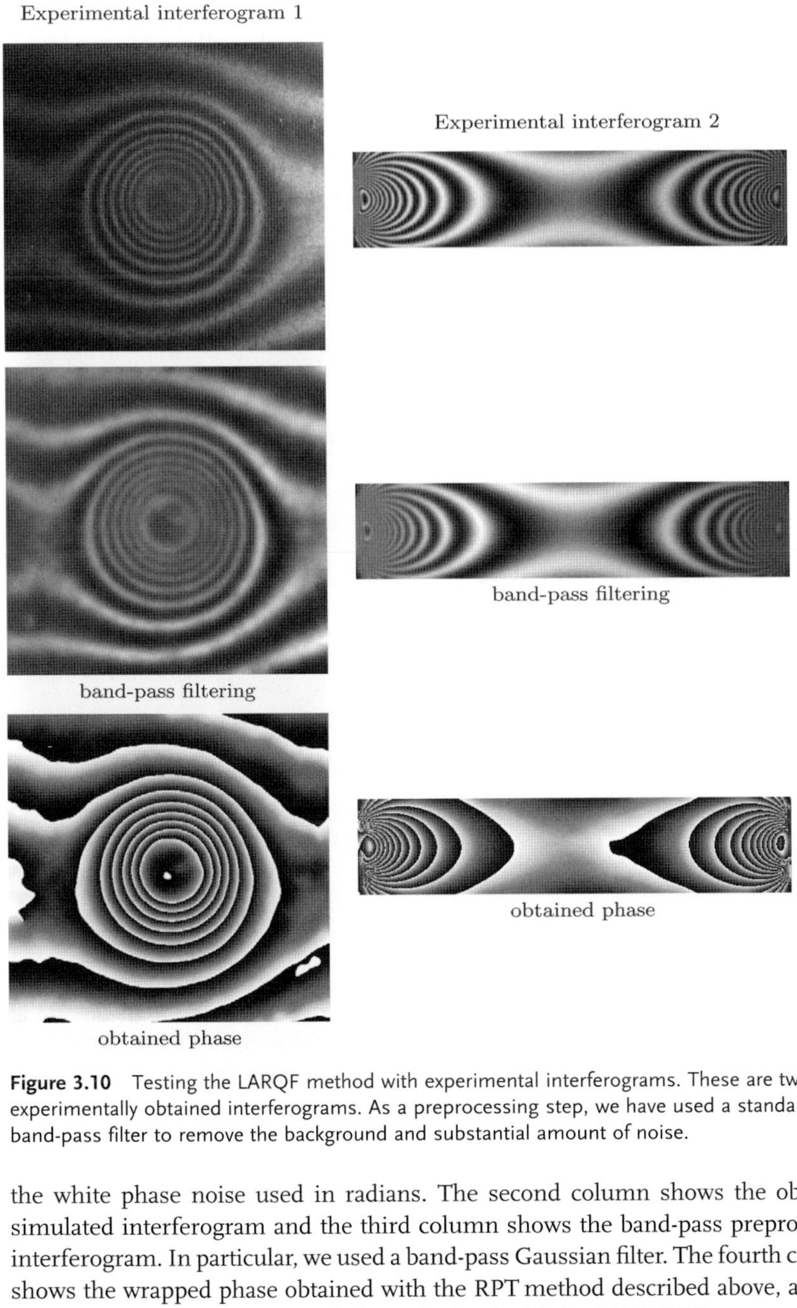

Figure 3.10 Testing the LARQF method with experimental interferograms. These are two experimentally obtained interferograms. As a preprocessing step, we have used a standard band-pass filter to remove the background and substantial amount of noise.

the white phase noise used in radians. The second column shows the obtained simulated interferogram and the third column shows the band-pass preprocessed interferogram. In particular, we used a band-pass Gaussian filter. The fourth column shows the wrapped phase obtained with the RPT method described above, and the fifth column shows the wrapped phase obtained with the LARQF. We can see that in most of the cases the RPTobtains a wrong modulating phase. This is because after the low-pass filtering the fringe pattern is not normalized as required by the RPT. We did this intentionally to show this drawback of the RPT method. On the other hand, the

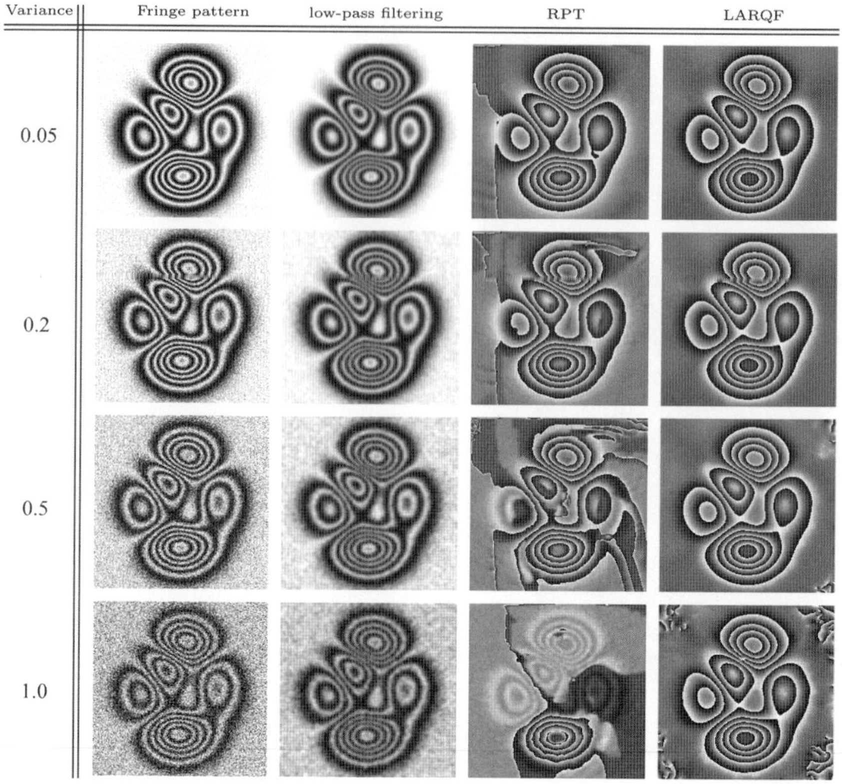

Figure 3.11 Comparison of the LARQF demodulation method with the RPT demodulation method. We can see that the LARQF technique outperforms the RPT when the fringe pattern is not normalized and it is severely noisy.

phase obtained with the LARQF corresponds to the expected modulating phase and it is obtained without fringe pattern normalization.

In conclusion, the LARQF is a very good alternative to the RPT technique. However, we must say that one of the main characteristics of the RPT method is that the RPT obtains the phase already unwrapped. When using the LARQF, it is a necessary phase unwrapping technique after demodulation [1].

3.7 Single Interferogram Demodulation Using Fringe Orientation

3.7.1 Orientation in Interferogram Processing

We are going to start this section by giving a global overview of the applications of the fringe orientation in interferogram processing. Some of these applications in phase demodulation will be discussed in detail later on.

In the field of digital image processing, orientation is a local feature linked to the spatial structure of an image. For example, regions with a border are highly structured and have a well-defined orientation. On the other hand, noisy or smooth areas are poorly oriented structures and consequently their orientation is not well defined. Therefore, orientation is a useful tool for corner detection [27] or directional filtering [28]. Specifically, orientation is a feature associated with locally simple images. For a locally simple image, the Fourier spectrum of a small neighborhood will be concentrated in a narrow spot oriented at the same angle as the local gradients [29].

Interferograms are the archetypical example of locally simple images. The usual model for an interferogram is a phase-modulated cosine irradiance signal, called the fringe pattern. Interferograms are locally highly structured due to the apparition of fringes; in them, the local orientation corresponds to the fringe orientation. For an interferogram excluding the phase extrema (maxima, minima, and saddle points), the fringe orientation can be computed from the image gradient. In the usual definition, the fringe orientation is parallel to the interferogram gradient. In addition, the fringe orientation can be interpreted as the local spatial frequency vector orientation in the Fourier space.

Historically, the first application of orientation in interferogram processing was directional filtering [30–35]. All these techniques assume that the noise is white, so the image is corrupted also by high-frequency noise. If the interferogram is processed in an arbitrary direction, it is difficult to separate the fringe signal from the noise. However, if we filter along the fringes, the signal becomes low frequency and it is easily separable from the noise.

In interferogram processing, the most important application of the fringe orientation is the phase demodulation from a single-image closed fringe interferogram. Research in recent years has demonstrated the central role that fringe orientation plays in robust phase demodulation from a single interferogram. As we will see later, the reason is that the two-dimensional demodulation process can be split into two stages: quadrature and fringe orientation computation.

Marroquin *et al.* [36] were the first to realize this possibility. They demonstrated how their original nonlinear regularized demodulation technique could be made more robust by dividing it into two iterative linear estimation processes, one for the orientation and another one for the spatial frequency.

At the same time, the original row-by-row phase tracker [22] was also improved thanks to the orientation information. The process of estimating both the spatial frequency and the local phase offset was enhanced using a fringe-following scanning strategy [37] implemented into the RPT. In this case, the orientation information was implicitly introduced by the fringe-following scanning strategy. Guiding the demodulation process as much as possible along isophasic lines decouples fairly well the local phase estimation from that of the spatial frequencies in its neighborhood, making the process far more robust.

This orientation information demonstrated its importance in single-image closed fringe interferogram demodulation; nevertheless, its exact function remained unclear. It was not until the work of Larkin *et al.* [38, 39] that the role of the fringe orientation was explicitly stated in phase demodulation. In his work, Larkin showed how the calculation of the modulating phase of a fringe pattern could be done only through the explicit calculation of the local fringe orientation using the so-called spiral phase transform. Although this work opened new ways for demodulation fringe patterns, the demonstration of the role of the fringe orientation was heuristic. As a result of this limitation, the mathematical origin of the orientation term and the spiral phase transform could not be clearly established.

Servin *et al.* [40] mathematically demonstrated the exact nature of the demodulation process with the so-called general *n*-dimensional quadrature transform. Starting with the standard interferogram mathematical model, Servin *et al.* obtained the general expression of the quadrature operator in the direct space (the space of the interferogram). This operator is divided into two parts. The first one is related to the fringe direction through the interferogram's phase. The second one is a nonlinear operator that under the approximation of local spatial monocromaticity (locally simple images) yields the spiral phase transform of Larkin. Finally, Quiroga *et al.* [41] demonstrated how the orientation information can be used to generalize any 1D spatial PSI algorithm to process general closed fringe interferograms.

With this brief introduction in mind, in the next sections we are going to explain in detail how the orientation can be computed, after which we will examine its application in phase demodulation from single-image interferograms.

3.7.2 Fringe Orientation and Fringe Direction

Graphically, orientation can be represented as a two-tip arrow, although *a priori* we cannot know the "true" direction of the arrow. To define the direction, we need to "eliminate" one tip and leave a single-tip arrow. In interferogram processing, we define the fringe direction as the vectorial field that points perpendicular to the fringes at every location. To be more specific, we are going to assume that the interferogram intensities can be described by the usual cosine formula

$$I(x,y) = b(x,y) + m(x,y)\cos[\phi(x,y)], \tag{3.32}$$

where I is the two-dimensional irradiance and b and m are low-frequency signals that represent, respectively, the background and the modulation of the fringe pattern. Finally, $\phi(x,y)$ is the modulating phase that we want to estimate. We want to make it clear that in this interferogram model we have the fringe pattern that has no spatial carrier. Also, for brevity we are going to eliminate the two-dimensional spatial dependence of our functions.

With this model, the fringes can be interpreted as isophase lines and the fringe direction at every point is given by

$$\mathbf{n}_\phi = \nabla\phi/|\nabla\phi| = (\omega_x, \omega_y)/|\nabla\phi|. \tag{3.33}$$

The vector $(\omega_x, \omega_y) = \nabla\phi$ represents the interferogram's spatial frequencies. This vector field always points perpendicular to the interferogram fringes and has a well-defined direction. An alternative representation of the fringe direction is the direction angle β subtended between the direction vector and the x-axis, given by

$$\beta = \arctan\frac{\omega_y}{\omega_x}. \tag{3.34}$$

The relation between the fringe direction vector and its angle is

$$\mathbf{n}_\phi = (\cos\beta, \sin\beta). \tag{3.35}$$

However, we do not have direct access to the interferogram phase. The phase is wrapped by the cosine function. Therefore, the orientation information must be obtained from the intensity of the interferogram. In this case, we cannot directly calculate the fringe direction vector, but we can do it for the fringe orientation vector. In effect, if we use Equation 3.33 with the intensities instead of the phase, we will obtain the fringe orientation vector given by

$$\mathbf{n}_I = \nabla I/|\nabla I| = \left(I_x, I_y\right)/|\nabla I|. \tag{3.36}$$

From now on, we are going to assume that the b and m signals are substantially smoother than the signal $\cos\phi$; with this condition, using our interferogram model Equation 3.32 in our last equation, we obtain

$$\mathbf{n}_I = -b\sin\phi(\nabla\phi/|\nabla I|) = -\text{sign}(\sin\phi)\cdot\mathbf{n}_\phi. \tag{3.37}$$

In case that b and m were not smooth, they would introduce an error in the fringe orientation [42]. One possible solution to this problem is to normalize the fringe pattern (background suppression and normalized modulation) [24, 43] or simply modify the experimental setup until the smoothness condition is fulfilled.

From Equation 3.37, the “two-tip” character of the fringe orientation vector $\mathbf{n}_I$ is clear. Every time the function $\sin\phi$ undergoes a sign change, $\mathbf{n}_I$ has a corresponding direction flip. Again, we can interpret $\mathbf{n}_I$ in angular terms and define the fringe orientation angle, θ, as the angle subtended by $\mathbf{n}_I$ with the x-axis.

$$\theta = \arctan\frac{I_y}{I_x}. \tag{3.38}$$

The relation between the fringe orientation vector and the fringe orientation angle is given by

$$\mathbf{n}_I = (\cos\theta, \sin\theta). \tag{3.39}$$

From their respective definitions, β and θ are defined modulo 2π and π, respectively. To clarify the meaning of the fringe direction and fringe orientation vectors, Figure 3.12 shows a 258×258 circular interferogram with the fringe direction and orientation vectors superposed; in addition, Figure 3.12 shows the fringe direction and orientation angles codified as a gray valued image. The fringe pattern is a radial chirp with spatial frequencies ranging from 0 to π, explicitly given by

$$I = \cos(0.5\pi(x^2 + y^2)/N), \quad x, y = -M, \ldots, 0, \ldots, M. \tag{3.40}$$

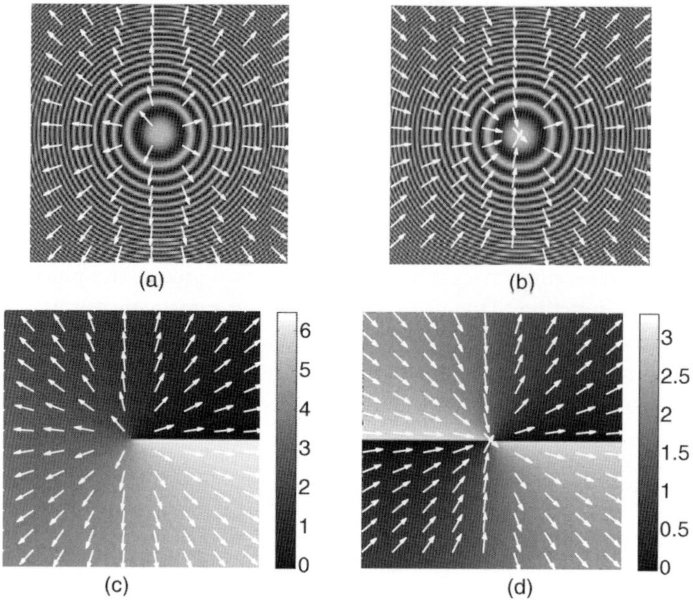

Figure 3.12 Fringe direction and fringe orientation. (a) Fringe direction vectors superposed onto a circular fringe pattern; (b) fringe orientation vectors using Equation 3.39; (c) fringe direction angle codified as a gray level image (black and white corresponds to 0 and 2π rad, respectively). The fringe direction vector is superposed. (d) Fringe orientation angle codified as a gray level image (black and white corresponds to 0 and π rad, respectively). The fringe orientation vector is superposed.

Figure 3.12 graphically depicts the two main properties that any orientation estimator must have. First, the estimator must be scale invariant, that is, independent of the local spatial frequency. Second, the estimator must be isotropic, that is, independent of the local fringe orientation.

Before we finish this section, we have two final caveats. The first refers to the orientation representations. At first glance, Equations 3.37 and 3.39 seem to be different. The orientation vector from Equation 3.37 has a direction flip every half fringe, while the orientation vector of Equation 3.39 smoothly varies in the range $[0, \pi]$ (as Figure 3.12 shows). However, both represent the same "double-tip arrow" character. Equation 3.37 is used in formal expressions, while Equation 3.39 is used to compute practically the orientation angle. The second warning refers to the digital calculation of fringe orientation and direction.

With the concepts of fringe orientation and direction in mind, we are going to proceed with the different techniques used for calculating fringe orientation.

3.7.3
Orientation Computation

In this section, we are going to introduce three numerical techniques based on the use of spatial derivatives, namely, gradient, plane fit, and the directional derivative techniques. A fourth Fourier transform-based method will be presented in Section 3.8.

3.7.3.1 **Gradient-Based Orientation Computation**

This is the most direct method for orientation calculation. It is based on the direct application of Equations 3.36 and 3.38 [42, 44]. A typical approximation for the gradient is

$$\nabla I = \frac{1}{2}(I(x-1,y)-I(x+1,y), I(x,y-1)-I(x,y+1)). \tag{3.41}$$

Figure 3.12 shows the orientation angle computed for a noiseless circular interferogram. However, for real interferograms, the direct gradient computation of the orientation is not feasible due to the signal noise. In Figure 3.13, we show graphically how the direct gradient of the orientation is affected by noise. Figure 3.13a and c shows the orientation vector and orientation angle obtained for the chirp fringe pattern of Figure 3.12a, with a Gaussian phase noise having a variance of 0.2 rad. The image sizes of Figure 3.13a and b are 258 × 258 and 303 × 289, respectively. As Figure 3.13b and d shows, for an experimental interferogram, the situation is even worse (compare qualitatively with Figure 3.12)

To alleviate the problem there are at least three possibilities: making the gradient estimation more robust, filtering the orientation angle, or using a combination of

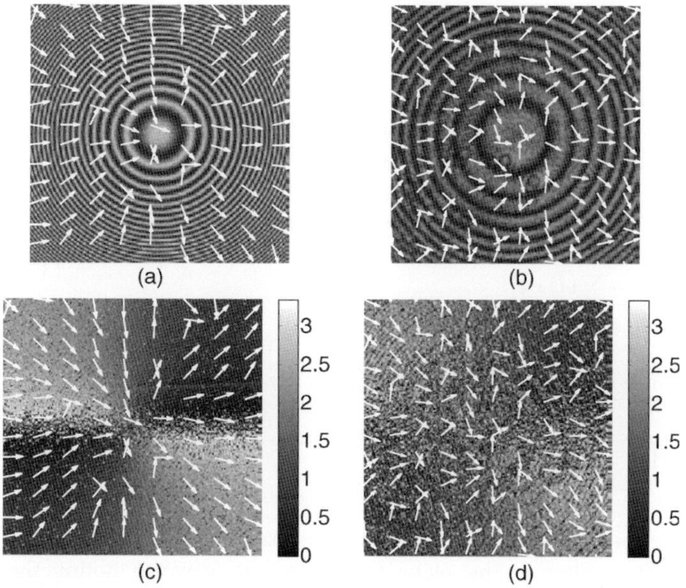

Figure 3.13 Fringe orientation in the presence of noise. (a) Chirp interferogram of Figure 3.12a with Gaussian phase noise with a variance of 0.2 rad. (b) Experimental interferogram. (c) Orientation angle obtained for (a) directly from the gradient information and (d) orientation angle obtained for (b) directly from the gradient information.

both. To make the gradient calculation more robust, a popular choice is the derivative of Gaussian (DoG) filter [42, 45].

The second possibility is to filter the orientation angle. However, for a general closed fringe interferogram, the orientation angle will present two π discontinuities for each fringe center, as can be seen in Figure 3.12. That means that conventional low-pass filtering techniques cannot be applied directly to the orientation angle. For wrapped signals, one possibility is to apply the sine–cosine filter [46]. This filter calculates the sine and the cosine of the wrapped signal generating two continuous signals that can be easily filtered. However, this sine–cosine filtering cannot be applied directly to the fringe orientation. The reason is that θ and $\theta+\pi$ rad represent the same orientation, but $\cos\theta = -\cos(\theta+\pi)$ and $\sin\theta = -\sin(\theta+\pi)$. If we average the sine and the cosine of the orientation, θ and $\theta+\pi$ orientations will cancel instead of summing up and the filtering will be incorrect. To solve this problem, the orientation must be doubled before filtering. In this case, the orientation averaging will annihilate perpendicular orientations and enforce parallel orientations. Mathematically, orientation sine–cosine filtering consists in computing first the sine and cosine of the double angle

$$\begin{aligned} S &= \sin 2\theta, \\ C &= \cos 2\theta, \end{aligned} \tag{3.42}$$

next both signals are filtered obtaining S_f and C_f, and finally the filtered fringe orientation is computed by

$$\theta_f = \frac{1}{2}\arctan\frac{S_f}{C_f}. \tag{3.43}$$

Typically, the filtering consists in a moving average over a small neighborhood of size N. Taking into account that $\sin 2\theta = 2\sin\theta\cos\theta$ and $\cos 2\theta = \cos^2\theta - \sin^2\theta$, we can express the filtering in terms of the irradiance gradient as

$$\begin{aligned} S_f &= \frac{\sum \sin 2\theta}{N^2} = \frac{\sum 2I_x I_y}{N^2}, \\ C_f &= \frac{\sum \cos 2\theta}{N^2} = \frac{\sum (I_y^2 - I_x^2)}{N^2}. \end{aligned} \tag{3.44}$$

And the orientation as

$$\theta = \frac{1}{2}\arctan\left(\frac{\sum 2I_x I_y}{\sum (I_y^2 - I_x^2)}\right), \tag{3.45}$$

where the irradiance derivatives can be directly estimated or using any robust alternative as the DoG filter or the plane fit method. It is worth noting that Equation 3.45 is also the orientation obtained from the structure tensor often used in computer vision [29].

In Figure 3.14, we show the performance of the DoG filter combined with a sine–cosine method. The DoG filter used in Figure 3.14 had $\sigma = 2$ pixels (px)

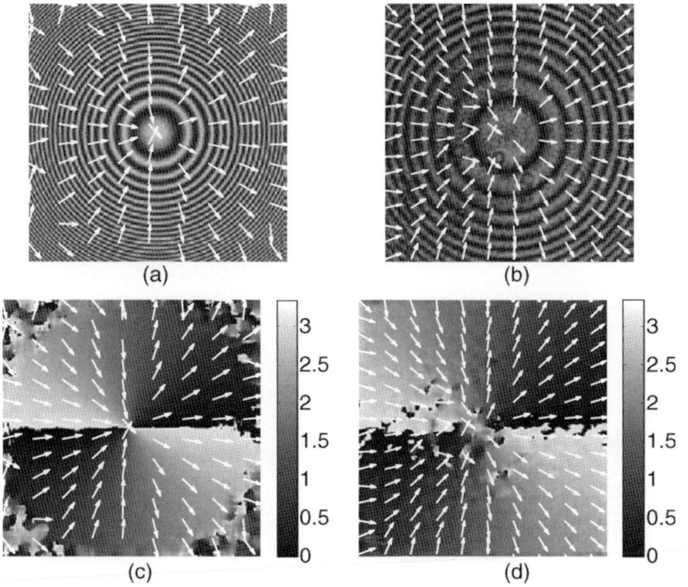

Figure 3.14 Fringe orientation in the presence of noise for the same images of Figure 3.13. In these images, we have used a DoG filter to estimate the gradients and a sine–cosine filter for the fringe orientation. (a) Chirp interferogram with Gaussian phase noise of 0.2 rad amplitude; (b) experimental interferogram; (c) orientation angle obtained for (a); and (d) orientation angle obtained for (b).

with a window size of 12 px. The sine–cosine filter used a moving averaging filter with a window of 10 px.

As Figure 3.14 shows, the combination of a DoG filter and the sine–cosine filtering is a powerful tool for fringe orientation computation. However, as both methods use local average over a small neighborhood, there is a dependence on the fringe's spatial frequency that must be taken into account [42, 44, 47]. This dependence can be observed in Figure 3.14c where the fringe orientation angle is not correctly computed at the high-frequency regions in the corners of the interferogram. In other words, the DoG method is not scale invariant; however, from the rotational symmetry of the fringe orientation, we can see that the method is isotropic.

3.7.3.2 **Plane Fit Orientation Calculation**

To reduce the influence of the noise in the gradient estimation, another possibility is the use of a local plane least square fit for every pixel of the interferogram [44, 47]. In this technique, the local irradiance at pixel (x, y) of a small neighborhood of size N is modeled as a plane $I(x, y) = a + bx + cy, \quad x, y = -N, \ldots, 0, \ldots, N$. The least square fit gives as parameters

$$a = \frac{\sum_{xy} I(x, y)}{(2N+1)^2}, \quad b = \frac{\sum_{xy} x \cdot I(x, y)}{\sum_{xy} x^2}, \quad c = \frac{\sum_{xy} y \cdot I(x, y)}{\sum_{xy} y^2}, \tag{3.46}$$

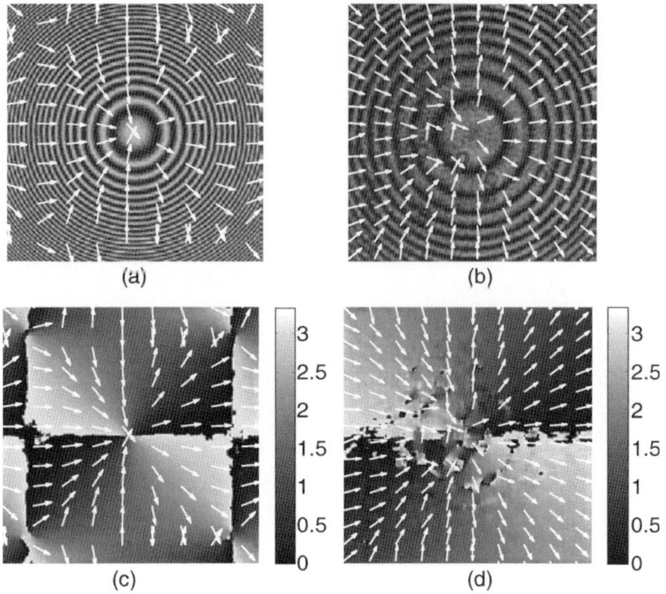

Figure 3.15 Orientation estimation by local plane fit and sine–cosine filtering. The images are the same as those in Figure 3.13. (a) Chirp interferogram with Gaussian phase noise of 0.2 rad amplitude; (b) experimental interferogram; (c) orientation angle obtained for (a); and (d) orientation angle obtained for (b).

And the fringe orientation angle is given by

$$\theta = \arctan(c/b) = \arctan\left(\frac{\sum_{xy} y \cdot I(x,y)}{\sum_{xy} x \cdot I(x,y)}\right), \tag{3.47}$$

where we have used that due to the window symmetry $\sum_{xy} x^2 = \sum_{xy} y^2$.

In Figure 3.15, we show the results obtained for a combination of gradient estimation by plane fit and sine–cosine filtering for the same interferograms of Figure 3.13. In this case, the plane fit used a 5×5 window and the sine–cosine filtering employed a 10×10 moving average. As in the case of DoG, filtering window size affects the performance of the orientation estimation for high spatial frequencies. However, the structure of the noise is very different. In this case, the recovery error has a quasiperiodic structure [44, 47] associated with the number of fringes that fit in the plane fit processing window. In the particular case of Figure 3.15a, the radial chirp interferogram has a variable spatial frequency. At the points with about 5 pixels/fringe, the 5×5 plane fit will incorrectly estimate the local gradient. In Figure 3.15c, this area can be seen as "extra" fringe center in the orientation. In this case, we can see that the plane fit method is not scale invariant and that the isotropic properties are not as effective as in the case of the DoG filter.

3.7.3.3 Minimum Directional Derivative

This method is based on the fact that the directional derivative has a minimum along the fringe. If the directional derivative is determined for 0, 45, 90, and 135° from the measured derivatives, it is possible to obtain the direction for which the directional derivative has a minimum [48]. The directional derivatives can be defined as

$$\begin{aligned} d_0 &= \sqrt{2}\cdot |I(x-1,y)-I(x+1,y)|,\\ d_{45} &= |I(x-1,y+1)-I(x+1,y-1)|,\\ d_{90} &= \sqrt{2}\cdot |I(x,y-1)-I(x,y+1)|,\\ d_{135} &= |I(x-1,y-1)-I(x+1,y+1)|. \end{aligned} \tag{3.48}$$

From these expressions, we can calculate the cumulative sum for every square neighborhood $D_i = \sum_{xy} d_i$, and from them the fringe orientation can be obtained as [48]

$$\theta = \frac{1}{2}\arctan\left(\frac{D_{45}-D_{135}}{D_0-D_{90}}\right) + \pi/2. \tag{3.49}$$

In principle, this technique seems very similar to the DoG or the plane fit methods. However, it combines seamlessly the sine–cosine averaging with first differences. In fact, the minimum directional derivative method is more isotropic and scale invariant than the DoG and plane fit methods, as we show in Figure 3.16. In this figure, we

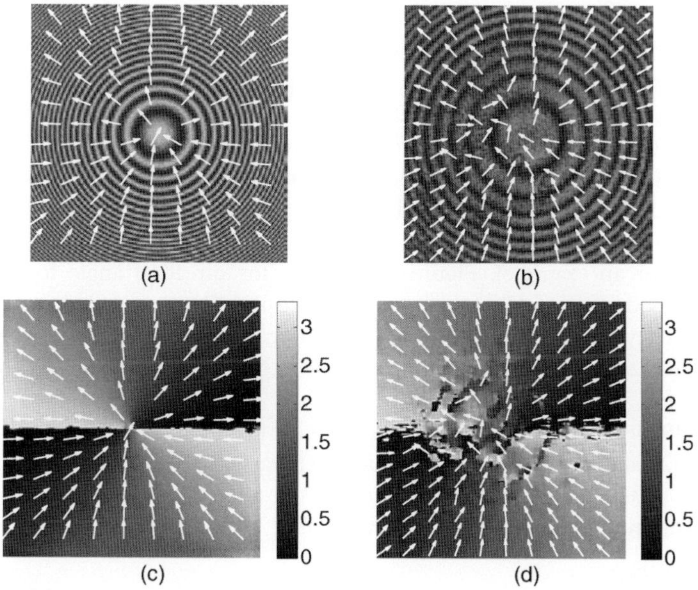

Figure 3.16 Orientation estimation by local plane fit and sine–cosine filtering for the same images as in Figure 3.13. (a) Chirp interferogram of equation with Gaussian phase noise of 0.2 rad amplitude; (b) experimental interferogram; (c) orientation angle obtained for (a); and (d) orientation angle obtained for (b).

show the orientation estimation for the same interferograms as the former examples. In this case, we have used the minimum directional derivative method with a cumulative sum window of 12×12 to make it comparable with the former results. As can be seen in the results of Figure 3.16, the isotropic and scale invariant properties of the minimum directional derivative method are better than the DoG or plane fit techniques.

3.7.4 Direction Computation

In this section, we are going to present two methods to solve the two-tip character of the orientation in order to obtain the fringe direction (one-tip arrows).

3.7.4.1 Regularized Phase Tracking Direction Estimation

In this technique, the direction estimation is presented as a phase unwrapping problem. From their definitions, the relation between the orientation and the direction angles is

$$\theta = \beta + k\pi, \tag{3.50}$$

where k is an integer such that $0 \leq \theta \leq \pi$; as a consequence,

$$W[2\theta] = W[2\beta + 2k\pi] = W[2\beta], \tag{3.51}$$

where W denotes the modulo 2π wrapping operator, that is, for any ϕ, $W[\phi] = \phi + 2k\pi$ with k being an integer such that $-\pi < W[\phi] \leq \pi$. Equation 3.51 states that the wrapped version of 2θ is equal to the wrapped version of 2β. Thus, by unwrapping $W[2\theta]$, it is possible to obtain 2β.

The unwrapping of $W[2\theta]$ is not a straightforward unwrapping problem since, for a closed fringe interferogram, the direction angle β is a discontinuous function [49]. If the center of a closed fringe is encircled, the direction angle will present a continuous variation from 0 to 2π rad, presenting a 2π discontinuity starting at the fringe center (see Figure 3.12c for an example). Figure 3.17 shows the direction angle for a complex-shaped closed fringe pattern generated using the peak function of MATLAB. Therefore, we can say that the direction angle is a circular magnitude "2π continuous." As a result, 2β will be a discontinuous signal for closed fringe patterns, but "4π continuous."

Quiroga *et al.* [49] proposed a regularized phase tracker [37] method for the phase unwrapping of $W[2\theta]$ adapted to the "4π continuous" character of 2β. This method is based on the RPT phase unwrapping procedure, but the regularization term takes into account discontinuous character of 2β. If we define

$$\begin{aligned} C &= \cos(W[2\theta]), \\ S &= \sin(W[2\theta]), \end{aligned} \tag{3.52}$$

the RPT method will consist in the minimization with respect to $(2\beta, \nu_x, \nu_y)$of the next cost function

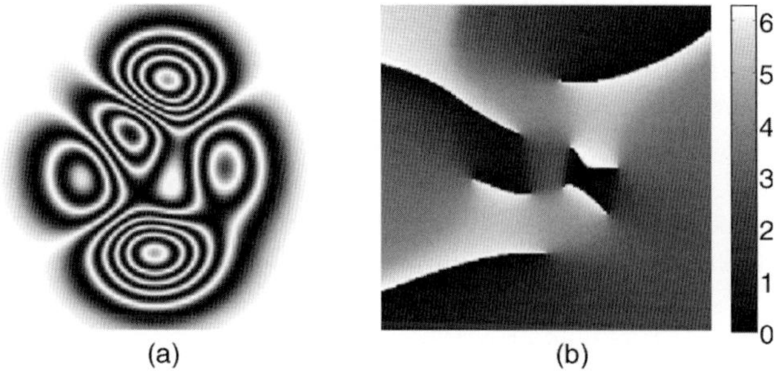

Figure 3.17 (a) Complex fringe pattern and (b) its direction angle. Observe the 2π discontinuities starting at the fringe centers and saddle points.

$$U_r(2\beta, \nu_x, \nu_y) = \sum_{\xi\eta \in N \cap L} |C(\xi,\eta) - \cos p(x,y,\xi,\eta)|^2 + |S(\xi,\eta) - \sin p(x,y,\xi,\eta)|^2 + \lambda |W_{4\pi}[2\beta(\xi,\eta) - p(x,y,\xi,\eta)]|^2 s(\xi,\eta), \tag{3.53}$$

where N is the size of the local neighborhood used and $p(x,y,\xi,\eta)$ is given by

$$p(x,y,\xi,\eta) = 2\beta(x,y) + \nu_x(x,y)(x-\xi) + \nu_y(x,y)(y-\eta), \tag{3.54}$$

where $W_{4\pi}$ denotes the modulo 4π wrapping operator, that is, for any ϕ, $W_{4\pi}[\phi] = \phi + 4k\pi$, with k being an integer such that $-2\pi < W_{4\pi}[\phi] \leq 2\pi$. As in the usual RPT methods, the cost function of Equation 3.53 is minimized at every spatial location and the image is scanned sequentially and/or guided by a quality map. In Equation 3.53, the regularization term is adapted to the 4π continuous character of 2β by using the 4π wrapping operator instead of the usual local difference. Due to the 4π continuous character of 2β, the RPT technique and any other direction estimation method depend on the scanning strategy used. The scanning strategy must guarantee that the critical orientation points, fringe centers, saddle points, and flat phase regions are encircled and processed last. Typical quality maps for guiding the path-following minimization of cost function (Equation 3.53) are the modulus of the fringe pattern gradient, $|\nabla I|$, or the amplitude of the phasor used to compute the orientation. For example, in the case of the minimum directional derivative method from Equation 3.49, the quality map for guiding the path-following minimization could be

$$B_\theta = \sqrt{(D_{45} - D_{135})^2 + (D_0 - D_{90})^2}. \tag{3.55}$$

The RPT direction estimation shares with the RPT unwrapping technique its performance with the noise. However, the nonlinear minimization of Equation 3.53 imposes a heavy computational load for standard image and regularizing neighborhood sizes. For a 512×512 image with a neighborhood size of 10 px on a modern computer, one should expect a processing time of about 10 min.

3.7.4.2 Vector Field-Regularized Direction Estimation

The nonlinear character of Equation 3.53 makes the local optimization a heavy computational task. To overcome this problem in this section, we are going to introduce a regularized local linear technique for direction estimation.

In this case, instead of calculating the direction angle directly, one can try to estimate the direction vector. Thus, the goal is to find a smooth vector field parallel to the measured orientation vector field [50]. This method is denominated the vector field-regularized direction estimation (VFR).

In 2D, this problem can be easily formulated as a local minimization. If $\mathbf{n}_I$ is the orientation vector (Equation 3.36), we can find a perpendicular smooth vector field $\mathbf{p} = (p_x, p_y)$ by minimizing at every location the cost function

$$U_{\mathbf{r}}(\mathbf{p}) = \sum_{\mathbf{r}' \in N \cap L} (\mathbf{p}(\mathbf{r}) \cdot \mathbf{n}_I(\mathbf{r}'))^2 + \lambda \|\mathbf{p}(\mathbf{r}) - \mathbf{p}(\mathbf{r}')\|^2 s(\mathbf{r}'), \tag{3.56}$$

where $s(\mathbf{r})$ is a Boolean function used to indicate if the site $\mathbf{r}$ has already been estimated, λ is the regularization parameter, and N indicates the neighborhood around site $\mathbf{r}$. Once $\mathbf{p}$ is obtained, the direction field is obtained as

$$\mathbf{n}_\phi = (-p_y, p_x) \tag{3.57}$$

and the fringe direction angle as

$$\beta = \arctan \frac{-p_x}{p_y}. \tag{3.58}$$

Details about the minimization of Equation 3.56 and the local lineal system that it generates can be found in Ref. [50]. This technique is fast, locally linear, and also very robust due to the regularization approach. In this case, for a 512×512 image with a neighborhood size of 10 px on a modern computer, one should expect a processing time of about 10 s.

To illustrate the performance of the RPT and VFR techniques, Figure 3.18 shows the direction angle obtained for the fringe pattern of Figure 3.17 using both methods. Compare the results obtained with the theoretical results shown in Figure 3.17. In both cases, the quality map used to guide the image scanning was the amplitude of the orientation phasor obtained using the minimum directional derivative method. Here, the processing neighborhood was 11×11 with regularization parameter $\lambda = 1$. In Figure 3.18, the fringe pattern is a 258×258 px image generated using the MATLAB peaks. For the RPT method, we used the MATLAB function *fminunc* for the local minimization of the cost function (Equation 3.53). Processing time was 470.5 s for the RPT and 3.3 s for the VFR. Although Figure 3.18 shows the results for a fringe pattern, it is a good example of the general behavior of both techniques. The RPT approach generates better estimations, but needs more processing time. In contrast, the VFR tends to generate a somewhat poorer direction estimation, although the processing time is much smaller.

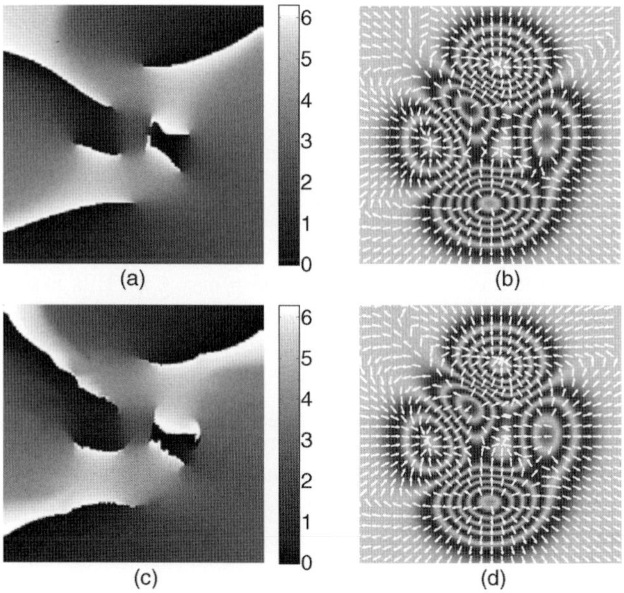

Figure 3.18 Direction estimation results obtained for the RPT and the VFR methods compare the results obtained with the theoretical direction shown in Figure 3.17. (a) and (b) direction angle and direction vector obtained using the RPT. (c) and (d) direction angle and direction vector obtained using the VFR.

3.8 Quadrature Operators

3.8.1 Phase Demodulation of 1D Interferograms

The 1D Hilbert transform, H_1, is a very useful mathematical tool to obtain the quadrature of a single-frequency sinusoidal signal $I(x) = \cos \omega_0 x$. According to Bracewell [51], the Hilbert transform of such signal is

$$H_1(I) = -\sin \omega_0 x, \quad \omega_0 \neq 0. \tag{3.59}$$

Because of this, we may have the (wrong) impression that the Hilbert transform H_1 always gives the quadrature of a cosine signal and that the phase can be demodulated by computing the phase of the analytic signal $A = I - iH_1(I)$ [51]. The transforming properties of the Hilbert operator become clear by looking at the form of the frequency response of H_1 as applied to a real function $I(x) = \cos \phi(x)$, which is [51]

$$\mathrm{FT}(H_1(I)) = H_1(u)I(u) = -i\,\mathrm{sign}\,u \cdot I(u) = -i\frac{u}{|u|}I(u), \tag{3.60}$$

where the Fourier transform is represented by FT( ), u is the frequency variable, and $I(u) = \mathrm{FT}(I)$. According to this, the Hilbert transform still renders the expected result (the quadrature signal) when it is applied to a more complicated carrier frequency signal, that is,

$$H_1(\cos(\omega_0 x + \psi)) = -\sin(\omega_0 x + \psi), \tag{3.61}$$

provided that the local frequency does not change its sign, that is,

$$\omega_0 + \mathrm{d}\psi/\mathrm{d}x > 0 \quad \forall x. \tag{3.62}$$

In this case, this cosine signal has its two spectral lobes well separated, and the H_1 operator may be used to recover the modulating phase $\phi = \omega_0 x + \psi$. This is the reason why the Fourier method is so widely used to find the modulating phase of carrier frequency interferograms.

Unfortunately, Equation 3.61 will not hold if condition (3.62) is not satisfied, that is, if the modulating phase is not a monotonically increasing function of x. Without a carrier, one needs to know the sign of the local frequency to obtain the expected quadrature signal. Therefore, to obtain the one-dimensional quadrature of $\cos\phi(x)$, one needs to use the following formula that relates the 1D quadrature transform Q_1 to the 1D Hilbert transform H_1:

$$Q_1(\cos\phi) = \frac{\mathrm{d}\phi/\mathrm{d}x}{|\mathrm{d}\phi/\mathrm{d}x|} H_1(\cos\phi) = \operatorname{sign}\omega_x \cdot H_1(\cos\phi) = -\sin\phi. \tag{3.63}$$

Applying Q_1 reduces them to H_1 only when the local frequency ω_x of the signal is positive everywhere, so that $\operatorname{sign}\omega_x = 1$ over the whole domain of interest. In general, to obtain the quadrature signal of the cosine of a nonmonotonic function $\phi(x)$, one needs to know the sign of the local spatial frequency to correct the sign of the signal obtained by the application of the Hilbert transform. Consequently, in 1D the demodulation of a general interferogram is a nonlinear process composed of two steps: the linear application of the Hilbert transform and the nonlinear estimation of $\operatorname{sign}\omega_x$.

3.8.2
Phase Demodulation from a Single Interferogram: the Vortex Transform

The extension H_1 to the 2D case is not direct. How can we define in 2D the equivalent of the 1D Hilbert transform? A straightforward generalization for the 1D Hilbert transform given by

$$H_2(u, v) = \operatorname{sign} u \quad \text{or} \quad H_2(u, v) = \operatorname{sign} v \tag{3.64}$$

was used by Kreis [3] to find the modulating phase of a closed fringe interferogram. However, as in the 1D case, unless the modulating phase is monotonic, with two well-separated spectral side lobes (as in spatial carrier interferometry), the Hilbert transforms (Equation 3.64) will not work [40]. In other words, if the Hilbert transforms (Equation 3.64) are used, the obtained phase will always be monotonic. In the case of closed fringe interferograms, where the phase is not spatially monotonic, the estimated phase obtained using the Hilbert operators (3.64) will not represent the actual nonmonotonic modulating phase. Thus, it will present serious spurious ringing effects along the path where this phase is stationary.

To solve this problem and extend the Hilbert transform to 2D, Larkin *et al.* [38, 39] proposed a 2D isotropic Hilbert transform based on the Spiral phase linear operator given by

$$S(u,v) = \frac{u+iv}{\sqrt{u^2+v^2}} = \exp(i\alpha(u,v)), \tag{3.65}$$

where $\alpha(u,v)$ is the polar angle in the frequency space. Following Larkin, the Spiral operator can be interpreted as the 2D equivalent of the sign function of Equation 3.60. In fact, any section through the origin of $S(u,v)$ is a sign function. We are going to analyze the application of the Spiral operator to a monochromatic interferogram given by

$$I(\mathbf{r}) = \cos(\omega_0 \cdot \mathbf{r}) = \cos(\omega_{0x}x + \omega_{0y}y). \tag{3.66}$$

The Fourier transform of $I(\mathbf{r})$ is

$$I(\mathbf{q}) = \frac{1}{2}(\delta(\mathbf{q}-\omega_0) + \delta(\mathbf{q}+\omega_0)), \tag{3.67}$$

where $\mathbf{q} = (u,v)$ denotes a point in the frequency space. In Fourier space, the application of the Spiral operator is equivalent to the product of its frequency response by the signal spectra.

$$\mathrm{FT}(S(I)) = I(\mathbf{q}) \cdot S(\mathbf{q}) = \frac{1}{2}(\delta(\mathbf{q}-\omega_0) + \delta(\mathbf{q}+\omega_0))\exp(i\alpha(\mathbf{q})). \tag{3.68}$$

Using the properties of the delta functions, we obtain

$$\mathrm{FT}(S(I)) = \frac{1}{2}(\delta(\mathbf{q}-\boldsymbol{\omega}_0) - \delta(\mathbf{q}+\boldsymbol{\omega}_0))\exp(i\alpha(\boldsymbol{\omega}_0)) \tag{3.69}$$

with

$$\alpha(\boldsymbol{\omega}_0) = \arctan(\omega_{0y}/\omega_{0x}) = \beta, \tag{3.70}$$

and finally,

$$\mathrm{FT}(S(I)) = \frac{1}{2}(\delta(\mathbf{q}-\boldsymbol{\omega}_0) - \delta(\mathbf{q}+\boldsymbol{\omega}_0))\exp(i\beta), \tag{3.71}$$

where β is the fringe direction angle defined in (3.34). If we make the inverse Fourier transform back to the direct space, we obtain

$$S(I) = i\exp(i\beta)\sin(\boldsymbol{\omega}_0 \cdot \mathbf{r}). \tag{3.72}$$

Equation 3.72 is a very important result. It states that the Spiral operator applied to a monochromatic fringe pattern returns the interferogram's quadrature with a phase factor given by the fringe direction angle. From Equation 3.72, Larkin proposes the next 2D quadrature operator:

$$V(I) = i\exp(-i\beta)S(I). \tag{3.73}$$

This linear quadrature operator is denominated “vortex transform” by Larkin. Although the result of Equation 3.73 has been derived for a monochromatic interferogram, it can be extended in general to a single-image closed fringe pattern with a nonmonotonic phase. If I is a general fringe pattern, DC filtered and with constant AC term [43] given by

$$I(\mathbf{r}) = m\cos\phi(\mathbf{r}), \tag{3.74}$$

the quadrature term can be obtained applying the vortex transform

$$V(I) = i\exp(-i\beta(\mathbf{r}))S(I) \approx -m\sin\phi(\mathbf{r}), \tag{3.75}$$

where $\beta(\mathbf{r})$ is the local fringe direction angle. The modulating phase can be obtained from the analytical signal

$$A(\mathbf{r}) = I(\mathbf{r}) - iV(I). \tag{3.76}$$

Equation 3.75 represents the first demodulation method in which the role of the fringe direction and the local magnitude of the spatial frequencies are clearly factorized. The demodulation process of an interferogram can be divided into two stages: the nonlinear fringe direction computation and the application of a linear filter. As Larkin [39] and Servin [40] demonstrate, Equation 3.75 is a very good approximation of the exact quadrature operator. As we have shown above, Equation 3.75 is exact only in the case of monochromatic phase. However, any smooth phase function ϕ can always be modeled locally as a plane. As a consequence, “locally” the interferogram spectrum will consist of two spectral side lobes instead of the two deltas of Equation 3.67. The application of the spiral transform will invert the sign of one of the side lobes, as in Equation 3.71, and the inverse Fourier transform will yield the quadrature signal. A formal stationary phase analysis can be found in Ref. [39]. A very important result of the stationary phase analysis is that the error in approximation (3.75) is proportional to the phase curvature. Another interesting point is that for a general nonmonochromatic interferogram, the vortex operator is no longer linear, as in Equation 3.73, due to the necessity of computing the local direction $\beta(\mathbf{r})$.

Figure 3.19 shows the demodulation results obtained using the vortex transform with two experimental interferograms. Figure 3.19a depicts a 430×474 Fizeau interferogram. Figure 3.19b shows the demodulated phase. In this case, a normalization and direction computation is necessary. The total processing time was 4.5 s. Figure 3.19c shows a 574×768 circular dark field isochromatic fringe pattern of a diametrally loaded disk. The demodulated phase obtained using the vortex is shown in Figure 3.19d. In this case, the total processing time for normalization, direction computation, and vortex transform was 6 s.

3.8.3 Vortex Transform-Based Orientation Computation

We are going to briefly return to the orientation computation presenting a Fourier transform method based on the vortex operator [52]. If I is a general fringe pattern,

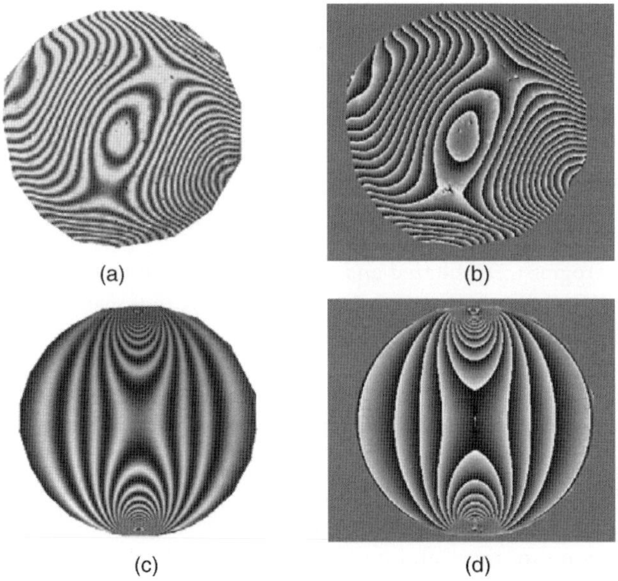

Figure 3.19 Demodulation results obtained using the vortex transform. (a) Fizeau interferogram, (b) demodulated phase, (c) photoelastic fringe pattern, (d) and demodulated phase (see text for details).

DC-filtered and with constant AC term [43] as given by Equation 3.74, we have seen that the application of the Spiral transform gives

$$S(I) = i\exp(i\beta(\mathbf{r}))\sin(\phi). \tag{3.77}$$

If we apply again the Spiral phase operator, we have

$$S(S(I)) = S^2(I) = i\exp(i2\beta(\mathbf{r}))\cos(\phi), \tag{3.78}$$

from which the orientation can be computed:

$$[iS(I)]^2 - iI(\mathbf{r})S^2(I) = m^2\exp(2i\beta(\mathbf{r})). \tag{3.79}$$

We say orientation because from the phasor signal of Equation 3.79 we obtain $W[2\beta(\mathbf{r})] = W[2\theta(\mathbf{r})]$. For replication periodic fringe patterns over rectangular regions of interest, Equation 3.79 is an interesting frequency–space isotropic alternative to the direct-space methods of Section 3.8.

3.8.4
The General n-Dimensional Quadrature Transform

In the previous sections, we have discussed the problem of the demodulation of 1D signals and concluded that for a general nonmonotonic modulating phase, the demodulation can be divided into two steps: the nonlinear estimation of the frequency sign and the linear application of the Hilbert transform. We continued analyzing the 2D case in the frequency space using the vortex transform. Again, for a general case of nonmonotonic phase, the demodulation is a nonlinear process with two steps: nonlinear direction estimation and the linear application of the Spiral phase operator.

In this section, we will present a general n-dimensional quadrature transform that works well both for closed fringes and for carrier frequency interferograms [40]. Moreover, the result of this section will also permit us to obtain as a special case the two-dimensional results stated in the previous sections.

Given a general interferogram, DC filtered and with constant modulation m, the expected result of an n-dimensional quadrature operator $Q_n(\,)$ is to transform the fringe pattern into its quadrature. This operation may be represented by

$$Q_n(I) = -m \sin \phi(\mathbf{r}), \tag{3.80}$$

where $\mathbf{r} = (x_1, \ldots, x_N)$ is the n-dimensional vector position. The modulating phase can be obtained from the angle of the analytical signal

$$A(\mathbf{r}) = I - iQ_n(I). \tag{3.81}$$

Following Servin *et al.* [40], the first step toward obtaining the quadrature signal will be to obtain the gradient of the fringe pattern, which is

$$\nabla I = -m\nabla \phi \sin \phi, \tag{3.82}$$

where we are assuming that the AC signal is smooth or constant. If we multiply both sides by $\nabla\phi$, we obtain

$$\nabla I \cdot \nabla\phi = -m|\nabla\phi|^2 \sin \phi. \tag{3.83}$$

Rearranging the terms, one obtains

$$\frac{\nabla I}{|\nabla\phi|} \cdot \frac{\nabla\phi}{|\nabla\phi|} = -m \sin \phi. \tag{3.84}$$

That finally gives place to the generalized quadrature transform (GQT):

$$Q_n(I) = \mathbf{n}_\phi \cdot \mathbf{H}_n(I) = -m \sin \phi, \tag{3.85}$$

where

$$\mathbf{n}_\phi = \frac{\nabla\phi}{|\nabla\phi|} \tag{3.86}$$

is the fringe direction vector and

$$\mathbf{H}_n(I) = \frac{\nabla I}{|\nabla\phi|} \tag{3.87}$$

is the generalized n-dimensional Hilbert transform. The GQT expression (Equation 3.85) factorizes the demodulation of an interferogram in two nonlinear operations: the first is the computation of the fringe direction vector and the second is the nonlinear Hilbert transform, which under local monochromaticity can be approximated by the vortex transform. In vectorial notation, the Spiral phase transform spectrum is given by

$$\mathbf{S}_n(\mathbf{q}) = \frac{i\mathbf{q}}{|\mathbf{q}|}, \tag{3.88}$$

where $\mathbf{q} = \sum_1^N q_i \mathbf{e}_i$ is the position vector in Fourier space. In this notation, the Fourier transform of a linear combination of vectors is given by

$$\mathrm{FT}\left(\sum_1^N a_i \mathbf{e}_i\right) = \sum_1^N \mathrm{FT}(a_i)\, \mathbf{e}_i \tag{3.89}$$

By applying the properties of the Fourier transform of the derivative, we get

$$\mathrm{FT}(\mathbf{S}_n(I)) = \frac{i\mathbf{q}}{|\mathbf{q}|}\mathrm{FT}(I) = \frac{1}{|\mathbf{q}|}\mathrm{FT}(\nabla I). \tag{3.90}$$

If the modulating phase is locally monochromatic, it can be demonstrated [40] that the application of a linear filter can be approximated as

$$T(\mathbf{q}) \cdot \mathrm{FT}(I) \approx \mathrm{FT}(T(\boldsymbol{\omega}(\mathbf{r})) \cdot I(\mathbf{r})), \tag{3.91}$$

where $\omega(\mathbf{r}) = \nabla\phi$ is the spatial frequency vector. In other words, if the phase is locally monochromatic (i.e., can be locally approximated by a plane), we can identify the Fourier space coordinates as spatial frequencies and approximate $T(\mathbf{q}) \approx T(\boldsymbol{\omega}(\mathbf{r}))$. In our case, the local monochromaticity makes it possible to approximate $|\mathbf{q}| \approx |\boldsymbol{\omega}(\mathbf{r})|$ and Equation 3.90 becomes

$$\mathrm{FT}(\mathbf{S}_n(I)) \approx \mathrm{FT}\left(\frac{\nabla I}{\left(\sum_1^N \omega_i^2\right)^{1/2}}\right) = \mathrm{FT}\left(\frac{\nabla I}{|\nabla\phi|}\right). \tag{3.92}$$

This finally gives

$$\mathbf{S}_n(I) \approx \frac{\nabla I}{|\nabla\phi|} = \mathbf{H}_n(I). \tag{3.93}$$

That is, the vectorial spiral transform (including the 2D complex version of Equation 3.65) is an approximation of the Hilbert transform (Equation 3.87), assuming local monochromaticity (smooth continuous modulating phase). In the case of a pure monochromatic phase (straight fringes), the approximations turn into equalities. One can see that the GQT reduces in 1D and 2D to the results previously presented and makes straightforward the extension to the n-dimensional case. Especially interesting are the spatiotemporal interferograms that can be treated as 3D signals. Equations 3.86 and 3.87 also tell us that the exact demodulation is possible only if $|\nabla\phi| \neq 0$, otherwise neither the fringe direction nor the Hilbert operator is defined at these sites. Practically, this means that at the closed fringe centers and the constant phase regions of a 2D interferogram, we will always obtain nonreliable results.

3.9
2D Steering of 1D Phase Shifting Algorithms

In addition to clarifying the role of the Hilbert transform and the fringe direction in the demodulation process, the GQT operator (Equation 3.85) give us the path to generalize in the direct space (the space of the interferogram image) all the existing 1D SPSI methods, extrapolating here to the general case of closed fringe interferograms [41].

We are going to present the 2D case, but the technique can automatically handle the n-dimensional case. Any SPSI filter can be interpreted as a 1D quadrature filter along direction k (rows or columns); this method, applied to a general interferogram (Equation 3.32), will generate an analytical signal given by

$$g_k(I) = f_k(I) + ih_k(I) = m \cdot g_k(\omega_k)(\cos\phi + i\sin\phi)k = \{x, y\}, \tag{3.94}$$

where $g_k(\omega_k) = \mathrm{FT}(g_k)$ is the frequency response of the filter. For nonlinear filters, it is also possible to compute $g(\omega_k)$ as shown in Ref. [41]. In Equation 3.94, f_k is a low-pass filter used to eliminate the background signal and h_k is a Hilbert filter that, for the selected range of frequencies, filters the background and generates the quadrature. For example, a typical SPSI filter is the five-step Hariharan method applied along rows

$$g_x(x) = (2\delta(x)-\delta(x-2)-\delta(x+2)) + 2i(\delta(x-1)-\delta(x+1)), \tag{3.95}$$

where

$$f_x(x) = 2\delta(x)-\delta(x-2)-\delta(x+2). \tag{3.96}$$

And

$$h_x(x) = 2(\delta(x-1)-\delta(x+1)). \tag{3.97}$$

If we apply the Hariharan SPSI to an interferogram, we will obtain

$$g_x(I) = g_x{}^*I = (2I(x,y)-I(x-2,y)-I(x+2,y)) + 2i(I(x-1,y)-I(x+1,y)), \tag{3.98}$$

where * denotes the convolution operation. From the phase of the analytical signal $g_x(I)$, we can obtain ϕ. As we have seen, in the case of a closed fringe interferogram, this filter will fail and the obtained phase will be monotonic in x-direction. The same SPSI applied along columns will have the same problem: it will fail and will generate a monotonic phase in y-direction.

$$g_y(I) = g_y{}^*I = (2I(x,y)-I(x,y-2)-I(x,y+2)) + 2i(I(x,y-1)-I(x,y+1)). \tag{3.99}$$

Thus, the question is how can we extend the 1D SPSI filters to the 2D case? The GQT offers a solution to this problem [41]. First, we must compose a vectorial Hilbert operator as

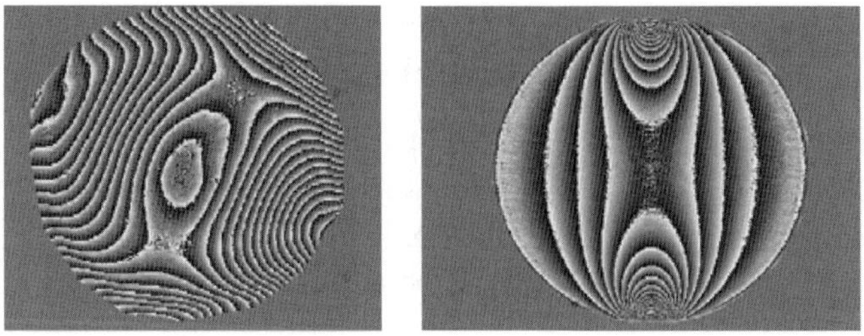

(a) (b)

Figure 3.20 Demodulation results obtained steering two 1D SPSI methods. (a) Demodulated phase of the Fizeau interferogram of Figure 3.19a. (b) Demodulated phase of the photoelastic pattern of Figure 3.19b (see text for details).

$$\mathbf{h}_2(I) = (h_x(I), h_y(I)). \tag{3.100}$$

And from it compute the quadrature as

$$q_2(I) = \mathbf{n}_\phi \cdot \mathbf{h}_2(I) = -mg(\omega_\mathbf{x}) + g(\omega_y)\sin\phi. \tag{3.101}$$

The low-pass part of the 2D SPSI can be constructed as

$$f_2(I) = f_x(I) + f_y(I) = m(g(\omega_x) + g(\omega_y))\cos\phi. \tag{3.102}$$

And finally, the phase can be computed from the analytical signal

$$A(\mathbf{r}) = f_2(I) - iq_2(I). \tag{3.103}$$

Equations 3.101 and 3.102 tell us how we can steer any 1D SPSI by composing the operations along rows and columns using the direction information.

Figure 3.20 shows the demodulation results obtained using two steered 1D SPSI methods. Figure 3.20a depicts the wrapped phase obtained for the interferogram of Figure 3.19a using the ultrawide band SPSI, as presented in Ref. [17]. The total processing time, including the direction computation, was 4 s. Figure 3.20b shows the demodulated phase of the photoelastic fringe pattern of Figure 3.19b using the SPSI method of [53]. In this case, the processing time was 14 s including the direction computation.

3.10 Conclusions

In this chapter, we have presented several techniques to deal with the problem of demodulating a single interferogram with a spatial carrier and without this phase inclination.

As seen in this chapter, demodulation of a single interferogram with spatial carrier was pioneered by Takeda *et al.* [2] with their Fourier analysis technique. Many improvements have been made to this basic technique, but the original method is now as useful as it was when proposed. After the landmark of Fourier interferometry, the use (among other carrier methods) of phase shifting algorithms in the spatial domain was proposed. This, in turn, helped to improve the understanding of the phase shifting algorithms in general to finally propose a comprehensive theory of phase shifting interferometry based on the existing theory of linear system engineering [9].

On the other hand, in the past 10 years or so the number of different techniques to phase demodulate a single interferogram with closed fringes has substantially grown. For limitation of space, we have not reviewed all the methods available in the literature, but the presented methods give an idea of the additional difficulties faced when solving this problem compared to the case when a spatial carrier is introduced.

References

1 Malacara, D., Servin, M., and Malacara, Z. (2005) *Interferogram Analysis for Optical Testing*, 2nd edn, CRC Taylor and Francis, Boca Raton.
2 Takeda, M., Ina, H., and Kobayashi, K. (1982) Fourier-transform method of fringe-pattern analysis for computer-based topography and interferometry. *J. Opt. Soc. Am. A*, **720** (1), 156–160.
3 Kreis, T. (1986) Digital holographic interference-phase measurement using the Fourier-transform method. *J. Opt. Soc. Am. A*, **3** (6), 847–855.
4 Fernandez, A., Blanco-Garcia, J., Doval, A.F., Bugarin, J., Dorrio, B.V., Lopez, C., Alen, A.J., Perez-Amor, M., and Fernandez, J.L. (1998) Transient deformation measurement by double-pulsed-subtraction TV holography and the Fourier transform method. *Appl. Optics*, **370** (16), 3440–3446.
5 Pedrini, G. and Tiziani, H.J. (1994) Double-pulse electronic speckle interferometry for vibration analysis. *Appl. Optics*, **330** (34), 7857–7863.
6 Freischlad, K. and Koliopoulos, C.L. (1990) Fourier description of digital phase-measuring interferometry. *J. Opt. Soc. Am. A*, **70** (4), 542–551.
7 Phillion, D.W. (1997) General methods for generating phase-shifting interferometry algorithms. *Appl. Optics*, **360** (31), 8098–8115.
8 Surrel, Y. (1996) Design of algorithms for phase measurements by the use of phase stepping. *Appl. Optics*, **350** (1), 51–60.
9 Servin, M., Estrada, J.C., and Quiroga, J.C. (2009) The general theory of phase shifting algorithms. *Opt. Express*, **170** (24), 21867–21881.
10 Proakis, J.G. and Manolakis, D.G. (1995) *Digital Signal Processing. Principles, Algorithms, and Applications*, 3rd edn, Prentice-Hall, New York.
11 Carré, P. (1966) Installation et utilisation du comparateur photoelectrique et interferentiel du bureau international des poids et measures. *Metrologia*, **2** (0), 13–23.
12 Bruning, J.H., Herriott, D.R., Gallagher, J.E., Rosenfeld, D.P., White, A.D., and Brangaccio, D.J. (1974) Digital wavefront measuring interferometer for testing optical surfaces and lenses. *Appl. Optics*, **130** (11), 2693.
13 Wyant, J.C. (1975) Use of an AC heterodyne lateral shear interferometer with real-time wavefront correction systems. *Appl. Optics*, **140** (11), 2622–2626.
14 Morgan, C.J. (1982) Least-squares estimation in phase-measurement interferometry. *Opt. Lett.*, **70** (8), 368–370.
15 Schwider, J., Burow, R., Elssner, K.E., Grzanna, J., Spolaczyk, R., and Merkel, K. (1983) Digital wave-front measuring interferometry: some systematic error sources. *Appl. Optics*, **220** (21), 3421–3432.
16 Hariharan, P., Oreb, B.F., and Eiju, T. (1987) Digital phase-shifting interferometry: a simple error-compensating phase calculation algorithm. *Appl. Optics*, **260** (13), 2504–2506.

17 Estrada, J.C., Servin, M., and Quiroga, J.A. (2009) Easy and straightforward construction of wideband phase-shifting algorithms for interferometry. *Opt. Lett.*, **340** (4), 413–415.

18 Servin, M., Estrada, J.C., and Quiroga, J.A. (2009) Spectral analysis of phase shifting algorithms. *Opt. Express*, **170** (19), 16423–16428.

19 Ransom, P.L. and Kokal, J.B. (1986) Interferogram analysis by a modified sinusoid fitting technique. *Appl. Optics*, **25** (22), 4199–4204.

20 Servin, M. and Cuevas, F.J. (1995) A novel technique for spatial phase-shifting interferometry. *J. Mod. Optic.*, **420** (9), 1853–1862.

21 Gomez-Pedrero, J.A., Quiroga, J.A., and Servin, M. (2004) Temporal evaluation of fringe patterns with spatial carrier with an improved asynchronous phase demodulation algorithm. *J. Mod. Opt.*, **51** (0), 97–109.

22 Servin, M., Marroquin, J.L., and Cuevas, F.J. (1997) Demodulation of a single interferogram by use of a two-dimensional regularized phase-tracking technique. *Appl. Optics*, **360** (19), 4540–4548.

23 Servin, M., Marroquin, J.L., and Quiroga, J.A. (2004) Regularized quadrature and phase tracking from a single closed-fringe interferogram. *J. Opt. Soc. Am. A*, **210** (3), 411–419.

24 Guerrero, J.A., Marroquin, J.L., Rivera, M., and Quiroga, J.A. (2005) Adaptive monogenic filtering and normalization of ESPI fringe patterns. *Opt. Lett.*, **30**, 3018–3020.

25 Nocedal, J. and Stephen, J.W. (2006) *Numerical Optimization, Springer Series in Operations Research*, Springer-Verlag, New York.

26 Marroquin, J.L., Figueroa, J.E., and Servin, M. (1997) Robust quadrature filters. *J. Opt. Soc. Am. A*, **140** (4), 779–791.

27 Harris, C. and Stephens, M. (1988) A combined corner and edge detector. Proceedings of the 4th Alvey Vision Conference, Aug. 31–Sept. 2, 1988, University of Manchester, pp. 147–151.

28 Freeman, W.T. and Adelson, E.H. (1991) The design and use of steerable filters. *IEEE Trans. Pattern Anal.*, **13** (9), 891–906.

29 Granlund, G.H. and Knutsson, H. (1995) *Signal Processing for Computer Vision*, Kluwer Academic Publishers.

30 Yu, Q., Sun, X., Liu, X., and Qiu, Z. (2002) Spin filtering with curve windows for interferometric fringe patterns. *Appl. Optics*, **41** (14), 2650–2654.

30 Yu, Q., Yang, X., Fu, S., and Sun, X. (2005) Two improved algorithms with which to obtain contoured windows for fringe patterns generated by electronic speckle-pattern interferometry. *Appl. Optics*, **44** (33), 7050–7054.

31 Zhang, F., Liu, W., Wang, J., Zhu, Y., and Xia, L. (2009) Anisotropic partial differential equation noise-reduction algorithm based on fringe feature for ESPI. *Opt. Commun.*, **282** (12), 2318–2326.

32 Tang, C., Han, L., Ren, H., Gao, T., Wang, Z., and Tang, K. (2009) The oriented-couple partial differential equations for filtering in wrapped phase patterns. *Opt. Express*, **17** (7), 5606–5617.

33 Wang, H., Kemao, Q., Wenjing, G., Lin, F., and Soon, S.H. (2009) Fringe pattern denoising using coherence-enhancing diffusion. *Opt. Lett.*, **34** (8), 1141–1143.

34 Villa, J., Quiroga, J.A., and De la Rosa, I. (2009) Regularized quadratic cost function for oriented fringe pattern filtering. *Opt. Lett.*, **34** (11), 1741–1743.

35 Marroquin, J.L., Rodriguez Vera, R., and Servin, M. (1998) Local phase from local orientation by solution of a sequence of linear systems. *J. Opt. Soc. Am. A*, **15** (6), 1536–1544.

36 Servin, M., Marroquin, J.L., and Cuevas, F.J. (2001) Fringe-follower regularized phase tracker for demodulation of closed-fringe interferograms. *J. Opt. Soc. Am. A*, **18** (3), 689–695.

37 Larkin, K.G., Bone, D.J., and Oldfield, M.A. (2001) Natural demodulation of two-dimensional fringe patterns I. General background of the spiral phase quadrature transform. *J. Opt. Soc. Am. A*, **18** (8), 1862–1870.

38 Larkin, K.G. (2001) Natural demodulation of two-dimensional fringe patterns II. Stationary phase analysis of the spiral phase quadrature transform. *J. Opt. Soc. Am. A*, **18** (8), 1871–1881.

39 Servin, M., Quiroga, J.A., and Marroquin, J.L. (2003) General *n*-dimensional quadrature transform and its application to interferogram demodulation. *J. Opt. Soc. Am. A*, **20** (5), 925–934.

40 Quiroga, J.A., Servin, M., Estrada, J.C., and Gomez-Pedrero, J.A. (2009) Steerable spatial phase shifting applied to single-image closed-fringe interferograms. *Appl. Optics*, **48**, 2401–2409.

41 Zhou, X., Baird, J.P., and Arnold, J.F. (1999) Fringe-orientation estimation by use of a Gaussian gradient filter and neighboring-direction averaging. *Appl. Optics*, **38** (5), 795–804.

42 Quiroga, J.A. and Servin, M. (2003) Isotropic *n*-dimensional fringe pattern normalization. *Opt. Commun.*, **224**, 221–227.

43 Yang, X., Qifeng, Y., and Sihua, F. (2007) A combined method for obtaining fringe orientations of ESPI. *Opt. Commun.*, **273** (1), 60–66.

44 Gonzalez, R.C. and Woods, R.E. (2002) *Digital Image Processing*, 3rd edn, Prentice Hall, New York.

45 Aebischer, H.A. and Waldner, S. (1999) A simple and effective method for filtering speckle-interferometric phase fringe patterns. *Opt. Commun.*, **162**, 205–210.

46 Fu, S., Hui, L., Jinsong, C., and Qifeng, Y. (2007) Influence of window size on the fringe orientation estimation. *Opt. Commun.*, **272** (1), 73–80.

47 Yang, X., Qifeng, Y., and Sihua, F. (2007) An algorithm for estimating both fringe orientation and fringe density. *Opt. Commun.*, **274** (2), 286–292.

48 Quiroga, J.A., Servin, M., and Cuevas, F.J. (2002) Modulo 2π fringe orientation angle estimation by phase unwrapping with a regularized phase tracking algorithm. *J. Opt. Soc. Am. A*, **19**, 1524–1531.

49 Villa, J., Ismael De la Rosa, I., Gerardo Miramontes, G., and Juan Antonio Quiroga, J.A. (2005) Phase recovery from a single fringe pattern using an orientational vector-field-regularized estimator. *J. Opt. Soc. Am. A*, **22** (12), 2766–2773.

50 Bracewell, R.N. (2000) *The Fourier Transform and its Applications*, McGraw-Hill Book Co., New York.

51 Larkin, K.G. (2005) Uniform estimation of orientation using local and non-local 2-D energy operators. *Opt. Express*, **13**, 8097–8121.

52 Gómez-Pedrero, J.A., Quiroga, J.A., and Servín, M. (2008) Adaptive asynchronous algorithm for fringe pattern demodulation. *Appl. Optics*, **47**, 3954–3961.

4
Phase Evaluation in Temporal Speckle Pattern Interferometry Using Time–Frequency Methods

Alejandro Federico and Guillermo H. Kaufmann

4.1
Introduction

Temporal speckle pattern interferometry (TSPI) is an optical metrology technique primarily developed for the measurement of a wide range of low-speed dynamic physical quantities, which provides full field of view quantitative information without making contact with the object under test. With the sustained advance of imaging sensor technology and computers, various speckle-based and related techniques naturally matured toward its own digital conception that is nowadays well established. The TSPI technique is not an exemption, and the investigation of the generated digital signal and the development of novel image processing algorithms contribute even more to stand this concept, which increases its range of applications [1, 2]. It is interesting to highlight the development of digital holographic interferometry, where the analysis and synthesis of wavefronts are made by means of a computer and its range of applications competes with the well-known and assessed speckle interferometry techniques [3]. The breakthrough of the digital technology undoubtedly encourages the development of speckle metrology techniques, which requires the application of novel digital algorithms for data processing.

The TSPI technique was introduced in the early 1990s and appeared as a digital technique for measuring a range of dynamic physical quantities produced by diffusely reflecting objects [4]. In the application of the TSPI technique, the dynamics of the rough object produces intensity fluctuations at all pixels belonging to the time series of speckle interferograms recorded during the temporal history of the phenomenon to be analyzed. The sequential acquisition of a large number of speckle interferograms and its digital postprocessing facilitate the recovery of the optical phase distribution, which is associated with the physical quantity to be measured. Therefore, the whole temporal evolution of the analyzed physical parameter can be determined by recovering the object phase distribution. In this framework, by adding a known phase change as a function of time that acts as a temporal carrier, the optical phase distribution is commonly recovered using a point-by-point

Advances in Speckle Metrology and Related Techniques. Edited by Guillermo H. Kaufmann

ISBN: 978-3-527-40957-0

phase shifting algorithm along the temporal axis. If a sufficiently high temporal sampling rate is used, each pixel can be unwrapped as a function of time. It is important to note that this phase unwrapping procedure involves one-dimensional temporal phase signal analysis and tends to avoid the implementation of two- and three-dimensional complex spatiotemporal unwrapping procedures, which minimizes the propagation of phase recovery errors. However, the weakness of the acquired TSPI signals lies in its low tolerance to unwanted effects such as nonmodulated pixels, modulation loss, and noise, which sometimes make the object phase recovery process quite difficult. Although several strategies aiming the reduction of the experimental errors introduced in the phase recovery process are available, one of the key issues in TSPI is the development of more robust phase measurement approaches using special processing algorithms.

This chapter reviews various algorithms for processing TSPI signals, which were developed mostly during the last decade. The selected topics are analyzed and discussed within the conceptual frame of the time–frequency representations of the temporal evolution of the intensity signal acquired at each pixel of the charge-coupled device (CCD) camera. This framework allows developing a systematic approach in the phase recovery procedure, which is also advantageous when other related optical metrology techniques such as digital speckle pattern interferometry, fringe projection, and moiré are used. These issues are naturally addressed if the temporal variable is considered a dummy variable that can represent a privileged spatial direction. Therefore, the reader should note that the processing algorithms analyzed in this chapter can also be applied to the previously mentioned techniques.

4.2 The Temporal Speckle Pattern Interferometry Signal

The TSPI technique is based on phase measurement interferometry, which can be used to directly measure dynamic physical quantities by means of optical systems such as Michelson or Mach–Zehnder interferometers [5]. A typical example is the measurement of dynamic mechanical deformations generated by objects with nonspecular surfaces (i.e., the wavelength of the light is negligibly small compared to the surface roughness). Using a detector array and a temporal phase shifting device as the relevant components of the optical setup, the recorded data can be stored in a computer system and subsequently analyzed to evaluate the object phase distribution, which contains the information on physical quantity. Recently, the introduction of powerful lasers and high-speed digital recording devices in the optical systems have allowed to expand the temporal measurement range to more than 100 000 frames per second. The main characteristics of the TSPI signals are discussed below and the interested readers should also read Chapters 2 and 3 of Ref. [4] for technical details of the TSPI technique.

Digital phase measurement interferometry is implemented in TSPI by the superposition of two coherent optical fields having intensities I_1 and I_2, which is determined in a given spatiotemporal coordinate by the well-known interferometric equation:

$$I = I_1 + I_2 + 2\sqrt{I_1 I_2}\cos(\phi_1 - \phi_2), \tag{4.1}$$

where I is the resulting intensity and ϕ_1 and ϕ_2 are the phases associated with I_1 and I_2, respectively. As the phase difference $\phi_1 - \phi_2$ in Equation 4.1 is in general impossible to recover univocally, a known phase function, named the carrier, is commonly added to the interferometric optical setup to solve the ambiguity. The resulting intensity I is usually sampled by means of a CCD camera and the corresponding three-dimensional intensity data are recorded to be subsequently analyzed. In the presence of speckle fields, the terms in Equation 4.1 are affected by the fine structure of the speckle grains. Therefore, the recovery of the phase difference $\phi_1 - \phi_2$ is not a simple matter, as it will be shown later. To gain insight into this issue, as a typical TSPI signal we will consider the one generated in the measurement of dynamic deformations of rough objects when an out-of-plane speckle interferometer is used.

In an out-of-plane speckle interferometer (see Section 3.3 of Ref. [4]), Equation 4.1 is commonly written as

$$\begin{aligned} I(m,n,t) &= I_0(m,n,t) + I_r + 2\sqrt{I_0(m,n,t) I_r} \\ &\quad \times \cos[\phi(m,n,t) + \phi_c(m,n,t) + \phi_r(m,n,t)], \end{aligned} \tag{4.2}$$

where $I_0(m,n,t)$ and I_r are the object and the reference beam intensities, respectively, $\phi(m,n,t)$ is the object phase to be recovered, $\phi_c(m,n,t)$ is the carrier phase introduced in the reference beam with a constant temporal amplitude, and $\phi_r(m,n,t)$ is the random speckle phase. The integer indices $m = 1, 2, \ldots, N_m$ and $n = 1, 2, \ldots, N_n$ represent the spatial sampling coordinates at the CCD camera, where N_m and N_n are the number of pixels along the horizontal and vertical directions, respectively. $t = 1, 2, \ldots, N_t$ is the temporal coordinate and N_t is the number of temporal frames. Equation 4.2 is generally expressed in terms of the modulation intensity I_M and the intensity bias I_B to obtain the well-known formula for the resulting intensity $I(m,n,t)$ given by

$$\begin{aligned} I(m,n,t) &= I_B(m,n,t) + I_M(m,n,t) \\ &\times \cos[\phi(m,n,t) + \phi_c(m,n,t) + \phi_r(m,n,t)], \end{aligned} \tag{4.3}$$

$$I_M = 2\sqrt{I_0 I_r}, \tag{4.4}$$

$$I_B = I_0 + I_r. \tag{4.5}$$

In a speckle interferometer, the accuracy of the phase measurement is ultimately limited by various decorrelation sources that appear in the speckle fields produced by the successive deformation states of the tested object. The speckle decorrelation is itself an effect of the random variation in I_B, I_M, and ϕ_r (see Section 4.4 in Ref. [4]).

To illustrate the influence of the bias and the modulation intensities in the phase recovery process, Figure 4.1 shows a typical temporal history of a speckle pattern interference signal at a given pixel of the CCD camera. The resulting intensity $I(t)$ at the analyzed pixel corresponds to an object that was rotated in such a way that the phase distribution is a linear function of time. Here, a constant phase shift was introduced between the successive temporal frames, named a temporal carrier, and the resulting intensity was discretized in 256 gray levels. The presence of noise is clearly observed in the TSPI signal shown in Figure 4.1. The influence of the temporal variability, which is associated with the bias intensity and the existence of regions with very low modulation intensity values, can also be noted in the same figure. As the argument of the cosine function in Equation 4.3 and also the bias and the modulation intensities are random variables, a lower bound is imposed to the variance of the recovered phase. Therefore, this randomness makes the object phase ϕ to be susceptible to the introduction of phase evaluation errors, even if additional error sources such as sampling and intensity considerations, phase shifter calibration, detector nonlinearities, vibration, and air turbulence are removed. For these reasons, TSPI signals contain not only the information on the physical quantity to be measured but also the changes introduced in the fine speckle structure given by various decorrelation sources [6].

The development of experimental techniques aiming the reduction of decorrelation effects in speckle fields is an active area of research. However, the approach adopted in this chapter is the estimation of the phase coded in TSPI signals using space and time–frequency representations recently introduced in the literature. It will be shown that this point of view allows measuring robustly the space and time variations of the object phase and the instantaneous frequencies contained in the TSPI signals.

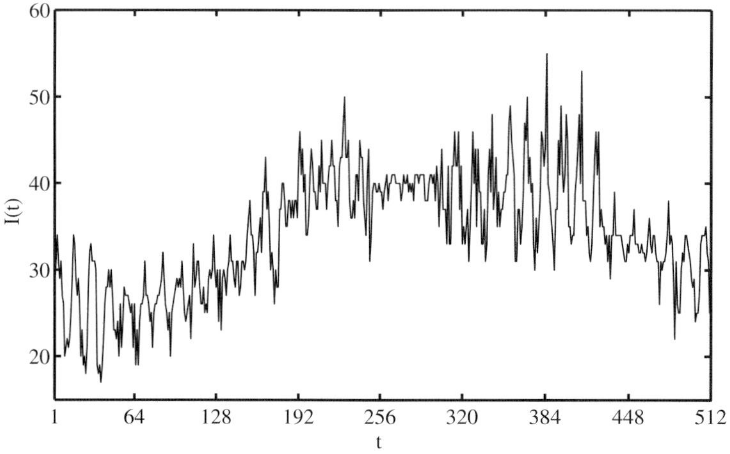

Figure 4.1 Temporal history of a speckle pattern interference signal at a given pixel of a CCD camera. (Reproduced with permission from Ref. [7], the Optical Society of America.)

4.3
The Temporal Fourier Transform Method

The Fourier transform FT and its inverse FT^{-1} establish a one-to-one relation between the time domain $\{t \in \mathbb{R}\}$ of a signal $s(t)$ and its frequency domain $\{\omega \in \mathbb{R}\}$, with $\mathbb{R}$ being the set of real numbers:

$$FT[s(t)](\omega) = \hat{s}(\omega) = \int_{-\infty}^{+\infty} dt e^{-i\omega t} s(t), \tag{4.6}$$

$$FT^{-1}[\hat{s}(\omega)](t) = s(t) = \frac{1}{2\pi}\int_{-\infty}^{+\infty} d\omega e^{i\omega t}\, \hat{s}(\omega). \tag{4.7}$$

It is assumed that $s(t)$ verifies the necessary conditions to guarantee the existence of the Fourier transform and its inverse [8]. Although the Fourier transform allows the passage from one domain to the other univocally, this does not allow a combination of the two domains. Time domain and frequency domain are two ways of looking a signal $s(t)$. The aim of introducing time–frequency signal representations is to characterize the TSPI signals in a time–frequency domain. This time–frequency combination yields a temporal localization of the spectral components of the signal, that is, to give information on the nonstationariness of the object phase. However, when time variations of the intensity signal are sufficiently slow and the frequency carrier is sufficiently high, Equation 4.1 can be separated in the frequency domain by recognizing the main contributions that appears in its spectrum (see Section 2.2.1.12 in Ref. [4]). Then, the multiplication of the spectrum by a given window function W in the frequency domain isolates one of the suitable frequency contributions. Therefore, by using the inverse Fourier transform, the object phase difference can be finally obtained if the user knows the frequency of the carrier introduced in the interferometer. With these reduced framework conditions, it is still possible to analyze a wide variety of measurement problems without considering the use of time–frequency algorithms (e.g., see Refs [9–14]). Below follows a discussion of the so-called Fourier transform phase evaluation method, which also introduces the necessity of using time–frequency algorithms in TSPI. This phase evaluation method is exemplified by analyzing the TSPI signal depicted in Figure 4.1.

The Fourier transform of Equation 4.3 results

$$\begin{aligned} FT[I(t)](\omega) &= FT[I_B(t)](\omega) + FT\left\{\frac{I_M(t)}{2} e^{i[\phi(t)+\phi_r(t)]} e^{i\phi_c(t)}\right\}(\omega) \\ &+ FT\left\{\frac{I_M(t)}{2} e^{-i[\phi(t)+\phi_r(t)]} e^{-i\phi_c(t)}\right\}(\omega). \end{aligned} \tag{4.8}$$

Note that as each pixel is analyzed independent of all the pixels in the image, the m and n spatial coordinates were dropped in Equation 4.8 for clarity. The success of the Fourier transform method is based on the quality obtained for the frequency splitting

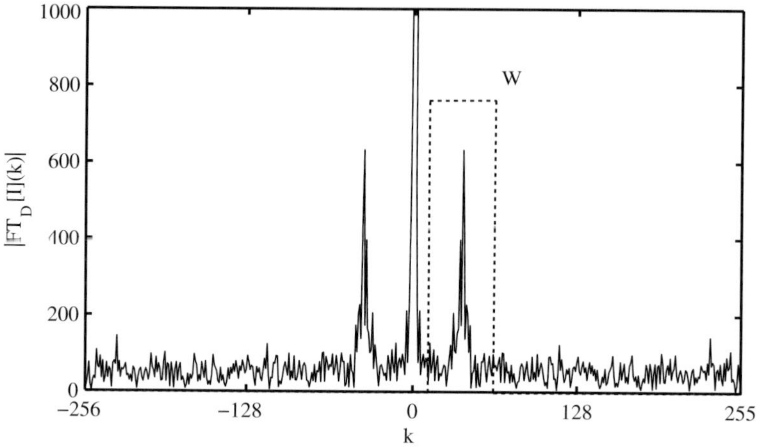

Figure 4.2 Fourier spectrum of the TSPI signal shown in Figure 4.1.

of the main contributions, which are associated with the terms $\exp(\pm i\phi_c)$. Using a numerical implementation of the discrete Fourier transform FT_D, the spectrum of the resulting intensity $|FT_D(I)|$ is illustrated in Figure 4.2. This figure displays the $k = 0$ contribution, commonly named the dc term, and the other two contributions corresponding to the side lobes centered on the positive and negative frequencies of the fixed frequency of the carrier $\omega_c = \partial\phi_c/\partial t$, which can be derived from Equation 4.8 using the translation property of the Fourier transform:

$$\begin{aligned} FT[I(t)](\omega) &= FT[I_B(t)](\omega) + FT\left\{\frac{I_M(t)}{2} e^{i[\phi(t)+\phi_r(t)]}\right\}(\omega-\omega_c) \\ &\quad + FT\left\{\frac{I_M(t)}{2} e^{-i[\phi(t)+\phi_r(t)]}\right\}(\omega+\omega_c). \end{aligned} \tag{4.9}$$

Applying the window filter function W in the frequency domain, the right lobe is isolated and translated by the constant carrier frequency ω_c via a rigid transformation $T_R(\cdot, \omega_c)$ toward the origin to remove the frequency of the carrier. Then, the following temporal intermediate term is obtained by applying the inverse Fourier transform

$$\alpha(t) = FT^{-1}\left\{T_R\left[FT\left\{\frac{I_M(t)}{2} e^{i[\phi(t)+\phi_r(t)]}\right\}(\omega-\omega_c), \omega_c\right]\right\}(t). \tag{4.10}$$

Therefore, the object phase difference is recovered from the equation

$$\Delta\phi(t) = \arctan\left\{\frac{\mathrm{Im}[\alpha(t)]}{\mathrm{Re}[\alpha(t)]}\right\} - \phi(t_0), \tag{4.11}$$

where $\phi(t_0)$ is defined as a reference state of the object phase at time t_0, which is given by the same expression

$$\phi(t_0) = \arctan\left\{\frac{\mathrm{Im}[\alpha(t_0)]}{\mathrm{Re}[\alpha(t_0)]}\right\}, \tag{4.12}$$

and where the functions Re[·] and Im[·] denote the real and imaginary parts, respectively. As the evaluated phase change difference values lie in the range $(-\pi, \pi)$, a temporal phase unwrapping procedure must be applied to remove the 2π phase discontinuities (see Section 2.4.1 of Ref. [4]).

A variation of the Fourier transform method can be used to improve the accuracy of the object phase evaluation. The common procedure consists in performing the inverse Fourier transform after the window function is applied in the frequency domain but without using the translation to the frequency origin. In this case, the object phase plus the phase associated with the frequency carrier are recovered. Therefore, the carrier phase must be removed in a subsequent step. This procedure allows to compensate for temporal nonlinearities produced by frequency variations of the carrier. Although this approach improves the accuracy in the recovered values of the phase difference, the selection of the window parameters is a very complex task due to the intrinsic limitation of the Fourier transform representation. The main contributions of the spectrum can be very broadened, so that the frequency splitting can be cumbersome. Figure 4.2 shows that a narrow window W can reject a substantive quantity of noise, but at the same time it degrades the temporal resolution of the analyzed TSPI signal.

Phase recovery based on the Fourier transform method is still an active area of research. Improvements in the carrier phase component removal [15, 16] and a comparison with the temporal phase shifting method [17] have been published. An analysis of the application of the Fourier transform method with and without introducing a temporal carrier was also carried out [18]. In Ref. [19], the zero order is removed using two complementary outputs of the interferometer to improve the spatial resolution by a factor of approximately 2.

The Fourier transform method behaves like a system that allows to pass certain frequencies and rejects all others by windowing. The Fourier spectrum $|\mathrm{FT_D}[I(t)](\omega)|$ shows all frequencies contained in the signal $I(t)$ and does not provide easy-to-interpret information on the frequencies contained in the object phase spectrum, which should be used to improve the filtering procedure by means of a well-adapted window design. This limitation can be overcome by using a different approach, named time–frequency representation, which will be described in the following section.

4.4 Time–Frequency Representations of the TSPI Signals

The Fourier spectrum does not explicitly show the time localization of the frequency components of the TSPI signals. The Fourier spectrum is an ideal tool for studying stationary signals, which are decomposed canonically into linear combinations of sines and cosines. However, as TSPI signals are primarily nonstationary functions, to analyze them we will need to apply techniques different from the Fourier transform. The other techniques, which are specific to the nonstationariness of the TSPI signal, include "decomposition in time–frequency atoms," which are waveforms well concentrated in time and in frequency. As typical examples, a windowed Fourier atom and a wavelet atom are two important classes of local time–frequency decompositions, which will be discussed later in this chapter. Other time–frequency representation techniques are based on the time–frequency energy density of the analyzed TSPI signal, which is computed by correlating the signal with a time and a frequency translation of itself, so that no atoms are required. Particularly, the decompositions used in this chapter are assumed to be complete. The readers interested in this topic will have to read the associated bibliography.

4.4.1 Preliminaries

4.4.1.1 The Asymptotic Signal and the Exponential Model

All the information contained in the TSPI signals is available by retrieving only the positive frequencies because they are real ones. The Fourier transform of a real signal is an even function, so that there is an unnecessary redundancy. Therefore, it is convenient to associate with the TSPI signal its analytic signal by means of a linear filtering that cancels its negative frequencies. A useful property of analytic signals is that they can be sampled at half the rate of the real signal, which follows from the fact that there are no negative frequencies present [20]. As it will be discussed in Section 4.4.4, another advantage of the analytic signal is that its time–frequency distribution contains fewer cross terms. More generally, given an arbitrary real signal $s(t)$, one can always obtain a representation in terms of its instantaneous modulus and argument in the form

$$s(t) = A(t)\cos[\phi(t)]. \tag{4.13}$$

This representation is not unique and therefore one can associate an infinity number of pairs (A, ϕ). However, one pair, named the canonical pair (A_s, ϕ_s), can be completely characterized by considering the Hilbert transform (HT). By definition, the analytic signal $z_s(t)$ associated with $s(t)$ is expressed as

$$z_s(t) = s(t) + i\mathrm{HT}[s](t), \tag{4.14}$$

where the Hilbert transform is defined as

$$\mathrm{HT}[s](t) = \frac{1}{\pi}\mathrm{PV}\int_{-\infty}^{+\infty} \mathrm{d}\tau \frac{s(\tau)}{t-\tau}. \tag{4.15}$$

This integral is improper in the sense of the Cauchy principal value denoted by PV [21]. Therefore, $z_s(t)$ is completely characterized by the pair (A_s, ϕ_s) by setting

$$z_s(t) = A_s(t)\mathrm{e}^{i\phi_s(t)}. \tag{4.16}$$

The canonical representation of the signal $s(t) = \mathrm{Re}[z_s(t)] = A_s(t)\cos[\phi_s(t)]$ allows the introduction of the instantaneous frequency $\nu_s(t)$ defined by

$$\nu_s(t) = \frac{1}{2\pi}\frac{\partial}{\partial t}\phi_s(t). \tag{4.17}$$

This definition always has sense mathematically, although its physical significance can be doubtful [22]. If the signal $s(t)$ is asymptotic and locally monochromatic,

$$\left|\frac{\partial}{\partial t}\phi(t)\right| \gg \left|\frac{1}{A(t)}\frac{\partial}{\partial t}A(t)\right|, \tag{4.18}$$

then by using the stationary phase method it is possible to obtain the analytic signal of an asymptotic locally monochromatic signal of the form given by Equation 4.13 close to the exponential model:

$$z_s(t) \approx A(t)\mathrm{e}^{i\phi(t)}. \tag{4.19}$$

There are conditions under which the HT exactly generates the identity

$$A(t)\cos[\phi(t)] + i\mathrm{HT}\{A(t)\cos[\phi(t)]\} = A(t)\mathrm{e}^{i\phi(t)}, \tag{4.20}$$

which are determined by Bedrosian's product theorem [23]. This theorem demonstrates that the identity is verified when the spectrum of the amplitude $\mathrm{FT}[A(t)](\omega)$ lies entirely in the region $|\omega| < \Omega$ and the spectrum associated with the cosine of the phase $\mathrm{FT}\{\cos[\phi(t)]\}(\omega)$ exists only outside Ω. The amplitude $A(t)$ and the phase $\phi(t)$ of a given TSPI signal may only be considered independently if the spectra of $\mathrm{FT}[A(t)](\omega)$ and $\mathrm{FT}\{\cos[\phi(t)]\}(\omega)$ are separated in frequency. In this case, the HT-based analytic signal can provide a better estimate of the instantaneous frequency. On the contrary, when the spectra are not separated, the Hilbert transform produces a result of overlapping and phase-distorted functions, although the analytic signal $z_s(t)$ is unique but with a questionable practical meaning.

The importance of the instantaneous frequency concept stems from the fact that the spectral characteristics of the TSPI signals are nonstationary functions. Therefore, a parameter needs to be used that accounts for the time-varying nature of the object phase. The instantaneous frequency associated with the TSPI signal is of significant practical importance, as it will be shown in the presentation of the different time–frequency representations described in this chapter.

4.4.1.2 **Fidelity Measures**

The mean square error (MSE) remains the standard criterion for the assessment of fidelity when novel phase recovery methods need to be compared. However, the MSE exhibits a weak performance. A review of the shortcomings of the MSE and proposed alternative methods can be seen in Ref. [24] and references therein. Particularly, by using easily computed statistics to evaluate the performance of the different object phase recovery methods, the introduction of a quality index Q is very convenient. The Q index is defined as [25]

$$Q = \frac{4\sigma_{EO}\bar{E}\bar{O}}{(\sigma_E^2+\sigma_O^2)[\bar{E}^2+\bar{O}^2]}, \tag{4.21}$$

where O is the original phase distribution and E is the one to be tested, respectively, and

$$\bar{E} = \frac{1}{N^2}\sum_{x,y=1}^{N} E(x,y), \quad \bar{O} = \frac{1}{N^2}\sum_{x,y=1}^{N} O(x,y), \tag{4.22}$$

$$\sigma_E^2 = \frac{1}{N^2-1}\sum_{x,y=1}^{N} [E(x,y)-\bar{E}]^2, \tag{4.23}$$

$$\sigma_O^2 = \frac{1}{N^2-1}\sum_{x,y=1}^{N} [O(x,y)-\bar{O}]^2, \tag{4.24}$$

$$\sigma_{EO} = \frac{1}{N^2-1}\sum_{x,y=1}^{N} [E(x,y)-\bar{E}][O(x,y)-\bar{O}], \tag{4.25}$$

for a square data set of size N^2.

The quality index Q has a dynamic range of $[-1,1]$ and allows to model any distortion as a combination of three different factors, commonly named in image processing: loss of correlation, luminance distortion, and contrast distortion, as it can be derived from Equation 4.21:

$$Q = \frac{\sigma_{EO}}{\sigma_E\sigma_O}\,\frac{2\bar{E}\,\bar{O}}{\bar{E}^2+\bar{O}^2}\,\frac{2\sigma_E\sigma_O}{\sigma_E^2+\sigma_O^2}. \tag{4.26}$$

The best value 1 is achieved when $O = E$. The lineal correlation coefficient between O and E is given by the first component in Equation 4.26 and its range is $[-1,1]$ with the best value 1. The second component in Equation 4.26 measures how close the mean is between O and E, with the best value 1 if $\bar{O} = \bar{E}$, and for intermediate values in the range $[0,1]$. Finally, the third component measures how similar the deviations are in the range of $[0,1]$ with the best value 1.

The overall quality index Q value is obtained by combining the statistical features corresponding to these three factors within local regions using a sliding window of size $S \times S$ pixels across the phase distribution. This window is moved pixel by pixel along the horizontal and vertical directions through all the rows and columns of the data, and finally the average quality index is evaluated as the mean value of the previously computed values.

4.4.2
The Windowed Fourier Transform

Time localization of frequency components can be obtained by suitable prewindowing the TSPI signal with a windowed Fourier atom $g_{b,\xi}(t)$, which is constructed with a real and a symmetric window $g(t) = g(-t)$ that is translated by b and modulated at the frequency ξ:

$$g_{b,\xi}(t) = e^{i\xi t} g(t-b). \tag{4.27}$$

The windowed Fourier transform of the signal $I(t)$ is defined by

$$\mathrm{SW}_I(b,\xi) = \int_{-\infty}^{+\infty} \mathrm{d}t\, e^{-i\xi t} g(t-b) I(t), \tag{4.28}$$

where the windowed Fourier atom is normalized $\|g_{b,\xi}\| = 1$ for any $(b,\xi) \in \mathbb{R}^2$ [8]. The windowed Fourier transform is a linear time–frequency representation and complex valued in general. This transform is also called the "short-time Fourier transform" or "short-time spectrum" because the multiplication by the short window $g(t-b)$ effectively suppresses the signal $I(t)$ outside a neighborhood around the analysis time point $t = b$. Therefore, $\mathrm{SW}_I(b,\xi)$ is simply a "local spectrum" of the signal $I(t)$ around the "analysis time" b.

The windowed Fourier transform unfold the intensity signal $I(t)$ in the time–frequency plane in such a way that a decomposition in time–frequency atoms is carried out. The choice of these time–frequency atoms are guided by an energy density of the signal $I(t)$ in the time–frequency plane, named the spectrogram, defined by

$$|\mathrm{SW}_I(b,\xi)|^2 = \left| \int_{-\infty}^{+\infty} \mathrm{d}t\, e^{-i\xi t} g(t-b) I(t) \right|^2. \tag{4.29}$$

The spectrogram measures the energy of the intensity $I(t)$ in the time–frequency neighborhood of (b,ξ). Therefore, time variations of the "instantaneous frequencies" can be measured.

For practical applications of the windowed Fourier transform, it is necessary to sample the time–frequency plane. The intensity signal $I(t)$ is considered a discrete signal of period N defined by $I[n]$ for each $0 \le n < N$, $n \in \mathbb{N}$, $\mathbb{N}$ being the set of natural numbers. The window $g(t)$ is sampled in the same way as the intensity signal, although it is symmetric and with unit norm. The windowed Fourier atoms are sampled as

$$g_{m,k}[n] = g[n-m] \exp\left(\frac{i2\pi nk}{N}\right), \tag{4.30}$$

and therefore the discrete windowed Fourier transform $\mathrm{SW}_I[m,k]$ of the intensity signal $I[n]$ is given by

$$\mathrm{SW}_I[m,k] = \sum_{n=0}^{N-1} I[n] g[n-m] \exp\left(\frac{-i2\pi nk}{N}\right), \tag{4.31}$$

where $0 \le m < N$, $m \in \mathbb{N}$. Note that $\mathrm{SW}_I[m,k]$ is computed for $0 \le k < N$, $k \in \mathbb{N}$, using the discrete Fourier transform of $I[n]g[n-m]$, which is calculated with N fast Fourier transforms of size N.

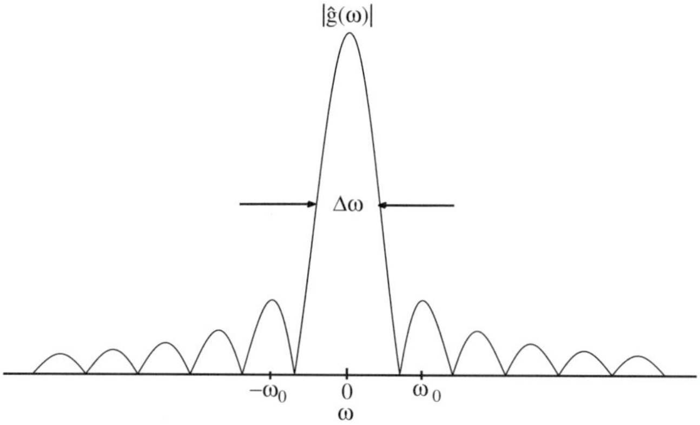

Figure 4.3 The energy spread of $\hat{g}(\omega)$ is characterized by its bandwidth $\Delta\omega$ and the maximum amplitude of the first side lobes located at $\omega = \pm\omega_0$.

$SW_I(b, \xi)$ is influenced by the choice of the analysis window $g(t)$ of finite support. The resolution in time and in frequency of the windowed Fourier transform depends on the spread of the window $g(t)$ in time and frequency. The modulus of the Fourier transform $|\hat{g}(\omega)|$ is a symmetrical function with a main lobe centered at $\omega = 0$, which decays to zero with oscillations (see Figure 4.3 for a graphical representation). Note that by concentrating the energy of $\hat{g}(\omega)$ near of $\omega = 0$, the frequency resolution of the windowed Fourier transform is maximized. To evaluate more precisely the energy spread of the Fourier transform $\hat{g}(\omega)$, it is convenient to consider the root mean square bandwidth $\Delta\omega$, the maximum amplitude of the first side lobes located in $\omega = \pm\omega_0$, and the asymptotic decay of $|\hat{g}(\omega)|$ for large frequencies, as it is shown in Figure 4.3. The windowed Fourier transform method is exemplified below by analyzing the TSPI signal shown in Figure 4.1 and an interpretation of the frequency parameters is also given.

With the introduction of the frequency carrier in the interferometer, a broadband around the frequency carrier in the Fourier spectrum is highlighted (see Figure 4.2). As an approximation, we could consider the analysis of a frequency tone given by the frequency carrier $\exp(i\omega_c t)$. If $\Delta\omega$ is small, from Equation 4.29 it is seen that $|SW_I(b, \xi)|^2 = |\hat{g}(\xi - \omega_c)|^2$, so that the spectrogram shows the energy concentrated near $\xi = \omega_c$. The $\hat{g}(\omega)$ side lobes cause problems by introducing shadows at $\xi = \omega_c \pm \omega_0$, which could be neglected if the maximum amplitude of its side lobes were small. As the frequency carrier is embedded in the TSPI signal, which has more frequency components with the possibility of much higher energy, the tone could still be detected if $|\hat{g}(\omega)|$ has a rapid decay when $|\omega|$ increases.

Figure 4.4 shows the spectrogram of the TSPI signal depicted in Figure 4.1, which was computed using Equation 4.31 and a Hanning window with a half window length of 128 pixels. The dark points indicate a large amplitude of the windowed Fourier coefficients, which were sampled into a $[0, 255]$ gray scale, although they are shown in the interval $[0, 10]$ for a better visualization of the temporal evolution of the frequency components. In this figure, the dc contribution $k = 0$ and the other main

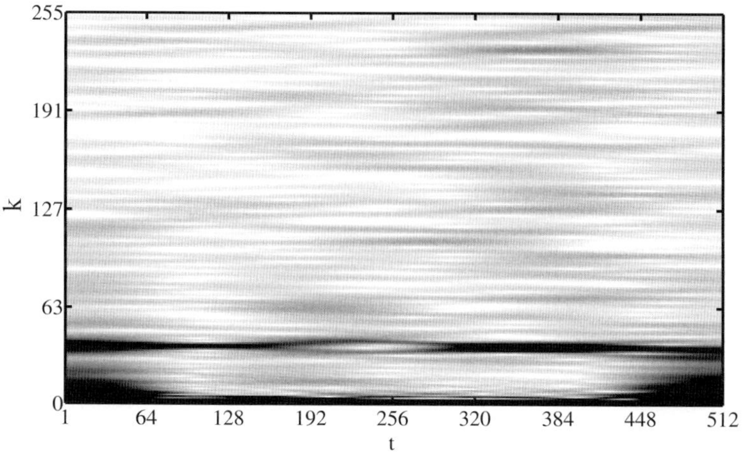

Figure 4.4 Spectrogram obtained by applying the windowed Fourier transform to the TSPI signal shown in Figure 4.1.

contribution corresponding to the broadband positive frequency of the carrier are observed. In the middle region of the spectrogram, an unexpected behavior at the broadband frequencies around the frequency carrier is clearly detected. This new information shows the detrimental effects in the phase recovery process due to the low levels of the intensity modulation values that are observed in Figure 4.1.

The windowed Fourier transform of the TSPI signal at the carrier frequency allows obtaining most of the standard temporal phase shifting algorithms (see Chapter 2 of Ref. [4]). This result can be shown by analyzing the TSPI signal given by Equation 4.3 and by assuming the phase shift between successive samples as $\Delta\phi_c = 2\pi/N$, where N is the integral number of samples per 2π cycle of the carrier, which in practice is commonly generated by introducing a Pockels cell in the reference beam of the speckle interferometer. From Ref. [4], the phase change with respect to the frame $t = 0$ can be estimated by

$$\Delta\phi(t) = \arctan\left[\frac{N(t)D(0) - D(t)N(0)}{D(t)D(0) + N(t)N(0)}\right], \quad t = 0, 1, 2, \cdots, N_t - M, \tag{4.32}$$

where $N(t)$ and $D(t)$ are determined by the complex variable

$$Z(t)e^{i\Delta\phi_c t} = \sum_{\tau=0}^{M-1} [a(\tau) + ib(\tau)] I(t+\tau), \tag{4.33}$$

with $N(t) = \mathrm{Im}[Z(t)]$ and $D(t) = \mathrm{Re}[Z(t)]$. The coefficients $a(\tau)$ and $b(\tau)$ depend on the particular phase shifting algorithm (see Table 2.1 in Ref. [4] for the values that have the most common phase shifting algorithms). The exponential term $\exp(i\Delta\phi_c t)$ eliminates the modulation introduced by the temporal carrier. By defining a window $w(\tau)$ of length M as

$$a(\tau) + ib(\tau) = w(\tau)e^{-i\Delta\phi_c \tau}, \quad \tau = 0, 1, 2, \cdots, M-1, \tag{4.34}$$

the right-hand side of Equation 4.33 is equivalent to the discrete windowed Fourier transform.

The technique of high-speed phase shifting speckle interferometry using temporal phase unwrapping allows to measure dynamic displacement fields, even for objects containing global discontinuities. Using numerical simulations to generate the data produced by a dynamic high-speed phase shifting speckle interferometer, it was demonstrated that longer sampling windows and higher framing rates are preferred in the presence of mechanical vibration in order to reduce phase change errors [26]. However, shorter sampling windows and higher framing rates increase the unwrapping success rate [27]. Due to the small phase errors introduced at each temporal frame are accumulated along the time axis, phase values strongly depend on the speckle rereference rate [28]. Reference [29] shows an analysis of the errors introduced in the phase evaluation by three sources: intensity errors, velocity errors, and speckle decorrelation. In addition, as the displacement of the object during image acquisition may cause the peak of the windowed Fourier transform to shift away from the carrier frequency, leading to the generation of phase errors and unwrapping failure, a method that searches for the peak of the windowed Fourier transform and evaluates the phase at that frequency instead of at the carrier frequency was presented in Ref. [30]. This last work suggests the necessity to take into consideration a neighborhood of frequencies to obtain a more accurate estimation of the object phase (see also Ref. [31]). This topic will be discussed in the following section.

4.4.3
The Wavelet Transform

TSPI signals have different time–frequency structures that depend on the problem at hand, so in practice windowed Fourier atoms cannot be sufficient to recover object phase distributions with low errors. To take advantage of the windowed Fourier atom approach described in the previous section, a possible extension consists in the use of time–frequency atoms by adding different time supports and in this way to obtain another parameter that favors the robustness of the phase recovery process.

A wavelet atom is a square-integrable complex-valued function $\psi(t)$ with a zero average, normalized $\|\psi\| = 1$, and centered in the neighborhood of $t = 0$ [8]. The set of time–frequency atoms is obtained by scaling ψ by $a \in \mathbb{R}_{>0}$ and by translating it by $b \in \mathbb{R}$:

$$\psi_{b,a}(t) = \frac{1}{\sqrt{a}}\psi\left(\frac{t-b}{a}\right). \tag{4.35}$$

The continuous wavelet transform W_I decomposes the TSPI signal $I(t)$ over dilated and translated wavelets $\psi_{b,a}(t)$

$$W_I(b,a) = \int_{-\infty}^{+\infty} \mathrm{d}t \frac{1}{\sqrt{a}}\psi^*\left(\frac{t-b}{a}\right) I(t), \tag{4.36}$$

where $(*)$ denotes complex conjugation. $W_I(b,a)$ measures, in a certain sense, the fluctuations of the signal $I(t)$ around the point b at the scale given by a. This time–scale representation involves the use of a wide range of scales for the analysis of $I(t)$, which is done by calculating the change from one scale to the next. This algorithmic scheme is called "multiresolution analysis."

A square-integrable function is said to be analytic if its Fourier transform is zero for negative frequencies (see Section 4.4.1). To measure the time evolution of frequency transients in the TSPI signal, a complex analytic wavelet that separates the amplitude and the phase components can be used. An analytic wavelet can be constructed using the frequency modulation of a real and a symmetric window, for example, a modulated Gaussian $g(t) = \exp(-t^2/2)$, being this the Morlet wavelet [32]:

$$\psi(t) = \pi^{-1/4}(e^{i\omega_0 t} - e^{-\omega_0^2/2})e^{-t^2/2}. \tag{4.37}$$

Its Fourier transform is a shifted Gaussian

$$\hat{\psi}(\omega) = \pi^{-1/4}[e^{-(\omega-\omega_0)^2/2} - e^{-\omega^2/2}e^{-\omega_0^2/2}], \tag{4.38}$$

adjusted so that $\hat{\psi}(0) = 0$ (zero average). In practice, one often takes $\omega_0 = 5$ and the second term in Equation 4.37 can be neglected. Therefore, it is a common procedure to consider the Morlet wavelet as $\psi_M(t) = \exp(-t^2/2)\exp(i\omega_0 t)$ and absorb the constant factors outside the integral.

The time–frequency resolution of an analytic wavelet transform depends on the time–frequency spread of the wavelet atoms $\psi_{b,a}(t)$. Supposing that $\psi(t)$ is centered at $t = 0$, then by means of the change of variable $\alpha = (t-b)/a$, it can be shown that

$$\int_{-\infty}^{+\infty} \mathrm{d}t (t-b)^2 |\psi_{b,a}(t)|^2 = a^2\sigma_t^2, \tag{4.39}$$

where

$$\sigma_t^2 = \int_{-\infty}^{+\infty} \mathrm{d}t\, t^2 |\psi(t)|^2. \tag{4.40}$$

In the frequency domain, if η is the center frequency of $\hat{\psi}$, then

$$\eta = \frac{1}{2\pi}\int_0^{+\infty} \mathrm{d}\omega\, \omega |\hat{\psi}(\omega)|^2. \tag{4.41}$$

The Fourier transform of $\psi_{b,a}(t)$ is a dilation of $\hat{\psi}(\omega)$ by $1/a$, then its center frequency is η/a and

$$\hat{\psi}_{b,a}(\omega) = \sqrt{a}\hat{\psi}(a\omega)e^{-i\omega b}. \tag{4.42}$$

The energy spread of $\hat{\psi}_{b,a}(\omega)$ around η/a is given by

$$\frac{1}{2\pi}\int_0^{+\infty} \mathrm{d}\omega \left(\omega - \frac{\eta}{a}\right) |\hat{\psi}_{b,a}(\omega)|^2 = \frac{\sigma_\omega^2}{a^2}, \tag{4.43}$$

where

$$\sigma_{\omega}^{2} = \frac{1}{2\pi}\int_{0}^{+\infty} d\omega (\omega - \eta)^{2} |\hat{\psi}(\omega)|^{2}. \qquad (4.44)$$

It is well known that the time and frequency energy concentrations are restricted by the Heisenberg uncertainty principle. According to Equations 4.39 and 4.43, the energy spread of a wavelet time–frequency atom $\psi_{b,a}(t)$ corresponds to a Heisenberg box centered at $(b, \eta/a)$ with sizes $a\sigma_t$ and σ_ω/a along the time and frequency axes, respectively. The area of the Heisenberg box in the time–frequency plane is $\sigma_t\sigma_\omega$ for all scales a, although the time and frequency resolution depends on the scale a. Larger scales increase the time spread, although the frequency support decreases [8].

The local time–frequency energy density of the analytic wavelet transform of a TSPI signal $I(t)$ is defined by

$$|W_I(b,a)|^2 = \left| \int_{-\infty}^{+\infty} dt \frac{1}{\sqrt{a}} \psi^{*}\left(\frac{t-b}{a}\right) I(t) \right|^2, \qquad (4.45)$$

which is called a scalogram and measures the energy of the signal $I(t)$ in the Heisenberg box centered at $(b, \eta/a)$ of each wavelet atom $\psi_{b,a}(t)$.

In the time domain, the correlation between the intensity signal $I(t)$ and the wavelet atoms can be written as the inverse Fourier transform of the product of the conjugate Fourier transforms of the wavelets and the Fourier transform of the signal:

$$W_I(b,a) = \frac{\sqrt{a}}{2\pi} \int_{-\infty}^{+\infty} d\omega\, e^{i\omega b} \hat{\psi}^{*}(a\omega) \hat{I}(\omega). \qquad (4.46)$$

Therefore, the wavelet transform is a bank of wavelet transform filters with different scales $\sqrt{a}\hat{\psi}(a\omega)$. This approach suggests an implementation for computing the discrete algorithm for the continuous wavelet transform by using fast Fourier transforms. The intensity signal $I(t)$ is considered a discrete signal of period N defined by $I[n]$ for each $0 \leq n < N,\ n \in \mathbb{N}$. The scale parameter is discretized as $a = 2^{m/M},\ m = 1, 2, \cdots, M$ with M being voices per octave $[2^j, 2^{j+1}],\ j \in \mathbb{N}$. Reference [33] provides several algorithms for computing various types of wavelet transforms and develops guidelines for an efficient implementation. Note that an octave-by-octave computation in the scale parameter a is insufficient, as it is usually the case considered in the discrete wavelet transform algorithm. For signal analysis, it is appropriate to generalize the octave-by-octave grid to a denser set of points. Therefore, the sampling is oversampled to obtain M voices per octave. Also, the computation of Equation 4.46 requires the extension of the signal at its left-hand and right-hand edges to avoid border effects. Symmetrical or zero padding extension techniques are commonly used in practice. Other techniques such as linear predictive extrapolation can also be used, in which the phase and the frequency of the TSPI signal are maintained [34]. After calculating the wavelet transform of the extended data, the set of wavelet coefficients is truncated appropriately.

4.4.3.1 Evaluation of the Ridge of a Wavelet Transform

From Equations 4.38 and 4.46, it can be seen that the negative frequencies and the dc contribution are neglected when the TSPI signal given by Equation 4.1 is analyzed. However, it is more convenient to eliminate the dc contribution by a specific previous filtering method and only to apply the wavelet analysis to the signal that corresponds to the intensity modulation term. It can be shown that this procedure improves the accuracy of the phase recovery process. When the dc-filtered TSPI signal $s(t) = I_M(t)\cos\phi(t)$ can be represented by its analytic signal (see Equation 4.16), the amplitude spectrum lies entirely outside the spectrum of the cosine of the phase. In this case, the Hilbert transform generates exactly the quadrature component of the signal and in consequence the analytic signal allows the definition of the instantaneous frequency of the phase $\phi(t)$ (see Section 4.4.1). Therefore, its physical meaning as the frequency associated with the wavefront phase is conserved.

The signal $z_s(t)$ is called asymptotic with respect to the Gaussian window corresponding to the Morlet wavelet, when $s(t)$ oscillates fast with respect to its amplitude variations, which are supposed to be locally constant on the support of the Gaussian window $g(t)$:

$$\left|\frac{\partial}{\partial t}\phi_s(t)\right| \gg \left|\frac{1}{A_s(t)}\frac{\partial}{\partial t}A_s(t)\right|, \tag{4.47}$$

$$\left|\frac{1}{A_s(t)}\frac{\partial}{\partial t}A_s(t)\right| \ll \left|\frac{1}{g(t)}\frac{\partial}{\partial t}g(t)\right|. \tag{4.48}$$

In the analytic asymptotic limit, Equation 4.36 can explicitly show the behavior of the continuous wavelet coefficients when the stationary phase approximation is used [35]. Below follows a description of this approximation and also the procedure to determine the instantaneous frequencies of the signal from the knowledge of the wavelet coefficients.

The expression for the wavelet transform of $z_s(t)$ is given by

$$W_I(b,a) = \frac{1}{2a}\int_{-\infty}^{+\infty} \mathrm{d}t\, A_s(t)\, g\left(\frac{t-b}{a}\right) \mathrm{e}^{i\,\left[\phi_s(t)-\omega_0\frac{t-b}{a}\right]}, \tag{4.49}$$

where the factor 1/2 comes from the analytic transformation of the signal $s(t)$ and where a different normalization was used for the wavelets in Equation 4.35:

$$\psi_{b,a}(t) \equiv \frac{1}{a}\psi\left(\frac{t-b}{a}\right). \tag{4.50}$$

In the analytic asymptotic limit, the amplitude of the analytic signal $A_s(t)$ is approximated up to the first order in a Taylor series around a point of the stationary phase $t_s = t_s(b,a)$. At this point, the phase $\phi_s(t)$ verifies the stationary phase condition $\phi_s'(t_s) = \omega_0/a$, where the prime ($'$) means the derivative with respect to the variable t. The wavelet coefficients in this approximation are described for the zero- and first-order contribution of the wavelet transform, and can be obtained by direct integration as [36]

$$|W_I(b,a)|^{(0)} = \sqrt{\frac{\pi}{2}} \frac{A_s(t_s)\exp\left[-(1/2)\left((t_s-b)^2\phi_s''(t_s)^2\right)/\left(a^2\phi_s''(t_s)^2+a^{-2}\right)\right]}{[1+a^4\phi_s''(t_s)^2]^{1/4}}, \tag{4.51}$$

$$\arg[W_I(b,a)]^{(0)} = \phi_s(t_s) - \omega_0\frac{t_s-b}{a} + \frac{1}{2}\frac{(t_s-b)^2\phi_s''(t_s)}{1+a^4\phi_s''(t_s)^2} + \frac{1}{2}\arctan\left[a^2\phi_s''(t_s)\right], \tag{4.52}$$

$$|W_I(b,a)|^{(1)} = \sqrt{\frac{\pi}{2}} \frac{A_s'(t_s)\exp\left[-(1/2)\left((t_s-b)^2\phi_s''(t_s)^2\right)/\left(a^2\phi_s''(t_s)^2+a^{-2}\right)\right]}{[1+a^4\phi_s''(t_s)^2]^{3/4}}|t_s-b|, \tag{4.53}$$

$$\arg[W_I(b,a)]^{(1)} = \phi_s(t_s) - \omega_0\frac{t_s-b}{a} + \frac{1}{2}\frac{(t_s-b)^2\phi_s''(t_s)}{1+a^4\phi_s''(t_s)^2} + \frac{3}{2}\arctan\left[a^2\phi_s''(t_s)\right]. \tag{4.54}$$

From Equation 4.53 note that the first-order term contribution is proportional to $|t_s-b|$.

The ridge of the wavelet transform is defined as the set of points $t_s(b,a)=b$. It follows from the definition of stationary phase points that on the ridge $a_r(b)=\omega_0/\phi_s'(b)$. Therefore, the knowledge of the ridge gives information about the instantaneous frequency.

The skeleton of the wavelet transform, defined as the restriction of the transform to the associated ridge, is obtained from Equations 4.51 through 4.54 as

$$W_I(b,a_r(b)) \approx \sqrt{\frac{\pi}{2}} \frac{\exp\left\{(i/2)\arctan[a_r(b)^2\phi_s''(b)]\right\}}{[1+a_r(b)^4\phi_s''(b)^2]^{1/4}} z_s(b). \tag{4.55}$$

The skeleton reproduces the analytic signal $z_s(t)$ of $s(t)$ up to a corrective function, which can be evaluated by the expression $\phi_s'(b) = -a'_r(b)\omega_0/a_r(b)^2$ [35]. The extraction of the ridge can be performed from Equations 4.51 and 4.52 by tracking the local maximum at $t_s=b$ when the influence of the corrective function in the denominator can be neglected.

Figure 4.5 shows the obtained scalogram corresponding to the analyzed TSPI signal depicted in Figure 4.1 when the intensity bias was removed by using a conventional Fourier filtering technique. The dark points indicate large values of the wavelet coefficients, which were sampled into a scale composed by [0, 255] gray levels. The obtained local maxima are pointed out by means of the dark continuous line. It should be noted that the local tracking of the maximum cannot be successfully

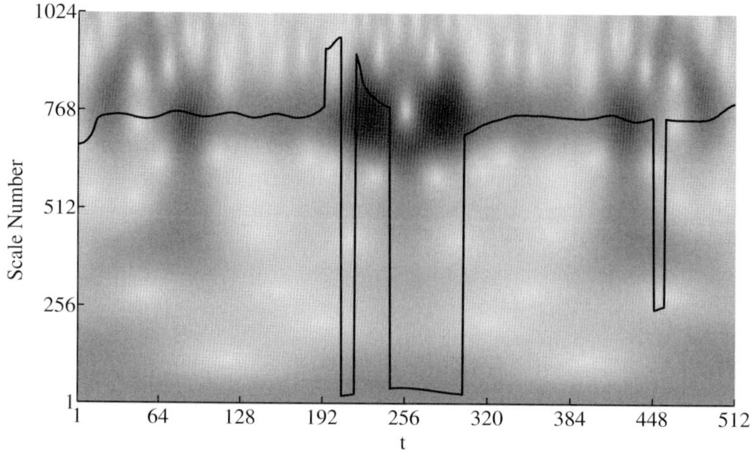

Figure 4.5 Scalogram corresponding to the TSPI signal shown in Figure 4.1, where the continuous dark line points out the local maxima.

done due to the existence of several isolated local maximum points. This problem is better described using the associated phase of the wavelet coefficients and the local maxima representation that is shown in Figure 4.6. In this figure, the phase of the wavelet coefficients, which lies in the interval $(-\pi, \pi)$, is discretized into $[0, 255]$ gray levels. As the phase recovery process depends on the ridge extraction procedure, new methods are proposed to improve the accuracy of the recovered object phase. A reliable ridge extraction algorithm based on a cost function was introduced in Refs [37, 38], where it was demonstrated that this new approach outperforms the maximum ridge extraction algorithm.

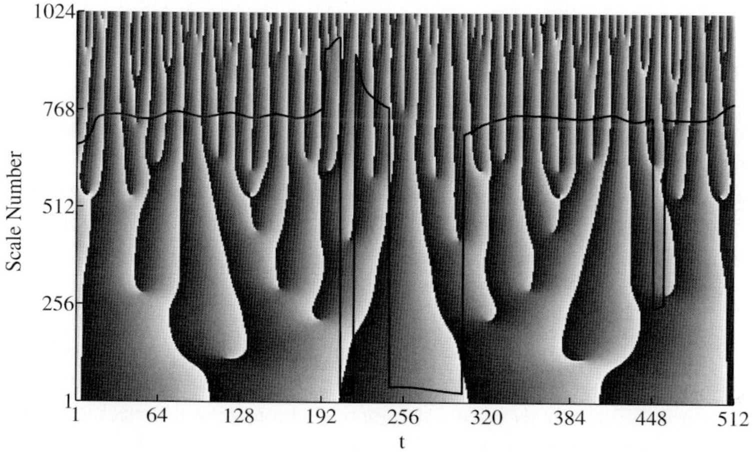

Figure 4.6 Phase of the wavelet coefficients obtained from Figure 4.5 with the corresponding line of local maxima.

The ridge can also be obtained from the knowledge of the argument of the wavelet transform by calculating

$$\frac{\partial}{\partial b}\{\arg[W_I(b,a)]\} = \frac{\omega_0}{a}, \tag{4.56}$$

being unnecessary to compute the whole wavelet transform [35, 39]. To numerically solve Equation 4.56, a fixed point algorithm is generally used. Equation 4.56 improves the accuracy of the ridge extraction and requires the application of a phase unwrapping technique. The use of parallel computing algorithms to determine the ridge has been tested in Refs [40, 41], a procedure that improves the computational time. However, the step of phase unwrapping can be avoided when the local maximum tracking method is applied. The phase map is obtained by integration of the evaluated instantaneous frequency up to one additive constant for each pixel, which has to be known in advance.

4.4.3.2 Applications of the Morlet Transform Analysis in TSPI and Other Related Techniques

Wavelet analysis has become a standard tool used by many researchers working in signal and image processing. Many industrial applications, which denote the interest of this approach, can be seen in a recent review that includes more than 190 referenced papers [42]. In TSPI, the pioneering work performed at the Swiss Federal Institute of Technology at Lausanne [43, 44] marked a turning point in phase recovery using wavelet analysis. It is worth to mention that with the introduction of wavelet analysis, the processing of data generated by other related optical metrology techniques such as holographic interferometry, digital speckle pattern interferometry, white light interferometry, fringe projection, and moiré, have also been improved.

Interferometric techniques were widely used to characterize different properties of a variety of media. For example, laser plasma interferograms are currently analyzed by using these techniques. The application of the Morlet wavelet transform is much more accurate and produces phase maps considerably less noisy than those recovered with the Fourier transform method [45]. The Morlet wavelet transform has been evaluated for the phase measurement in digital speckle pattern interferometry [46] and other fringe patterns [47, 48] from the ridge of the wavelet transform. More recently, a novel method based on the Morlet wavelet transform has been developed to evaluate phase maps from single fringe patterns containing closed fringes [49].

In the measurement of displacement and velocity fields of continuous deforming objects, the results obtained using the temporal Morlet wavelet analysis also show that it minimizes the influence of noise and improves those given by the temporal Fourier transform [50]. Recently, a method based on the Morlet wavelet analysis was presented to measure out-of-plane displacement on vibrating objects by using digital speckle pattern interferometry with the introduction of a temporal carrier [51]. The absolute displacement of the vibrating object was obtained without the need of temporal or spatial phase unwrapping. Wavelet analysis characterizes the optimized frequency at each instant and extracts the instantaneous frequencies, thus being an advantage in comparison with the temporal Fourier transform method. The transient curvature and twist of a continuously deforming object have been characterized from a series of speckle interferograms by using a digital shearography technique with the Morlet wavelet analysis and introducing a temporal carrier [52]. This approach also limits the influence of various noise sources and improves the results obtained in the phase recovery process. Consequently, wavelet analysis allows numerical spatial differentiation to obtain good-quality transient curvature and twist results. Recently, the results obtained with the Morlet wavelet analysis to process sheared speckle interferograms recorded with the introduction of a temporal carrier were also compared with those given by the temporal Fourier transform [53].

Strain analysis and three-dimensional shape measurement have also been tackled using moiré methods [54, 55]. Strains are the instantaneous frequencies of the moiré fringe pattern and the ridge of the Morlet wavelet transform extracts them locally. The effectiveness of the wavelet approach is illustrated in Ref. [56] by comparing it with other traditional methods. It was shown that strain extraction was insensitive to noise, so that accurate results could be obtained. Moreover, the sensitivity of the moiré method was improved to obtain a pixel-by-pixel in-plane strain distribution by introducing a cost function for the adaptive selection of the ridge of the wavelet transform [37]. Recently, the shadow moiré technique with the addition of Morlet wavelet analysis was proposed for velocity, displacement, three-dimensional surface profile measurements of continuously deforming objects [57] and also for profiling of objects with height steps [58]. An improvement in the phase recovering process is found in Ref. [59] by introducing a statistical procedure to search the phases of the Morlet wavelet transform in the whole scaling range. The phase map associated with the deformation field was obtained by the difference of the modulation phases of the wavelet transform coefficients (see Equation 4.55).

Fringe projection profilometry is one of the most used techniques for noncontact three-dimensional shape measurement for its high accuracy, fast speed, and easy implementation. Particularly, the use of the Morlet wavelet transform has been implemented in measurement problems with different characteristic length scales, for example, human body shape [60] and microcomponents [61]. In dynamic measurements, the transient phase changes that are produced by a vibrating object using the Morlet wavelet analysis with the introduction of a temporal carrier have also been investigated [62]. Due to the recent interest in phase recovery methods using the Morlet wavelet analysis, a comparison of three ridge extraction algorithms is provided in Ref. [63]. In addition, the Morlet wavelet analysis has been compared in fringe projection profilometry [64] and particularly with the Fourier transform and the windowed Fourier transform methods for phase extraction from single fringe patterns [65].

Vertical scanning interferometry for the three-dimensional profile measurement of microcomponents is considered a standard technique [66]. It is important to note that the Morlet wavelet analysis was also tested in three-dimensional profiling using a white light interferometric technique [67]. The wavelet method accurately determines local fringe peak and improves the vertical resolution of the measurement.

4.4.3.3 The Chirped Wavelet Transform

The expression of the wavelet transform of $z_s(t)$ given by Equation 4.49 takes the form of a rapidly oscillating integral. The most important contribution to $W_I(b,a)$ is provided by the stationary points $t_s = t_s(b,a)$ of the argument in the integrand $\phi_s'(t_s) = \omega_0/a$. Moreover, the integration bounds do not contribute to the asymptotic expansion of $W_I(b,a)$ due to the fast decay and regularity of the Gaussian window $g(t)$. Clearly, a better fit could be obtained if a new nonlinear term $\exp(i\beta t^2/2), \beta \in \mathbb{R}$, were introduced in the analyzing wavelet. This novel approach was proposed in Ref. [39] within the framework of the characterization of acoustic signals.

The chirped wavelet transform W_I^c of an analytic signal $z_s(t)$ is given by

$$W_I^c(b,a) = \frac{1}{2a}\int_{-\infty}^{+\infty} \mathrm{d}t\, A_s(t)\, g\left(\frac{t-b}{a}\right) \mathrm{e}^{i\left[\phi_s(t)-\omega_0\frac{t-b}{a}\right]}\, \mathrm{e}^{-i\frac{\beta}{2}(t-b)^2}, \tag{4.57}$$

where the parameter β is used to finely adjust the frequency variation rate. In the analytic asymptotic limit, Equation 4.57 can explicitly show the behavior of the chirped wavelet coefficients when the stationary phase approximation is used. Below follows a description of this approximation and also the procedure to determine the instantaneous frequencies of the TSPI signal from the knowledge of the chirped wavelet coefficients, which are analogous to the analysis presented in Section 4.4.3.1.

Evaluation of the Ridge of the Chirped Wavelet Transform The amplitude $A_s(t)$ and phase $\phi_s(t)$ of the analytic signal are approximated by its series expanded around the analysis time b:

$$\phi_s(t) = \phi_s(b) + \phi_s'(b)(t-b) + \frac{1}{2}\phi_s''(b)(t-b)^2, \tag{4.58}$$

$$A_s(t) = A_s(b) + A'_s(b)(t-b). \tag{4.59}$$

Replacing Equations 4.58 and 4.59 in Equation 4.57 and arranging the terms of the integrand in powers of $(t-b)$, the following identity is obtained:

$$W_I^c(b,a) = \frac{\mathrm{e}^{i\phi_s(b)}}{2a}\int_{-\infty}^{+\infty} \mathrm{d}t[A_s(b) + A'_s(b)(t-b)] \times \mathrm{e}^{i\left[\phi_s'(b)-\frac{\omega_0}{a}\right](t-b)}\, \mathrm{e}^{-\left\{\frac{1}{a^2}-i[\phi_s''(b)-\beta]\right\}\frac{(t-b)^2}{2}}, \tag{4.60}$$

which can also be integrated by means of the formula given in Ref. [36].

It should be noted that in Equation 4.60, the new factor β only contributes to the term corresponding to $\phi_s''(b)$. Therefore, Equations 4.51 through 4.54 are again obtained by replacing $\phi_s''(b)$ with the term $[\phi_s''(b)-\beta]$. Therefore, since the ridge is defined as the set of points $t_s(b,a) = b$, from the definition of stationary phase points it follows that $a_r(b) = \omega_0/\phi_s'(b)$ along the ridge points, so the information on the instantaneous frequency can be obtained. The skeleton of the chirped wavelet transform is then described by

$$W_I^c(b, a_r(b)) \approx \sqrt{\frac{\pi}{2}} \frac{\exp\left((i/2)\arctan\{a_r(b)^2[\phi_s''(b)-\beta]\}\right)}{\{1+a_r(b)^4[\phi_s''(b)-\beta]^2\}^{1/4}} z_s(b). \tag{4.61}$$

The ridge extraction algorithm consists in the estimation of both parameters a_r and β at the same time. In the model described by Equation 4.56, the ridge is determined from the knowledge of the instantaneous frequency of the analytic signal by measuring the phase variation of the Morlet wavelet transform in a small neighborhood of time b. In the chirped wavelet transform, the proposed ridge extraction algorithm should also measure the phase variation of the transform in a neighborhood around the pair (b, a) belonging to the time–scale domain. The determination of the ridge and β (window-matching condition) is performed from the following set of equations:

$$\left(\frac{\partial}{\partial b} + \beta\frac{\partial}{\partial \omega}\right)\arg\left[W_I^c(b, a, \beta)\right] = \frac{\omega_0}{a}, \tag{4.62}$$

$$\left(\frac{\partial}{\partial b} + \beta\frac{\partial}{\partial \omega}\right)^2 \arg\left[W_I^c(b, a, \beta)\right] = \beta, \tag{4.63}$$

which are solved by means of a fixed point algorithm with two variables [39]. The importance of the ridge extraction method is clearly demonstrated by the fact that is the key tool to recover the object phase within the dynamic phase shifting framework. Below follows an outline of the algorithm proposed in Ref. [44] to solve Equations 4.62 and 4.63, which also includes the procedure for the evaluation of the Morlet wavelet ridge. To simplify the notation, the phase of the chirped wavelet transform is defined as $\arg[W_I^c(b, a, \beta)] = \varphi(b, a)$ and the β dependence in φ is explicitly omitted.

Two arbitrary initial values a_0 and β_0 are introduced in Equations 4.62 and 4.63 to estimate the new ones a_1 and β_1:

$$\frac{\partial}{\partial b}\varphi(b, a_0) + \beta_0\frac{\partial}{\partial \omega}\varphi(b, a_0) = \frac{\omega_0}{a_1}, \tag{4.64}$$

$$\begin{aligned} &\frac{\partial}{\partial b}\left[\frac{\partial}{\partial b}\varphi(b, a_0) + \beta_0\frac{\partial}{\partial \omega}\varphi(b, a_0)\right] \\ &+\beta_0\frac{\partial}{\partial \omega}\left[\frac{\partial}{\partial b}\varphi(b, a_0) + \beta_0\frac{\partial}{\partial \omega}\varphi(b, a_0)\right] = \beta_1, \end{aligned} \tag{4.65}$$

where $\varphi(b, a)$ is the phase calculated at the frequency $\omega = \omega_0/a$. The derivatives of the phase are computed with respect to ω instead of a. The advantage of this procedure is that the frequency variation $\Delta\omega$ used to compute the phase differences is set to β. Therefore, to compute the transform at the analysis frequency $\omega + \Delta\omega$, a scale factor $\tilde{a} = a\omega_0/(\omega_0 + a\Delta\omega)$ must be used. The algorithm starts with a rough estimation of the parameters and stops once the convergence is obtained for the final values a_k and β_k. Then, the algorithm proceeds to the next time sample $b + 1$, using

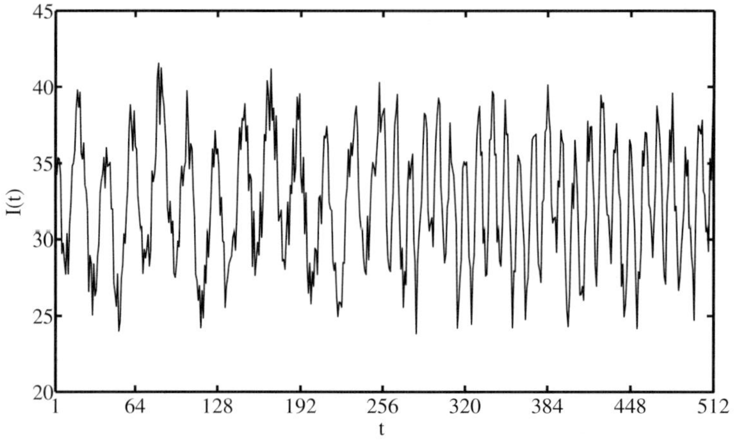

Figure 4.7 TSPI signal with an abrupt frequency variation at time $t_c = 256$.

$a_0 = a_k\omega_0/(\omega_0 + a_k\beta_k)$ and $\beta_0 = \beta_k$. In practice, each ridge iteration requires to compute 13 values of the phase $\varphi(b, a)$ and the different derivatives in Equations 4.64 and 4.65 are estimated as its finite differences.

Below we illustrate the performance of the chirped wavelet transform approach by analyzing the TSPI signal shown in Figure 4.7. This TSPI signal presents an abrupt frequency variation at time $t_c = 256$, which is characterized by doubling its frequency at this characteristic time. Figure 4.8 depicts the characterization of the corresponding local maxima of the chirped wavelet coefficients, which were calculated by solving Equations 4.64 and 4.65. In this figure it is important to note that the abrupt

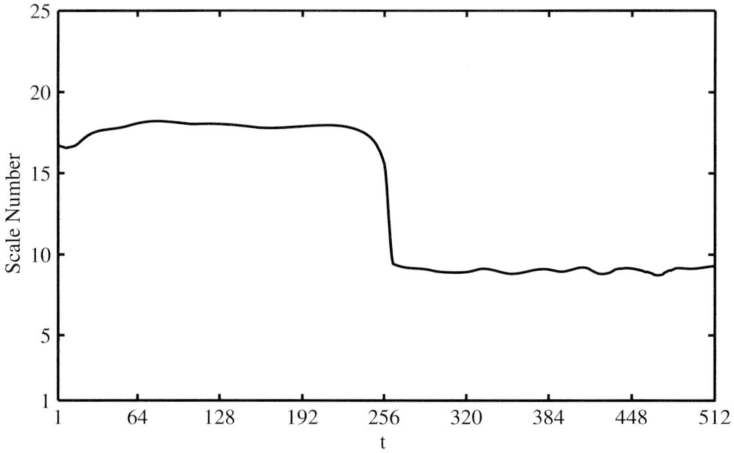

Figure 4.8 Tracking of the local maxima using Equations 4.64 and 4.65 corresponding to the TSPI signal shown in Figure 4.7.

frequency transition is properly determined. The use of an additional parameter β improves the ridge characterization when the corresponding chirped wavelet is compared with the Morlet wavelet and an abrupt frequency variation needs to be estimated.

4.4.3.4 Other Wavelet Transforms

The use of the one-dimensional continuous wavelet transform to recover object phase information shows many advantages over the Fourier transform method, as previously discussed. Several analytic wavelets can be used and also specifically designed with suited properties, for example, Fresnelets that are a wavelet-like bases that are well adapted for the reconstruction and processing of optically generated Fresnel holograms recorded on CCD arrays [68].

Another example is the Paul wavelet transform, which was originally applied to the analysis of fringe patterns [69] and is defined as

$$\psi_{\mathrm{P}}(t) = \frac{2^n n!(1-it)^{-(n+1)}}{2\pi\sqrt{(2n)!/2}}, \tag{4.66}$$

where $n \in \mathbb{N}$ is the order of the wavelet. In the analytic asymptotic limit around a stationary phase, the Paul wavelet transform coefficients can be described for the zero-order contribution, as seen in Section 4.4.3.1. Using Equation 4.46 and observing that the Fourier transform of the Paul wavelet is

$$\hat{\psi}_{\mathrm{P}}(\omega) = \frac{2^n}{\sqrt{n(2n-1)!}}\,\omega^n \mathrm{e}^{-\omega} H(\omega), \tag{4.67}$$

where $H(\omega)$ is the Heaviside function, the instantaneous frequency of the phase is given by the expression

$$\phi'(b) = \frac{2n+1}{2a_{\mathrm{r}}(b)} - \phi_{\mathrm{c}}'(b) \tag{4.68}$$

along the ridge $a_{\mathrm{r}}(b)$ of the wavelet transform. In Ref. [70], a second-order Paul wavelet was used to recover the phase map by introducing a modulation carrier and applying a preprocessing wavelet-based denoising technique. The Paul wavelet method determines local phase gradients, so that the phase map is obtained by integration of Equation 4.68 without applying an unwrapping procedure.

Another example is the Gabor wavelet, which was used to analyze spatial carrier fringe patterns for three-dimensional shape measurement and its advantages over the Fourier transform method were shown in Refs [71, 72]. The Gabor wavelet is defined as

$$\psi_{\mathrm{GA}}(t) = \frac{1}{\pi^{1/4}}\left(\frac{2\pi}{\gamma}\right)^{1/2} \mathrm{e}^{-\left(\frac{\sqrt{2}\pi}{\gamma}\right)^2 t^2 + i2\pi t}, \tag{4.69}$$

where $\gamma = \pi\sqrt{2/\ln 2}$. More recently, numerical fringe pattern demodulation strategies to be applied in interferometry are given in Ref. [73], where a specialized fast algorithm based on a modified Gabor wavelet outperforms Fourier-based strategies for array sizes up to a few thousands of data points.

A thorough discussion on the application of wavelet transforms in order to retrieve phase information in temporally and spatially tilted fringe patterns was recently given in Ref. [74]. Paul, Gaussian, Morlet, B-spline, and Shannon wavelets were tested and compared with Fourier transform, phase stepping, and the windowed Fourier transform methods. Several ridge extraction algorithms and two methods for extending image edges to minimize distortions were also evaluated.

It is important to note that in order to obtain reproducible work [75], a collection of wavelet-related software tools can be found at the Internet. The access to these algorithms allows not only reproducing the results of different authors but also adapting to and developing novel methods. This approach is also an important matter in the analysis of TSPI signals, specially when the computational processing time needs to be reduced. With the fast development of multicore computers, numerical computations can be parallelized using high-level language tools, thus avoiding the typical embarrassing problems associated with parallel implementations [76]. In Ref. [77], a multicore CPU approach using the Matlab software is presented for parallel computing in fringe analysis. More recently, a parallel Matlab toolbox (known as pMatlab) has been introduced containing a set of Matlab scripts that allows any Matlab program to be run on a parallel computer (called MatlabMPI). This software was developed at the Lincoln Laboratory of the Massachusetts Institute of Technology to reduce the computational efforts, so that the user can quickly produce well-performing parallel programs [78].

4.4.4
The Quadratic Time–Frequency Distribution

Wavelet transforms and the windowed Fourier transform can be interpreted as time–frequency representations in terms of correlation functions. These transforms are computed by correlating the TSPI signal with different sets of time–frequency atoms. Therefore, the time–frequency resolution of the transform is limited by the time–frequency resolution of the atom's family involved. The quadratic structure associated with these transforms (scalogram and spectrogram) allows interpreting the corresponding time–frequency representation as a time–frequency energy distribution or instantaneous power spectrum.

The energetic approach seeks to combine the concepts of the instantaneous power and spectral energy density of the signal to devise a joint function of time and frequency, called a distribution, which describes the instantaneous power spectrum. The Wigner–Ville distribution is a time–frequency local energy that is computed by correlating the signal with a time and frequency translation of itself, which avoids any loss of time–frequency resolution as no atom's family is required. More generally, the spectrogram and scalogram can be described as a time–frequency averaging of the Wigner–Ville distribution [8].

The Wigner–Ville distribution $Q_z(t, \omega)$ of an analytic signal $z_s(t)$ is defined as a quadratic form that measures a local time–frequency energy [79]

$$Q_z(b,\omega) = \int_{-\infty}^{+\infty} \mathrm{d}t\, \mathrm{e}^{-it\omega} z_s\left(b+\frac{t}{2}\right) z_s^*\left(b-\frac{t}{2}\right). \tag{4.70}$$

$Q_z(b,\omega)$ is real due to the fact that the Fourier transform of $z_s(b+t/2)z_s^*(b-t/2)$ has Hermitian symmetry in t.

The interpretation of $Q_z(t,\omega)$ as a local time–frequency energy is suggested due to the fact that the time and frequency energy densities of the signal can be obtained as marginal integrals. Explicitly, the energy of the signal is computed in time or in frequency by the well–known relations

$$\|z_s(t)\|^2 = \int_{-\infty}^{+\infty} \mathrm{d}t |z_s(t)|^2 = \frac{1}{2\pi}\int_{-\infty}^{+\infty} \mathrm{d}\omega |\hat{z}_s(\omega)|^2, \tag{4.71}$$

where the identity is justified by the Plancherel formula. The square values $|z_s(t)|^2$ and $|\hat{z}_s(\omega)|^2/2\pi$ can be interpreted as energy densities in time and frequency, respectively, and it can be demonstrated that

$$\begin{aligned} \int_{-\infty}^{+\infty} \mathrm{d}t\, Q_z(t,\omega) &= |\hat{z}_s(\omega)|^2, \\ \int_{-\infty}^{+\infty} \mathrm{d}\omega\, Q_z(t,\omega) &= 2\pi |z_s(t)|^2. \end{aligned} \tag{4.72}$$

It is important to note that the Wigner–Ville distribution satisfies the following property. For a Gaussian windowed linear chirp

$$x_{\mathrm{LC}}(t) = \mathrm{e}^{-ct^2}\, \mathrm{e}^{i(\alpha_0+\alpha_1 t+\alpha_2 t^2)}, \tag{4.73}$$

with $\{c > 0,\ \alpha_i : i = 1,\ 2,\ 3\} \in \mathbb{R}$, the Wigner–Ville distribution concentrates the energy along the instantaneous frequency of the signal [79]:

$$Q_{x_{\mathrm{LC}}}(t,\omega) = \mathrm{e}^{-2ct^2}\, \mathrm{e}^{-\frac{1}{2c}(\omega-\alpha_1-2\alpha_2 t)^2}, \tag{4.74}$$

which can be determined by detecting the local maximum of the distribution. Therefore, the Wigner–Ville distribution should be a promising tool to analyze TSPI signals. Unfortunately, this is not the case due to the quadratic property of $Q_z(t,\omega)$. If a sum of two signals $z_x(t)$ and $z_y(t)$ were analyzed by means of the Wigner–Ville distribution, this will be the addition of the individual distributions plus "cross terms" that complicate the analysis. The spectrogram contains cross terms although they are attenuated, while those in the Wigner–Ville distribution are not. Cross terms also occur within a single component signal.

The Wigner–Ville distribution of a TSPI signal satisfies

$$Q_z(t,\omega) = Q_z(t,-\omega), \tag{4.75}$$

because the signal is real and therefore it has an even spectrum. Its Wigner–Ville distribution will have cross terms between the identical components in the positive and negative half planes. Then, the application of the analytic signal corresponding to the TSPI signal not only removes the redundant information but also many of the cross terms that complicate the interpretation of the Wigner–Ville distribution. The cross terms in the Wigner–Ville distribution take negative values and this fact shows that the distribution is not an energy density function. To recover the positivity property, it is necessary to average the Wigner–Ville distribution, although some time–frequency resolution is lost.

Smoothing $Q_z(t,\omega)$ with a kernel θ, the cross terms can be partially attenuated. The resulting smoothed distribution, also named as smoothed time–frequency distribution (STFD), is defined as

$$SQ_z(t,\omega) = \int_{-\infty}^{+\infty}\int_{-\infty}^{+\infty} \mathrm{d}t'\mathrm{d}\omega'\, Q_z(t',\omega')\,\theta(t,t',\omega,\omega'). \tag{4.76}$$

The time–frequency resolution of this distribution depends on the spread of the kernel θ. In addition, when the translation invariance properties of the kernel are imposed, $\theta(t,t',\omega,\omega') = \theta(t-t',\omega-\omega')$, this distribution can be defined as a member of a Cohen's class [80]. This class contains the quadratic time–frequency distributions that satisfy the time translation and frequency modulation invariance properties. Therefore, if a signal is translated in time or frequency, its distribution is translated by the corresponding amount.

The properties of smoothed distributions can be thought as constraints on the kernel. Therefore, different properties of the distribution can be obtained by modifying the kernel definition. This advantage allows adapting the distribution to the characteristics of the analyzed signal. In Ref. [81], it is shown that Gaussian kernel in a Wigner–Ville distribution can be used to estimate the instantaneous frequency. The first-order moment with respect to the frequency is a definition of the instantaneous frequency, which is consistent with the use of the analytic signal $z_s(t)$ [80]. This estimation is carried out by computing the first conditional frequency moment of the smoothed distribution, which means

$$\left\langle \frac{\partial}{\partial t}\phi_s(t) \right\rangle = \frac{\int_{-\infty}^{+\infty} \omega SQ_z(t,\omega)\mathrm{d}\,\omega}{\int_{-\infty}^{+\infty} SQ_z(t,\omega)\mathrm{d}\,\omega}. \tag{4.77}$$

Although some distributions can recover the instantaneous frequency from its first conditional moment of the frequency, the implementation of this calculation is often computationally expensive. Moreover, the use of this mathematical property generally leads to a degradation of the estimate due to the presence of noise. Since the smoothed distribution concentrates the energy around the instantaneous frequency in the time–frequency domain, the peak detection of the smoothed distribution can be used as an alternative procedure to estimate the instantaneous frequency. For the analytic signal $z_s(t)$, this means that an ideal SQ_{z_s} can be expressed by means of a relation of the form

$$SQ_z(t,\omega) = A_s^2(t)\delta[\omega-\langle\omega_{z_s}(t)\rangle], \tag{4.78}$$

where δ is the delta Dirac function. Note that when Equation 4.78 is satisfied, Equation 4.77 is verified automatically and the energy of the distribution SQ_z is concentrated along the instantaneous frequency of the signal. Although an estimate of the instantaneous frequency by using the local maximum of the distribution is appropriated, the existence of additional nonlinearities in the frequency dependence of the analyzed signal produces a biased estimate. The bias depends on the time–frequency resolution of the kernel, so its optimal selection can be very difficult to find in practice. To tackle this problem, the development of kernels adapted to the signal structure is a matter of constant investigation. Reference [81] presents an overview of several quadratic time–frequency distributions. Below follows the case when a Gaussian kernel is used.

The implementation of a Cohen's class for discrete signals is not straightforward. The frequency concept is associated with the Fourier transform, which is computationally implemented according to the properties of the analyzed signal [82]. There are several methods to compute time–frequency distribution of discrete signals [83–86]. Particularly, to compute the discrete Wigner–Ville distribution smoothed by the Gaussian kernel defined by

$$\theta(\tau, \nu) = \exp(-\tau^2/\eta_\tau) \exp(-\nu^2/\eta_\nu), \tag{4.79}$$

where η_ν and η_τ control the kernel spread, Ref. [8] provides some details on the numerical implementation, which are analyzed next.

From Equation 4.70, the Wigner–Ville distribution is the Fourier transform of $z_s(b+t/2)\, z_s^*(b-t/2)$. Therefore, for a discrete signal $z_s(n)$ defined over the interval $0 \leq n < N$, $n \in \mathbb{N}$, the integral in Equation 4.70 can be approximated by the discrete sum:

$$Q_z(n, k) \simeq \sum_{p=-N}^{N-1} z_s\left(n+\frac{p}{2}\right) z_s^*\left(n-\frac{p}{2}\right) \exp\left(-\frac{2\pi ikp}{N}\right). \tag{4.80}$$

Note that this last expression requires to know the values of z_s at half integers. Therefore, the use of an interpolation process is necessary to avoid aliasing. Substituting the interpolating function z_{int} in the discrete sum, it results

$$Q_z(n, k) \simeq \sum_{p=0}^{2N-1} z_{\text{int}}(2n+p-N) z_{\text{int}}^*(2n-p+N) \exp\left[-\frac{2\pi i(2k)p}{2N}\right], \tag{4.81}$$

where $0 \leq k < 2N$. This new discrete sum consists of N fast Fourier transforms of size $2N$, which requires $O(N^2 \log N)$ operations.

To obtain the smoothed distribution, the convolution of the discrete Wigner–Ville distribution Q_z with the discrete kernel θ is computed. The convolution is carried out by multiplying the two-dimensional Fourier transforms of the discrete distribution Q_z and the discrete kernel θ, and then transforming backward. This calculation also requires a total of $O(N^2 \log N)$ operations.

To find the phase distribution by means of the quadratic time–frequency distribution method, first the instantaneous frequency along the time axis is determined.

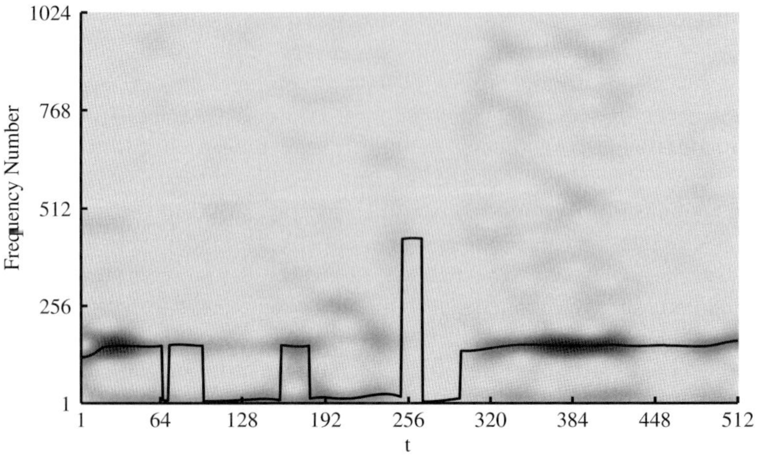

Figure 4.9 Smoothed time–frequency distribution of the TSPI signal shown in Figure 4.1, where the dark line points out the local maxima.

The modulus of the instantaneous frequency is estimated by tracking the local maximum of the smoothed distribution in the time–frequency plane. Afterward, the temporal evolution of the object phase is calculated by integration of the estimated instantaneous frequency knowing its sign and using an appropriate boundary condition.

Figure 4.9 depicts the smoothed time–frequency distribution of the TSPI signal shown in Figure 4.1 that is calculated using the discrete algorithm given by Equation 4.81. The dark points in this figure indicate large values of the smoothed distribution coefficients, which were sampled into [0, 255] discrete gray levels. The obtained local maxima are pointed out by means of the dark continuous line. This figure shows that the local maxima do not follow the instantaneous frequency track corresponding to the analyzed TSPI signal. This same behavior is also obtained for a wide set of values of the parameters η_v and η_τ. Figures 4.5 and 4.9 show this characteristic limitation produced by the wavelet transform and the smoothed time–frequency distribution, respectively. However, this limitation can be partially overcome by adequately prefiltering the TSPI signal, a topic that will be presented in the following section.

4.4.5
The Empirical Mode Decomposition and the Hilbert Transform

The analytic method is a phase recovery technique commonly used in signal processing applications. This method is based on the analytic signal, which is generated using the Hilbert transform (see Section 4.4.1). When the dc contribution of the TSPI signal is filtered out and the signal $s(t)$ can be represented by its analytic signal $z_s(t)$, the Hilbert transform generates exactly the quadrature component of the signal (see Equation 4.20) and consequently the analytic signal allows the recovery of the phase $\phi(t)$ in mod$[2\pi]$:

$$\phi(t) = \arctan\left\{\frac{\text{HT}[s(t)]}{s(t)}\right\}. \tag{4.82}$$

The recovery of the phase $\phi(t)$ depends on the removal of the bias intensity. The analytic method fails when it is applied to real-valued signals with nonzero local mean [22]. The other condition to be verified is fulfilled using Bedrosian's product theorem (see Section 4.4.1), which imposes a frequency condition between $I_M(t)$ and $\cos\phi(t)$. The validity of both conditions guarantees the identification of the recovered phase with the object phase.

The HT may also be expressed via the FT by observing that

$$\text{FT}\{\text{HT}[s(t)]\}(\omega) = -i\,\text{sgn}(\omega)\text{FT}[s(t)](\omega), \tag{4.83}$$

where

$$\text{sgn}(\omega) = \begin{cases} 1, & \omega > 0, \\ 0, & \omega = 0, \\ -1, & \omega < 0. \end{cases} \tag{4.84}$$

Therefore, the HT can be calculated by applying the inverse Fourier transform FT^{-1} to Equation 4.83 [21]. In addition, this derivation of the HT by means of the FT shows that the Fourier spectrum of the analytic signal $z_s(t)$ verifies

$$\begin{aligned} \text{FT}[z_s(t)](\omega) &= 0, & \omega < 0, \\ \text{FT}[z_s(t)](\omega) &= 2\,\text{FT}[s(t)](\omega), & \omega \geq 0. \end{aligned} \tag{4.85}$$

This property of the analytic signal allows connecting the Hilbert analysis with the temporal Fourier transform method (see Section 1.3). In the Fourier method, FT is applied to the resulting intensity. The Fourier spectrum appears separated in the frequency domain due to the carrier frequency and the modulation intensity $I_M(t)$ is considered a smooth temporal function. Therefore, using a filtering window in the frequency domain, one side of the spectrum can be isolated and recuperated, so that the bias intensity $I_B(t)$ is automatically removed (see Figure 4.2). Afterward, the inverse Fourier transform of the recuperated spectrum is determined and the phase distribution can be obtained by means of the arctan function (see Equations 4.10 and 4.11). Therefore, the analytic method is equivalent to the Fourier transform approach. This result also shows that if a comparison between both methods were made, the performance and accuracy of the obtained results would depend mainly on the numerical implementations that are used in the analysis. The analytic method was used in Ref. [87] to process TSPI signals for the detection of defects, and more recently to investigate thermal expansion of ceramic–stainless steel joints by using both in-plane and out-of-plane speckle interferometers [88].

The TSPI signal $I(t)$ given by Equation 4.1 should be previously processed to remove the bias intensity and the hypotheses of the Bedrosian's product theorem should also be fulfilled to apply the analytic method. However, as it was shown in

Section 4.2, the temporal variability of the bias intensity $I_B(t)$ and the modulation intensity $I_M(t)$ can be quite large (see Figure 4.1). Then, these facts can make the phase recovery process more difficult if it is considered that each pixel of the CCD camera acts as an independent sensor and several thousands of them are commonly involved. Consequently, it is very difficult to know if the conditions required to apply the analytic method are fulfilled, so that a practical approach is usually adopted. It is worth to emphasize that the implementation of this approach has a preponderant rôle in the phase recovery process.

In the following section, it will be shown that the application of the empirical mode decomposition (EMD) method can minimize the influence of the bias intensity on the phase recovery process when the analytic approach is used.

4.4.5.1 The Empirical Mode Decomposition Method

EMD is a nonlinear method for adaptively representing a nonstationary signal as a sum of zero-mean well-behaved fast and slow oscillation modes referred to as intrinsic mode functions (IMFs) [89]. The adapted decomposition is carried out through a sifting process that generates a fully data-driven method, so that no time–frequency atoms are required in the analysis process. Consequently, the frequency analysis does not correspond to a predetermined subband filtering, and the mode selection corresponds to an automatic and adaptive data-driven filtering.

In the empirical mode decomposition of a signal, the IMFs are defined by the signal itself using a sifting process. The features that define the IMFs are (i) over its entire length, the number of extrema and zero-crossings must either be equal or differ at most by one; and (ii) at any point, the mean value of the signal defined by the local maxima and the envelope defined by the local minima is zero. The IMFs result well defined for time–frequency analysis in a Hilbert spectrum. The signal is finally decomposed into a redundant set of IMFs and a residue without any loss of information.

The sifting process begins by identifying all local extrema of the signal $s(t)$. The maxima are then connected by cubic splines to form the upper envelope, and similarly for the lower envelope. An approximate first IMF h_1 is obtained by filtering out the mean of the envelopes, designated as m_1, from the signal s:

$$h_1 = s - m_1. \tag{4.86}$$

The amplitudes of the new local extrema are reduced by repeating the sifting process, treating h_1 as the input data. Denoting the mean of the envelopes of h_1 as m_{11}, an improved first IMF may be obtained as

$$h_{11} = h_1 - m_{11}. \tag{4.87}$$

This sifting process is repeated k times, until h_{1k} can be considered as the first IMF c_1 of the original data s:

$$c_1 = h_{1k} = h_{1(k-1)} - m_{1k}. \tag{4.88}$$

The residual signal r_1, which is given by

$$r_1 = s - c_1, \tag{4.89}$$

contains information of lower frequency components and is treated as a new signal to be subjected to the same sifting process as described above. The entire procedure is repeated until the nth residue

$$r_n = r_{n-1} - c_n \tag{4.90}$$

is a monotonic function and no more IMFs can be extracted. Finally, the decomposition of $s(t)$ can be expressed as

$$s(t) = \sum_{j=1}^{n} c_j(t) + r_n(t). \tag{4.91}$$

Numerical experiments tend to confirm that an increase in the required number of sifting iterations produces a signal overdecomposition by spreading out their components over adjacent modes. To terminate the sifting process, a criteria based on the evaluation of the normalized standard deviation SD between two consecutive sifts is commonly used, which is calculated as

$$\mathrm{SD} = \frac{\sum_{t=0}^{N-1} |h_{1k-1}(t) - h_{1k}(t)|^2}{\sum_{t=0}^{N-1} h_{1k-1}^2(t)}, \tag{4.92}$$

where N is the number of samples of the signal $s(t)$.

The SD value for terminating the sifting process is selected by analyzing several simple harmonic signals and the signal corresponding to the problem at hand. Typical values for SD can be set between 0.2 and 0.3 [89]. Another issue that also has to be taken into account when the EMD method is applied is related to boundary conditions of the analyzed signal. The spline fit at the beginning and at the end of the signal can result in large swings in the envelope, which can eventually propagate inward and corrupt especially the low-frequency components. The end effects can be overcome by adding data at the beginning and at the end of the signal to be analyzed through padding.

Figure 4.10 shows the IMFs and the residue obtained from the application of the empirical mode decomposition to the TSPI signal displayed in Figure 4.1. Note that the repeated sifting process gives a total of seven IMFs and a residue, which shows the high efficiency of the EMD method when it is compared with the traditional Fourier expansion. The decomposition also suggests that the bias intensity can be associated with last IMF plus the residue. Also, the first IMF usually contains the fine structure of the speckle and the high-frequency noise of a TSPI signal. The usefulness of this approach will be described in the following section.

4.4.5.2 **The Hilbert Transform**

The improvement produced by the EMD method in the phase recovery process of TSPI data by using the analytic approach was demonstrated in Refs [90–94]. As

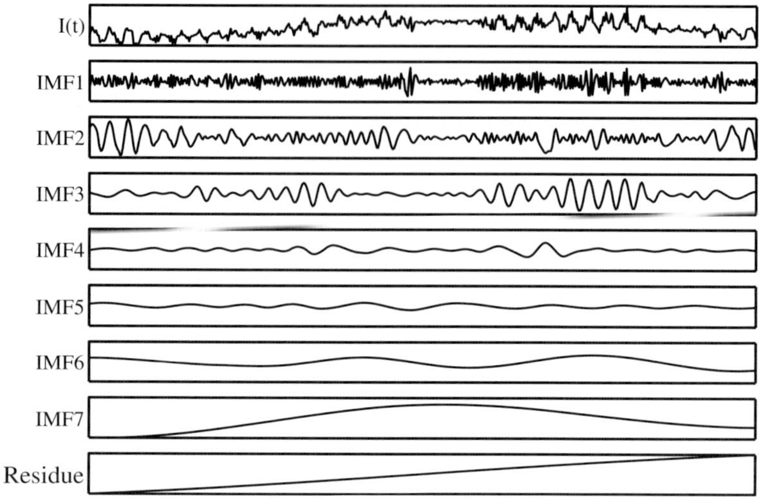

Figure 4.10 Empirical mode decomposition of the TSPI signal shown in Figure 4.1.

a typical example, Figure 4.11a and b illustrates the performance of the EMD method to remove the variation of the bias intensity and also to filter out the high-frequency noise of the signal displayed in Figure 4.1, respectively.

The Hilbert transform can be applied to the IMFs given by the empirical mode decomposition of the TSPI signal and it is known as the Hilbert–Huang transform [89]. In this case, the EMD method acts as a dyadic filter [95] and the object phase is recovered by analyzing a small number of IMFs, which admit well-behaved HTs [96]. This approach is very useful when the temporal variability of the bias and the modulation intensity are not sufficiently large, due to the fact that the phase information is mainly contained in the first IMFs of the decomposition.

Figure 4.12 shows the unwrapped phase $\phi_t(t)$ that is recovered using the Hilbert transform and applying a previous prefiltering through the EMD method. The recovered phase was obtained by applying Equation 4.82 to the TSPI signals depicted in Figure 4.11a and b, dashed and continuous lines, respectively. It should be noted that the improvement obtained in the recovered phase is through the application of the EMD method. The Hilbert transform approach by itself, that is, without the application of the EMD prefiltering, cannot follow the phase contained in the argument of the cosine function $\phi_t = \phi + \phi_c + \phi_r$ given by Equation 4.2.

The reader should note that Equation 4.83 is used in the analytic method and also in the EMD approach. Therefore, when the HT is numerically implemented by means of the inverse Fourier transform, the analytic method, that is, the HT with EMD approach described here, and the Fourier transform technique are essentially equivalent if the bias intensity is filtered by using the previously described procedure. Consequently, the relevant differences between these methods will be contained mainly in the numerical implementation of the HT as it was mentioned in the description of the analytic method, in the selection of the filtering window used in

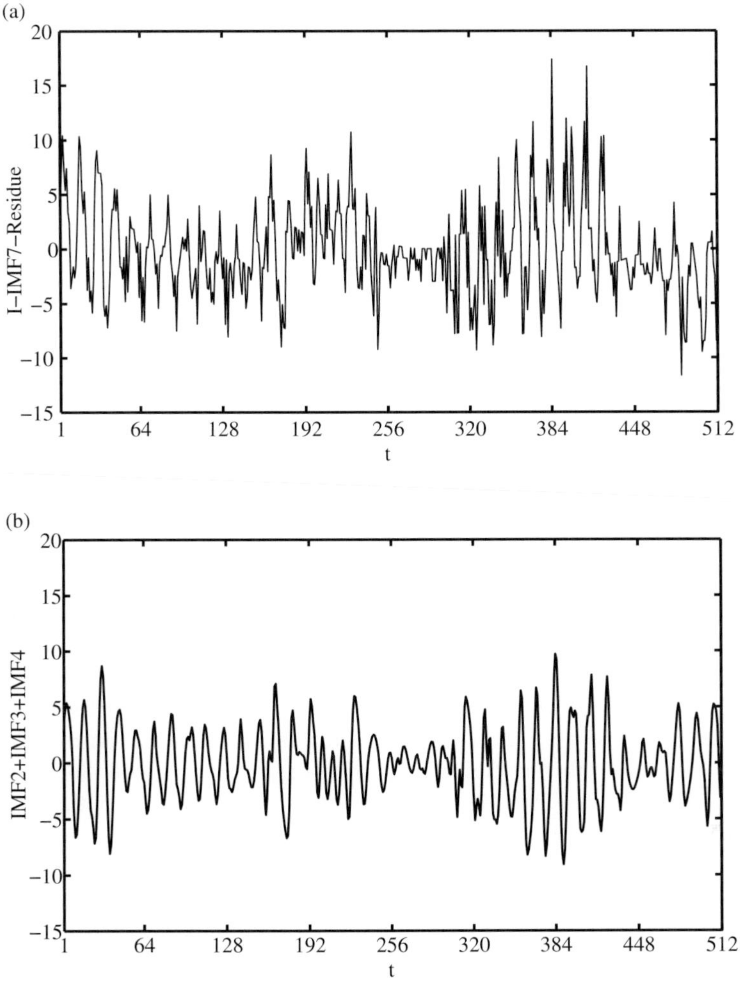

Figure 4.11 Application of the EMD method to the TSPI signal shown in Figure 4.1. (a) Removal of the variation of the bias intensity. (b) Filtering of the high-frequency noise. (Reproduced with permission from Ref. [7], the Optical Society of America.)

the Fourier transform method (i.e., rectangular, Hanning, etc.), and also in the involved IMFs.

Several algorithms can be numerically implemented to evaluate the HT. To mention some of them, the HT may be calculated by using Hermite polynomials, a convolution with an appropriated design of a filter with finite impulse response, and the inverse Fourier transform of Equation 4.83 [21]. This last procedure was implemented in Ref. [93] by using a FFT algorithm, which is faster than the decomposition on a basis of Hermite polynomials. No significant differences in performance and accuracy are obtained using a convolution of the intensity signal

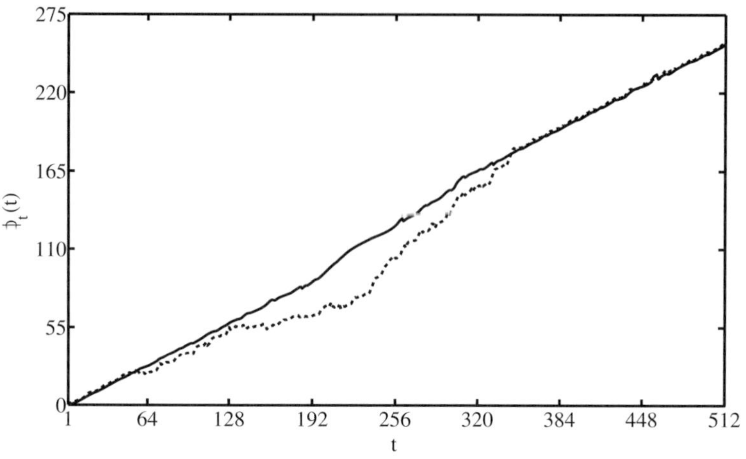

Figure 4.12 Unwrapped phase retrieved with the Hilbert transform when the signals shown in Figure 4.11a and b, dashed and continuous lines respectively, are processed.

with the finite impulse response filter or a convolution algorithm implemented with a FFT procedure.

At this point, the previous TSPI signal conditioning given by the EMD method should favor the application of time–frequency representation techniques such as those based on the wavelet transforms and smoothed Gaussian Wigner–Ville distributions. Therefore, this combination can seem to be an ideal tool for phase recovery processing of TSPI signals. However, in practice this is not the case because it is difficult to satisfy a suitable combination of IMFs for each pixel and several thousands of them are commonly involved. Therefore, a practical approach is usually adopted taking into account that the EMD improves the bias intensity filtering, so that a robust phase estimate can then be applied.

4.4.6 The Generalized S-Transform

The S-transform (ST) produces a time–frequency representation of the TSPI signal that uniquely combines a frequency-dependent resolution with simultaneously localizing the real and imaginary spectra (see Ref. [97] and references therein). The ST can be derived from the short-time Fourier transform (STFT) and the continuous wavelet transform (CWT). Therefore, the ST is a conceptual hybrid of the STFT and the CWT with their advantages combined within the ST framework. The ST is derived from the STFT (see Equation 4.28) by defining a normalized Gaussian window function $g_f(t)$ of width proportional to the inverse of the frequency f:

$$g_f(t) = \frac{1}{\sigma\sqrt{2\pi}} e^{-\frac{t^2}{2\sigma^2}} \quad \text{and} \quad \sigma(f) = \frac{1}{|f|}, \tag{4.93}$$

so that a better resolution in the phase space can be obtained. Replacing Equation 4.93 in Equation 4.28, the S-transform is obtained as

$$\mathrm{ST}_I(b,f) = \frac{|f|}{\sqrt{2\pi}} \int_{-\infty}^{+\infty} \mathrm{d}t \mathrm{e}^{-i2\pi ft} \mathrm{e}^{-\frac{1}{2}(t-b)^2 f^2} I(t). \tag{4.94}$$

The ST can be also associated with a CWT multiplied by a factor phase. Here, the mother wavelet is defined by a Gaussian window function, the scale being the inverse of the frequency and multiplied by an oscillatory exponential term:

$$\psi_{b,f}(t) = \frac{|f|}{\sqrt{2\pi}} \mathrm{e}^{-\frac{1}{2}(t-b)^2 f^2} \mathrm{e}^{i2\pi f(t-b)}. \tag{4.95}$$

Therefore,

$$\mathrm{ST}_I(b,f) = \mathrm{e}^{-i2\pi fb} W_I(b,f). \tag{4.96}$$

Note that the adopted normalization agrees with Equation 4.49, although the dilation factor a described in Section 4.4.3.1 is the inverse of the frequency. This approach splits the mother wavelet into the Gaussian window function that is localized in time and an oscillatory exponential kernel that selects the frequency being localized. As it is usual in the wavelet transform framework, the Gaussian window function is translated along the signal, but in this case the oscillatory exponential kernel remains stationary. Consequently, the ST combines a frequency-dependent resolution with a simultaneous localization of the real and imaginary spectra. Clearly, the mother wavelet does not satisfy the admissibility condition and therefore it is not strictly a CWT. However, the nonfulfillment of the admissibility condition gives an extra degree of freedom that is used for better estimation of the signal frequency. Therefore, the ST turns into a promising analysis tool to process nonstationary signals. In addition, the linear property of the ST reduces the presence of cross terms when it is compared with quadratic representations such as the Cohen's class distributions (see Section 4.4.4).

The generalized S-transform $S_I(b,f,\mathbf{p})\theta$ of a TSPI signal $I(t)$ is defined from the original ST by replacing in Equation 4.94 the Gaussian window with a generalized one $w_G(t,f,\mathbf{p})$:

$$S_I(b,f,\mathbf{p}) = \int_{-\infty}^{+\infty} \mathrm{d}t \mathrm{e}^{-i2\pi ft} w_G(b-t,f,\mathbf{p}) I(t). \tag{4.97}$$

The generalized window is translated by b along the time axis, and its width and shape are given by a function of the frequency f and a set of parameters denoted by $\mathbf{p}$ [98]. The generalized window w_G satisfies the normalization condition:

$$\int_{-\infty}^{+\infty} \mathrm{d}b\, w_G(b-t,f,\mathbf{p}) = 1. \tag{4.98}$$

By integrating Equation 4.97 over b and using the normalization condition Equation 4.98, it is obtained

$$\int_{-\infty}^{+\infty} \mathrm{d}b\, S_I(b,f,\mathbf{p}) = \mathrm{FT}[I(t)](f). \tag{4.99}$$

Therefore, $S(b,f,\mathbf{p})$ is invertible because $I(t)$ can be obtained from its Fourier transform. Then, the generalized S-transform (GST) can also be thought as a generalization of the Fourier transform to nonstationary signals.

A simple generalized window can be defined as

$$w_G(b-t,f,p_0) = \frac{|f|}{p_0\sqrt{2\pi}} \exp\left[\frac{-f^2(b-t)^2}{2p_0^2}\right], \tag{4.100}$$

where $p_0 \in \mathbb{R}_{>0}$ is the only element of $\mathbf{p}$ and clearly $p_0 = 1$ reproduces the ST. Note that the p_0 value can be increased for increasing frequency resolution, although the resolution in time is lost. The standard deviation of the Gaussian w_G contains p_0 wavelengths of the Fourier sinusoid at all frequencies, thus allowing the time and frequency resolution of the transform in the time–frequency space to be specified by the user.

The analyzing function is the product of a fixed Fourier sinusoid with a generalized window that have a frequency-dependent functional form, as previously mentioned. $S_I(\tau,f,p_0)$ is not a CWT because the Fourier sinusoid is not translated along the time axis with w_G. However, a multiresolution analysis while retaining the absolute phase of each frequency can be obtained.

Several generalized window functions have been reported in the literature to be used in the GST [98–100]. The selection of a Gaussian window function allows that the uncertainty principle reaches its minimum value with minimum computational requirements [98]. It should be noted that the saving in computational time is an important issue in the phase recovery process because huge data sets need to be processed. Recently, an improvement in the energy concentration of the ST has been proposed by using a frequency Kaiser window [101], although with an additional computational burden. Therefore, in some cases the implementation of a frequency Kaiser window could be a competitive choice with respect to the use of the Gaussian window function.

In Section 4.4.3.1, it was shown that the CWT method is based on the knowledge of the wavelet coefficients in the analytic asymptotic limit through the evaluation of the scalar products using the stationary phase method. This knowledge allows the specification of the ridge in the time–scale plane, which have the property to describe the temporal evolution of the phase or its gradients. As the GST can be formally derived from the continuous Morlet wavelet transform by interpreting the wavelet scale as $a = p_0/|f|$, the Morlet wavelet frequency as $\omega_0 = 2\pi p_0$, and by introducing an additional phase term $\exp[-2\pi i f b]$, the following identity is obtained:

$$S(b,f,p_0) = \frac{\mathrm{e}^{-2\pi i f b}}{\sqrt{2\pi}} W_I[b, a(p_0,f), \omega_0(p_0)]. \tag{4.101}$$

Therefore, the analytical expression of the GST at the ridge can be obtained from Section 4.4.3.1, where the stationary phase approximation was introduced. Note that

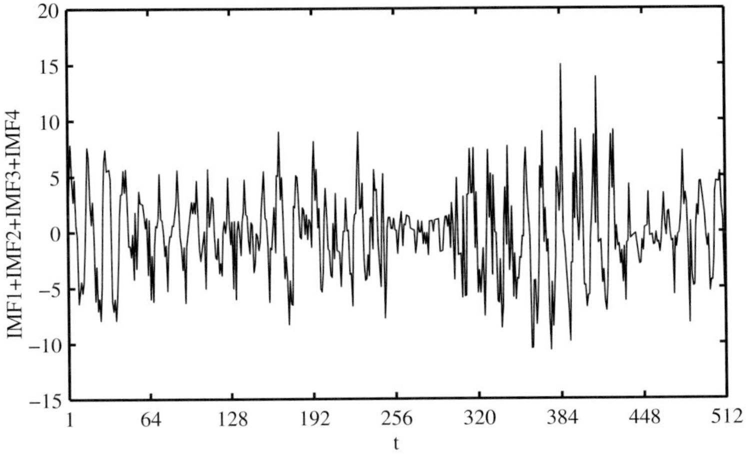

Figure 4.13 EMD removal of the bias intensity applied to the TSPI signal shown in Figure 4.1. (Reproduced with permission from Ref. [7], the Optical Society of America.)

the computational complexity of the GST is similar to the one produced by the CWT technique. An application of the GST is presented below and its performance is compared with the methods based on the CWT, HT, and Wigner–Ville distribution with a smoothing Gaussian kernel (STFD).

In practice, when the temporal variability of the bias and the modulation intensity are not sufficiently large, the phase information is mainly contained in the first IMF of the decomposition. Therefore, it is not convenient in the phase recovery process to remove the first IMF as several analyzed pixels in an experiment are mainly affected by noise. Figure 4.13 illustrates the removal of the intensity bias applied to the TSPI signal shown in Figure 4.1.

Figures 4.14 and 4.15 show the temporal evolution of the estimated instantaneous frequency obtained from the signal displayed in Figures 4.11a and 4.13, respectively, using the GST (continuous curve), the STFD (dashed curve), and the CWT (thin dashed curve) approaches. Figure 4.16 shows the integrated values of the estimated phase obtained from Figure 4.13. These values were determined by identifying the phase of the GST at the ridge (continuous curve) and also by applying the HT method (thin dashed curve). As the phase values were wrapped, a temporal one-dimensional unwrapping algorithm was applied. These last results were also compared with those determined by using the STFD approach (dashed curve) and by integrating the obtained instantaneous frequency values. In Figure 4.14, the GST is applied using a single value $p_0 = 15\pi$, although smaller values such as $p_0 = 2\pi$ can also produce accurate results. These results show that the proposed GST method is the most robust phase retrieval approach between CWT, HT, and STFD, and also for a wide variation of the parameter p_0 [7]. Also, it should be noted that the HT approach is the most sensible technique to the phase fluctuations and therefore it should be discouraged for its use in complex TSPI applications.

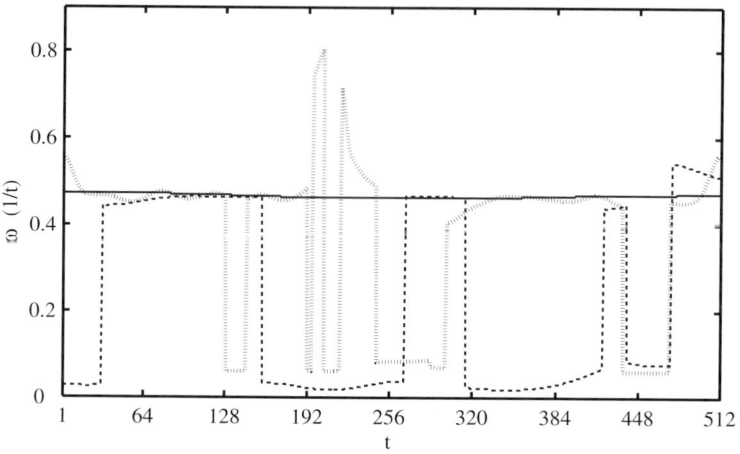

Figure 4.14 Instantaneous frequency values obtained from the signal shown in Figure 4.11a, GST being the continuous curve, STFD the dashed curve, and CWT the thin dashed curve. (Reproduced with permission from Ref. [7], the Optical Society of America.)

In Figures 4.17 and 4.18, the performance between the GST, CWT, and STFD approaches is compared when the TSPI signal contains very low phase gradient values and also regions with abrupt changes of the phase gradient. Figure 4.17 shows the temporal history of a TSPI signal $I(t)$ generated by the object mentioned in Section 4.2 when it returned to its initial position with a variable angular speed. To minimize the influence of the bias intensity, the EMD method was used to remove the first two IMFs. The phase gradients of the filtered signal were determined using

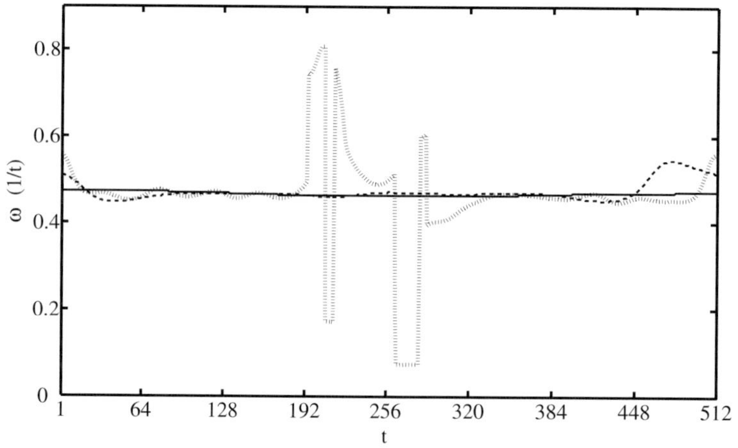

Figure 4.15 Instantaneous frequency values obtained from the signal shown in Figure 4.13, GST being the continuous curve, STFD the dashed curve, and CWT the thin dashed curve. (Reproduced with permission from Ref. [7], the Optical Society of America.)

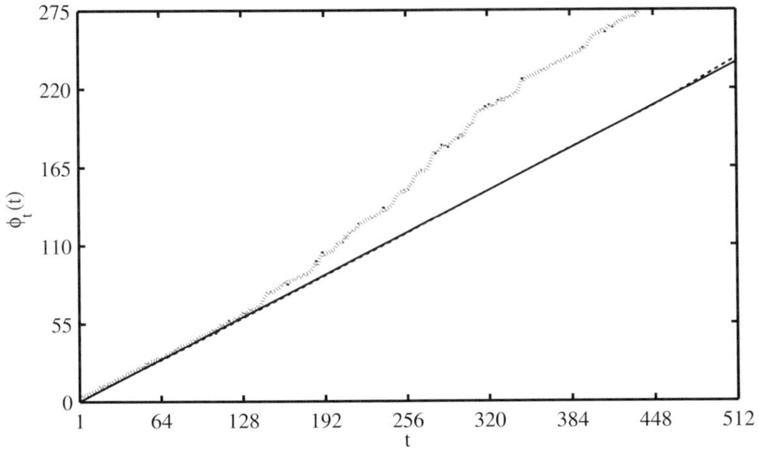

Figure 4.16 Integrated phase obtained from the signal shown in Figure 4.13 using the GST (continuous curve), the HT (thin dashed curve), and the STFD (dashed curve). (Reproduced with permission from Ref. [7], the Optical Society of America.)

the GST, CWT, and STFD approaches. The evaluated results are shown in Figure 4.18. It is observed that the CWT method fails near the null phase gradient due to the nonfulfillment of the analytic asymptotic limit. The STFD approach improves the detection of very slow phase variations and this behavior was already discussed in Ref. [102]. However, the STFD cannot detect abrupt phase gradient changes when large regions with very low phase gradient values are present. This result is obtained because only one external smoothing parameter is fixed by the operator that generates

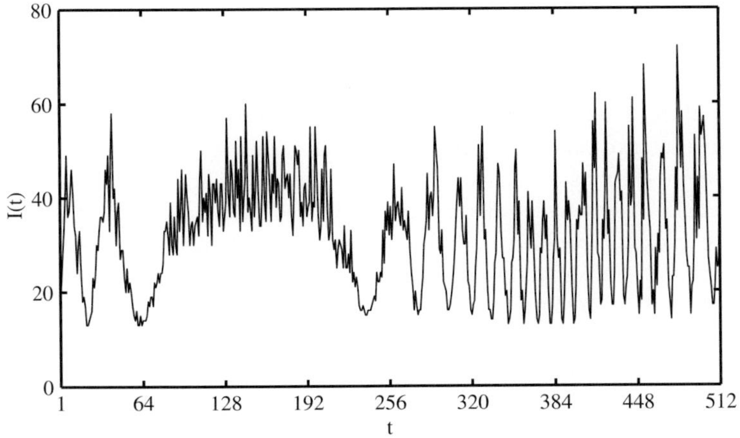

Figure 4.17 TSPI signal generated by an object that returns to its initial position. (Reproduced with permission from Ref. [7], the Optical Society of America.)

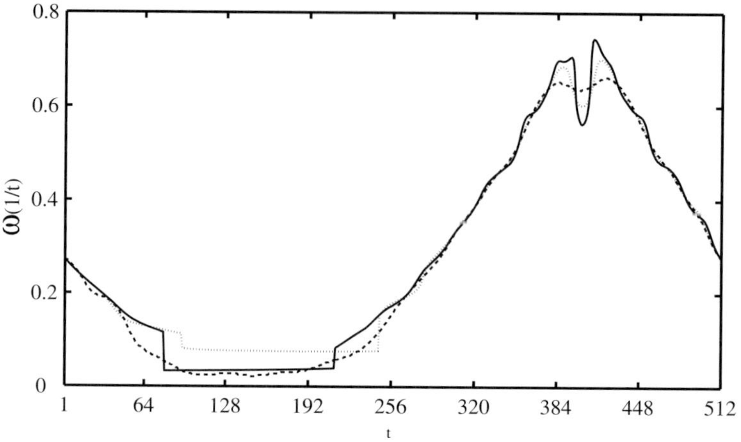

Figure 4.18 Instantaneous frequency obtained from Figure 4.17 using the GST (continuous curve), STFD (dashed curve), and the CWT (dotted curve). (Reproduced with permission from Ref. [7], the Optical Society of America.)

a trade-off. The GST approach reproduces quite well the performance given by the STFD method near the null phase gradients. In addition, the GST method generates an important improvement in the region that shows the abrupt phase gradient change, which is also detected by the CWT method by using the parameter value $p_0 = \pi/10$. Note that this single value can simultaneously estimate abrupt changes of the frequency in short time and also low phase gradient values in long times.

4.4.7
Two- and Three-Dimensional Approaches

In previous sections it was shown that the sequential acquisition of a large number of speckle interferograms and its postprocessing facilitate the recovery of the object phase distribution. The temporal object phase recovery involves only one-dimensional analysis of the TSPI signals. This procedure can be easily implemented and prevents the use of two- and three-dimensional phase unwrapping procedures. However, these one-dimensional methods of time–frequency analysis fail when a significant number of sets with adjacent nonmodulated pixels are present in the recorded data. Clearly, this problem is extremely difficult to solve using the described techniques because the information is absent in the pixel set, although this limitation could be significantly overcome by considering the information belonging to a neighborhood of pixels.

Speckle averaging is a simple way to take into account the variations in the reliability of phase estimates by assigning more weight to pixels with higher modulation (see Section 2.3.1 of Ref. [4]). This procedure was performed using a smoothing spatial kernel prior to the one-dimensional application of the phase recovery technique. More recently, a three-dimensional piecewise processing method was introduced, which is based on the removal of regions within each temporal pixel history where the phase is likely to be unreliable according to a predefined criterion. The corresponding regions of missing data are filled by the information of well-modulated neighboring pixels using an interpolation procedure based on the Delaunay triangulation [103].

Two- and three-dimensional Fourier transform methods have also been used to analyze dynamic displacements [13, 104]. The two-dimensional Fourier technique requires the recording of a time–sequence of fringe patterns, which are analyzed individually. It is worth noting that the two-dimensional Fourier transform can also be applied to the spatiotemporal slides (x, t) and (y, t). However, the two-dimensional approach cannot use any relation between consecutive slides. Therefore, a three-dimensional approach could be better adapted to the phase recovery process as more sampling points are analyzed in a three-dimensional kernel. To illustrate the use of the Fourier transform method in more than one dimension, the one-dimensional Fourier transform approach introduced in Section 4.3 is extended into three dimensions below.

The interferometric signal given by Equation 4.1 is modified by the introduction of the new frequency carriers ω_{cx} and ω_{cy} in the horizontal and vertical spatial domains, respectively. Therefore, the cosine argument in Equation 4.3 is modified by a new carrier phase:

$$\phi_c(m, n, t) = \omega_{cx}m + \omega_{cy}n + \omega_c t. \tag{4.102}$$

Applying the three-dimensional Fourier transform FT_{3D} to the modified Equation 4.3 and following a procedure analogous to the one described in Section 4.3, the phase information is encoded in the high-frequency components of the three-dimensional spectrum. Therefore, by band-pass filtering the frequency components of interest using a three-dimensional filter, the new Equation 4.10 can be expressed as

$$\begin{aligned}\alpha(m, n, t) = \mathrm{FT}_{3D}^{-1}\Bigg\{ T_{R3D}\Bigg[\mathrm{FT}_{3D}\Bigg\{\frac{I_M(t)}{2}\mathrm{e}^{i[\phi(t)+\phi_r(t)]}\Bigg\} \\ (\omega_x-\omega_{cx}, \omega_y-\omega_{cy}, \omega-\omega_c), (\omega_{cx}, \omega_{cy}, \omega_c)\Bigg]\Bigg\}(m, n, t),\end{aligned} \tag{4.103}$$

and then the phase difference can be recovered using Equation 4.11. Although these processing methods can improve the accuracy of the retrieved phase by introducing innovative band-pass filter in the frequency domain [105], the following section presents the description of other two- and three-dimensional approaches. These last approaches are introduced with the aim to overcome the difficulties produced by windowing, which is a very critical operation such as was previously discussed.

4.4.7.1 The Windowed Fourier Transform Method

The windowed Fourier transform method described in Section 4.4.2 can be generalized to two and three dimensions. A generalized three-dimensional windowed Fourier transform method was proposed in Ref. [106] for fringe analysis and it includes the two- and one-dimensional windowed Fourier transform methods as special cases [107, 108]. It is worth to mention that the two-dimensional windowed Fourier transform approach can not only be used in the spatial domain but can also be applied to spatial–temporal slides. The generalized windowed Fourier transform $S_3 W_I$ is an extension of Equations 4.27 and 4.28 by introducing the parameters $\{\sigma_i > 0;\ \xi_i : i = 1, 2, 3.\} \in \mathbb{R}$ for a Gaussian function $g(\mathbf{x})$ with $\mathbf{x} = (x_1, x_2, x_3) \in \mathbb{R}^3$,

$$g(\mathbf{x}) = C_g e^{-[(x_1/\sigma_1)^2 + (x_2/\sigma_2)^2 + (x_3/\sigma_3)^2]/2}, \tag{4.104}$$

where C_g is a normalization constant and $\|g(\mathbf{x})\| = 1$. Then,

$$S_3 W_I(\mathbf{b}, \xi) = \int_{-\infty}^{+\infty} d^3\mathbf{x}\, e^{-i\xi\cdot x} g(\mathbf{x}-\mathbf{b})\, I(\mathbf{x}). \tag{4.105}$$

The spectrogram $|S_3 W_I(\mathbf{b}, \xi)|^2$ measures the energy density of the intensity in a time–frequency space, as described in Section 4.4.2, and the instantaneous frequencies can be obtained by determining the local maximum of the spectrogram. Therefore, the phase can be determined by integrating the instantaneous frequencies or by using the arctan function procedure:

$$\phi(\mathbf{b}) = \arctan\left\{\frac{\mathrm{Im}[S_3 W_I(\mathbf{b}, \xi_r)]}{\mathrm{Re}[S_3 W_I(\mathbf{b}, \xi_r)]}\right\}, \tag{4.106}$$

where ξ_r is the phase gradient at the ridge of Equation 4.105 [106].

The two-dimensional windowed Fourier transform can be used for the determination of the phase distributions and its derivatives. Applications include strain determinations, fault and edge detections, and fringe segmentations with the implementation of the algorithms that appears in Ref. [109]. Deformation measurements by spatial–temporal analysis of interferogram sequences by means of the Fourier transform, windowed Fourier transform, and wavelet transform with algorithms extended to two- and three-dimensional processing were also presented in Ref. [110]. In this reference it is shown that the windowed Fourier transform analysis is better adapted than the Fourier transform approach to two-dimensional spatial and three-dimensional spatial–temporal analyses.

4.4.7.2 **Wavelet Transform Methods**

Two-Dimensional Directional Wavelet Transform In Section 4.4.3, the application of several wavelet transforms to process TSPI signals. Clearly, the wavelet function that is best adapted to the problem at hand should be selected. However, the introduction of a frequency carrier in the interferometer produces a directional feature in the recorded data that favors the application of directional wavelets, such as the Morlet wavelet.

The two-dimensional directional wavelet transform $U_I(a, \theta, \mathbf{b})$ of a two-dimensional TSPI signal $I(\mathbf{r})$, $\mathbf{r} = (x_1, x_2) \in \mathbb{R}^2$, with respect to the fixed wavelet $\psi(\mathbf{r})$, is the scalar product of $I(\mathbf{r})$ with the transformed wavelet $\psi_{a,\theta,\mathbf{b}}(\mathbf{r})$:

$$U_I(a, \theta, \mathbf{b}) = C_\psi \frac{1}{a}\int_{-\infty}^{+\infty} d^2\mathbf{r}\psi^*\left[\frac{1}{a} r_{-\theta}(\mathbf{r}-\mathbf{b})\right] I(\mathbf{r}), \tag{4.107}$$

where $r_{-\theta}$ denotes the usual rotation matrix in the plane, C_ψ is a normalization constant, and ψ is normalized to the unit [111]. The two-dimensional directional wavelet transform is characterized by the following parameters: the rotation $\theta \in [0, 2\pi)$, the translation $\mathbf{b} = (b_1, b_2) \in \mathbb{R}^2$, and the dilation $a \in \mathbb{R}_{>0}$.

A two-dimensional directional Morlet wavelet $\psi(\mathbf{r})$ is given by

$$\psi(\mathbf{r}) = e^{i\mathbf{k}_0 \cdot \mathbf{r}} e^{-\frac{1}{2}|A\mathbf{r}|^2} + \text{correction term}, \tag{4.108}$$

where $\mathbf{k}_0 = k_0(\cos\theta, \sin\theta)$ denotes a wavevector and R is the 2×2 anisotropy matrix:

$$R = \begin{bmatrix} 1/\sqrt{\varepsilon} & 0 \\ 0 & 1 \end{bmatrix}, \tag{4.109}$$

with $\varepsilon \geq 1 \in \mathbb{R}$. Usually, a correction term should be added in Equation 4.108 to enforce the admissibility condition, as discussed in Section 4.4.3, although this addition is numerically negligible for $k_0 \geq 2\pi$. With this condition, the modulus of the two-dimensional directional Morlet wavelet is a Gaussian function elongated in the x_1 direction when $\varepsilon > 1$ and its phase is constant along the direction orthogonal to $\mathbf{k}_0$. Note that the angular selectivity is incremented with the increase of the anisotropy ε and the modulus of the wavevector k_0. Therefore, the angular selectivity combines these two effects to obtain the best performance by selecting $\theta = \pi/2$, so that $\mathbf{k}_0$ becomes perpendicular to the large axis of the ellipse. Then, Equation 4.108 is given by

$$\psi(\mathbf{r}) = e^{ik_0 x_2} e^{-\frac{1}{2}(x_1^2/\varepsilon + x_2^2)}, \tag{4.110}$$

and the effective support of $\hat{\psi}(\mathbf{k})$ is contained in a convex cone centered at $\mathbf{k}_0$, where

$$\hat{\psi}(\mathbf{k}) = \sqrt{\varepsilon} e^{-\frac{1}{2}[\varepsilon k_1^2 + (k_2 - k_0)^2]}. \tag{4.111}$$

This cone becomes narrower as ε increases. The support of $\psi(\mathbf{r})$ and $\hat{\psi}(\mathbf{k})$ defines the localization properties that are compatible with the uncertainty principle. It must also be noted that the wavelet analysis is most efficient at small scales (high frequencies).

In wavelet analysis, $U_I(a, \theta, \mathbf{b})$ has an appreciable value only where $\psi_{a,\theta,\mathbf{b}}$ matches the features of the image local data. The tracking of the local maxima of $|U_I(a, \theta, \mathbf{b})|$ allows the specification of the ridge that describes the spatial evolution of the phase distribution or its gradients, as it was introduced in the one-dimensional case (see Section 4.4.3.1), although with a small variation.

To illustrate the use of the two-dimensional directional Morlet wavelet transform, below follows the recovery of the phase of a particular TSPI signal within a given spatiotemporal slice (y, t) corresponding to a TSPI data set $I(x, y, t)$. Figure 4.19 shows the spatiotemporal slice (y, t) corresponding to a simulated set of TSPI data $I(x, y, t)$. This slice contains the particular TSPI signal $I(120, 79, t)$ shown in Figure 4.20. Note the presence of anomalies in the TSPI signal in the middle of the temporal region. The presence of these anomalies complicate the ridge extraction

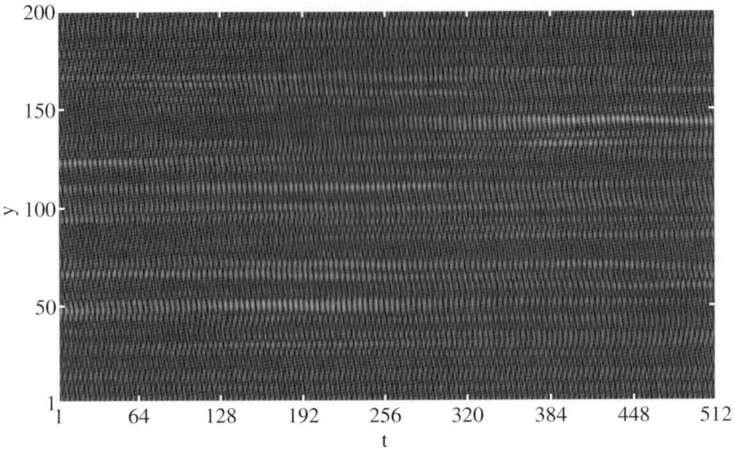

Figure 4.19 Spatiotemporal slice (y, t) corresponding to a simulated TSPI data set $I(x, y, t)$.

procedure, although the TSPI signal is prefiltered using the EMD approach. Applying an arbitrary discrete scale of numbers a, Figure 4.21 shows the application of the Morlet wavelet transform and also the ridge that is obtained. In this figure, the ridge is pointed out by means of the continuous black line and the phase wavelet coefficients belonging to the interval $(-\pi, \pi)$ are illustrated in a gray scale with $[0, 255]$ levels. As the two-dimensional directional wavelet transform takes into account the reliability variation of the phase estimates by using the information contained in the neighbor pixels, the extraction of the ridge can be consequently improved.

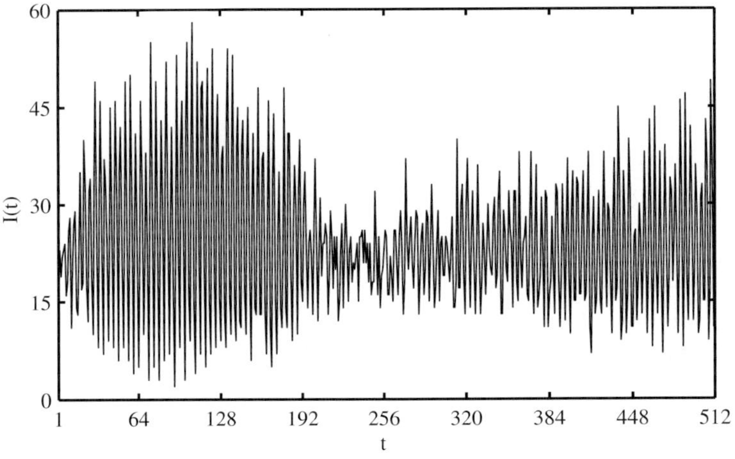

Figure 4.20 TSPI signal corresponding to the temporal history $I(120, 79, t)$ of the slice shown in Figure 4.19.

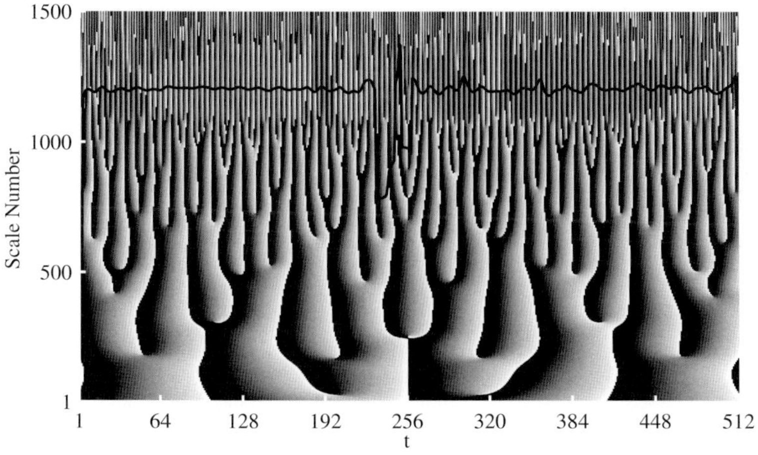

Figure 4.21 Phase of the Morlet wavelet coefficients obtained from the TSPI signal shown in Figure 4.20, where the dark continuous line points out the ridge of the transform.

Figures 4.22 and 4.23 show the results obtained when the two-dimensional directional wavelet transform is applied to the TSPI signal depicted in Figure 4.20 and an arbitrary scale of numbers a is used. Details of the ridge shown in Figure 4.22 are displayed in Figure 4.23. The readers should note that the estimation of the ridge values is more accurate when the information of the neighbor pixels is taken into account through the application of the two-dimensional wavelet transform.

The use of the described two-dimensional approach does not take into account the information located between the successive slides (x, t) and (y, t). Therefore, this

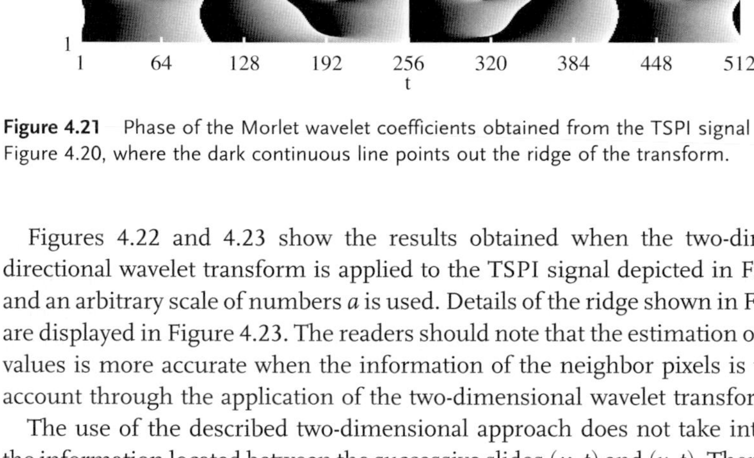

Figure 4.22 Phase of the two-dimensional directional wavelet coefficients obtained from the TSPI signal shown in Figure 4.20, where the dark continuous line points out the ridge of the transform.

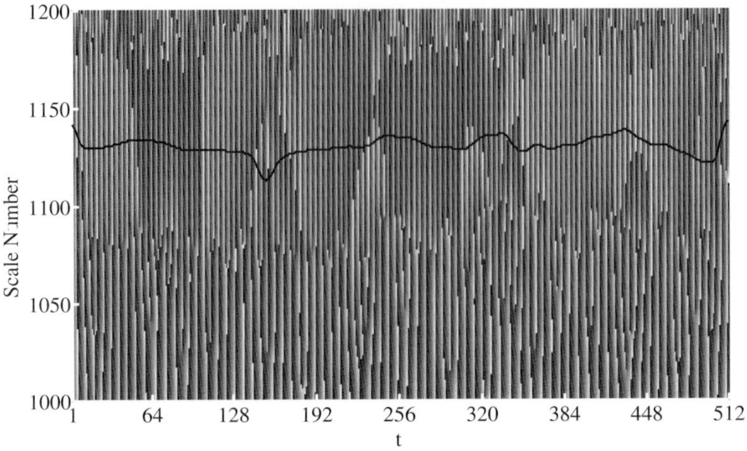

Figure 4.23 Details of the ridge shown in Figure 4.22.

approach suggests the application of the three-dimensional directional wavelet transform, which can actually consider the whole neighbor information.

Three-Dimensional Directional Wavelet Transform A three-dimensional wavelet is a square-integrable complex-valued function $\psi(x)$ with zero mean that is well localized both in the position and spatial frequency domains. The three-dimensional directional wavelet transform $D_I(a, \theta, \zeta, \mathbf{b})$ of a three-dimensional TSPI signal $I(x)$ is defined as [112]

$$D_I(a, \theta, \zeta, \mathbf{b}) = C_\psi \frac{1}{a} \int_{-\infty}^{+\infty} d^3\mathbf{x}\, \psi^* \left[\frac{1}{a} r_{-\theta\zeta}(\mathbf{x}-\mathbf{b}) \right] I(\mathbf{x}), \tag{4.112}$$

where $\mathbf{x} = (x_1, x_2, x_3) \in \mathbb{R}^3$, $a \in \mathbb{R}_{>0}$ is the dilation parameter, $\mathbf{b} \in \mathbb{R}^3$ is the displacement parameter, C_ψ is a normalization constant, and $r_{-\theta\zeta}$ acts as a rotation operator with θ around the x_3 axis and ζ around the x_2 axis. Note that the wavelet $\psi(a, \theta, \zeta, \mathbf{x})$ is translated by $\mathbf{b}$, rotated by (θ, ζ), and dilated by the scale a. Equation 4.112 is a convolution with a zero mean function $\psi(a, \theta, \zeta, \mathbf{x})$ and therefore its Fourier space representation

$$D_I(a, \theta, \zeta, \mathbf{b}) = C_\psi\, a \int_{-\infty}^{+\infty} d^3\mathbf{k} e^{i(\mathbf{b}\cdot\mathbf{k})} \hat{\psi}^* [a r_{-\theta\zeta}(\mathbf{k})]\, \hat{I}(\mathbf{k}), \tag{4.113}$$

shows that $D_I(a, \theta, \zeta, \mathbf{b})$ acts on the intensity $I(\mathbf{x})$ as a local filter in all variables: a, θ, ζ, and $\mathbf{b}$. Therefore, $D_I(a, \theta, \zeta, \mathbf{b})$ has an appreciable value only where it matches the features of the local intensity.

In a TSPI interferometer in which a temporal carrier is introduced, the output signal to be analyzed can be represented by the intensity $I(x, y, t)$ that is measured by a CCD detector at a specific pixel (x, y) and at time t (see Equation 4.3). Due to the symmetry imposed by the introduction of the temporal carrier, an appropriate selection of the rotation parameters should be $\theta = 0$ and $\zeta = 0$, so that $r_{-\theta\zeta}$ is the identity operator. In this configuration, the pair (x_1, x_2) is associated with the spatial pair (x, y), and the axis x_3 is associated with the time coordinate t. Although several analyzing wavelets could be proposed in this case, the extension of the Morlet wavelet given in Equation 4.37 is an obvious candidate.

The modified Morlet wavelet as the analyzing wavelet is defined as

$$\psi(x, y, t) = \mathrm{e}^{ik_0 t/\sigma_t} \mathrm{e}^{-[(x/\sigma_x)^2 + (y/\sigma_y)^2 + (t/\sigma_t)^2]/2}, \tag{4.114}$$

where $(\sigma_x, \sigma_y, \sigma_t) \in \mathbb{R}^3_{>0}$ are the parameters of anisotropy and $k_0 = 2\pi$. Usually, a correction term must be added in Equation 4.114 to enforce the admissibility condition, although this addition is numerically negligible for $k_0 \geq 2\pi$, as it was discussed in Section 4.4.3. Note that in the Fourier space representation, the Fourier transform of Equation 4.114 is a convex cone centered at k_0:

$$\hat{\psi}(k_x, k_y, k_t) = C\mathrm{e}^{-[\sigma_x^2 k_x^2 + \sigma_y^2 k_y^2 + \sigma_z^2 (k_t - k_0)^2]/2}, \tag{4.115}$$

where C includes the multiplicative constants. Therefore, the angular selectivity can be incremented with a suitable selection of the anisotropy parameters according to the problem at hand. All frequency changes due to abrupt conditions outside the cone determined by the selected anisotropy parameters are easily removed (see Equation 4.113).

To obtain a practical insight into the three-dimensional directional wavelet transform D_I, Equation 4.114 is replaced in Equation 4.112 and arranging the terms by identifying the two-dimensional spatial Gaussian dependence $\psi_G(x, y)$ and the one-dimensional temporal Morlet wavelet $\psi_M(t)$,

$$\psi(x, y, t) = \psi_G(x, y)\psi_M(t), \tag{4.116}$$

where

$$\psi_G(x, y) = \mathrm{e}^{-[(x/\sigma_x)^2 + (y/\sigma_y)^2]/2} \tag{4.117}$$

and

$$\psi_M(t) = \mathrm{e}^{ik_0 t/\sigma_t} \mathrm{e}^{-(t/\sigma_t)^2/2}. \tag{4.118}$$

Therefore, D_I can be seen as a spatial convolution of the temporal intensity frames with a two-dimensional Gaussian function and the CWT of the temporal history in each spatially convolved pixel by using the Morlet wavelet:

$$\begin{aligned} D_I(a, x, y, t) &= C_\psi \frac{1}{a} \int_{-\infty}^{+\infty} \mathrm{d}t' \, \psi_M^* \left[\frac{1}{a}(t' - t) \right] \\ &\times \int_{-\infty}^{+\infty} \int_{-\infty}^{+\infty} \mathrm{d}x' \mathrm{d}y' \, \psi_G \left[\frac{1}{a}(x' - x, y' - y) \right] I(x', y', t'). \end{aligned} \tag{4.119}$$

As it was discussed in Section 4.4.3.1, in the stationary phase approximation of the CWT, the tracking of the local maxima of $|D_I(a, \mathbf{b})|$ allows the characterization of the ridge a_r that has the property of describing the spatial evolution of the phase distribution as

$$\phi(\mathbf{b}) = \arctan\left\{\frac{\mathrm{Im}[D_I(a_r, \mathbf{b})]}{\mathrm{Re}[D_I(a_r, \mathbf{b})]}\right\}. \tag{4.120}$$

Note that this is a similar procedure to the one adopted by the CWT in stationary phase approximation for one-dimensional signals. Finally, as the phase values are wrapped, the one-dimensional phase unwrapping algorithm must be used. To avoid the process of phase unwrapping, an integration process of the obtained instantaneous frequencies should be carried out, as it was previously discussed.

To illustrate the performance of the phase recovery procedure using the three-dimensional directional wavelet transform approach, simulated TSPI data produced by an out-of-plane speckle interferometer will be analyzed [113]. This analysis is performed by introducing a temporal carrier and by generating interferograms having an average speckle size $s = 1, 3$, and 5 pixels. The phase corresponding to the simulated dynamic displacement field is selected as $\phi(x, y, t) = \phi_s(x, y)f(t)$, where

$$\phi_s(x, y) = 5(1-x^2)\mathrm{e}^{-x^2-(y+1)^2} - 20\left(\frac{x}{2} - x^3 - y^5\right) \times \mathrm{e}^{(-x^2-y^2)} - \mathrm{e}^{-(x+1)^2-y^2}, \tag{4.121}$$

with $-3.5 \leq x,\ y \leq 3.5$ discretized in 200×200 pixels, and $f(t) = 8 \times 10^{-3}(t-64)$ with $1 \leq t \leq 512$, $t \in \mathbb{N}$. The edges of the TSPI data were used in the calculation and removed from the results, as being the effective region of the object phase recovery evaluation: $32 \leq m,\ n \leq 168,\ m, n \in \mathbb{N}$, and $64 \leq t \leq 448$.

Figure 4.24 shows the selected phase distribution $\phi(m, n, t)$ corresponding to the maximum deformation ($t = 448$). Note that the phase distribution is a linear function of time and its minimum deformation corresponds to a plane surface

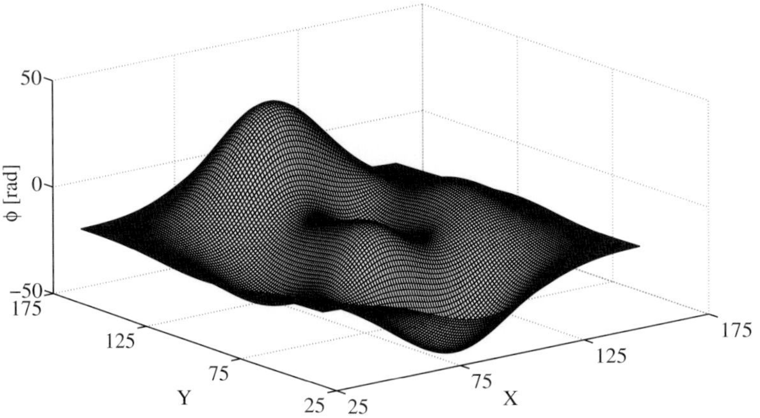

Figure 4.24 Phase distribution ϕ corresponding to the maximum deformation. (Reproduced with permission from Ref. [113], the Optical Society of America.)

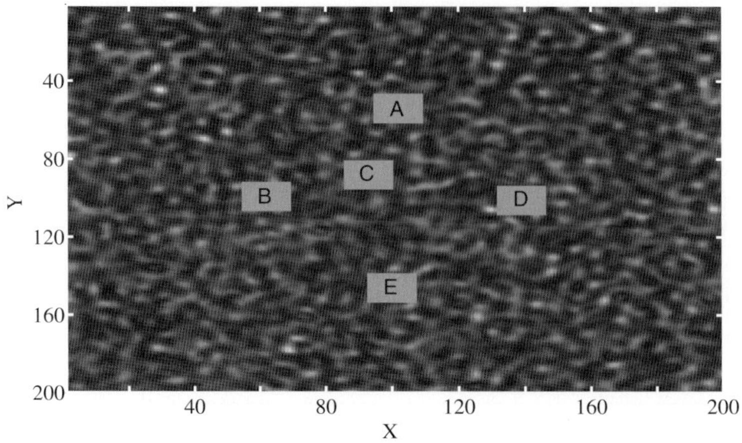

Figure 4.25 Spatial distribution of the zero blocks in the frame $t = 256$ when $s = 3$ pixels. (Reproduced with permission from Ref. [113], the Optical Society of America.)

$(t = 64)$. To simulate the modulation loss and noise, a random variable with a uniform distribution in the range $[-\pi/10, \pi/10]$ is added in the phase generation of the TSPI data. In addition, five blocks of zeros are also introduced in the TSPI data, namely, *A*, *B*, *C*, *D*, and *E* with dimensions $\Delta m = \Delta n = 15$ and $\Delta t = 33$ (7425 adjacent null pixels), temporally centered in $t = 256$ and spatially centered in the local maxima and minima of the phase ϕ. These blocks simulate the sets of adjacent nonmodulated pixels. Figure 4.25 displays the spatial distribution of the blocks in the frame $t = 256$ in the case of a time series of speckle interferograms having an average speckle size of 3 pixels. As a typical description of the TSPI signal, Figure 4.26 shows the temporal evolution of the modulation intensity corresponding to the pixel (93, 88) belonging to the block *C*.

The D_I can be computationally implemented as a convolution product in the Fourier space (see Equation 4.113). The TSPI data are first Fourier transformed and D_I is obtained via the inverse Fourier transform of the product of the transforms by using Equation 4.115. The tracking of the local maxima of $|D_I|$ can be carried out by using a common computational routine. Then, the phase distribution is retrieved using Equation 4.120 and finally a conventional one-dimensional unwrapping method along the time axis is applied.

To evaluate quantitatively the recovered object phase distributions, the quality index $Q(t)$ calculated at each temporal frame defined in Section 4.4.1 is used. The statistical features of each frame $\phi(t)$ by combining correlation, distortion, and deviation within local regions using a sliding window with a size 7×7 pixels are computed and finally the overall quality index is obtained as

$$Q = \frac{1}{M}\sum_{j=1}^{M} Q_j, \tag{4.122}$$

where M is the total steps required to analyze the frame data set by windowing.

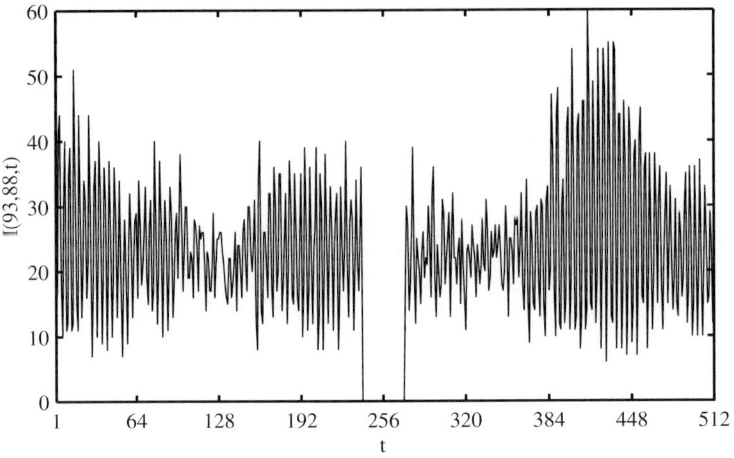

Figure 4.26 Evolution of the temporal intensity for a pixel belonging to the block C. (Reproduced with permission from Ref. [113], the Optical Society of America.)

Figure 4.27 shows the temporal evolution of the overall quality index Q for the retrieved phase obtained when $s = 1, 2$, and 3 pixels, with and without the application of a common smoothing procedure to remove the typical outliers produced in the phase recovery process. The lines with squares, rhombuses, and triangles-up represent the Q values obtained for $s = 1, 2$, and 3 pixels, respectively. These results were obtained by considering the presence of outliers and by using the parameters

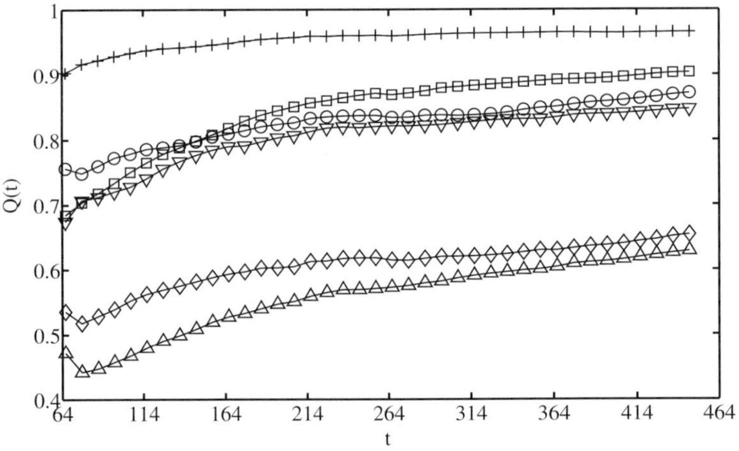

Figure 4.27 Temporal evolution of the quality index with squares, rhombuses, and triangles-up corresponding to $s = 1$, 2, and 3 pixels without smoothing and with dashes, circles and triangles-down corresponding to $s = 1$, 2, and 3 pixels with smoothing, respectively. (Reproduced with permission from Ref. [113], the Optical Society of America.)

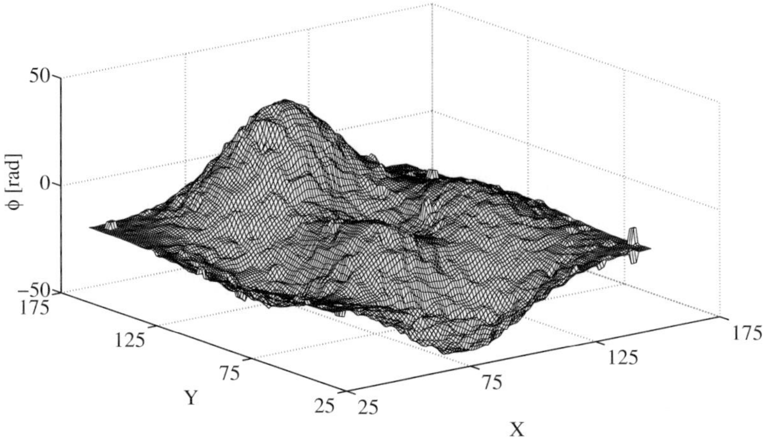

Figure 4.28 Retrieved phase distribution at the maximum deformation obtained with smoothing when $s = 3$ pixels. (Reproduced with permission from Ref. [113], the Optical Society of America.)

of anisotropy $\sigma_z = 6$, with $\sigma_x = \sigma_y = 0.1$, $\sigma_x = \sigma_y = 0.3$, and $\sigma_x = \sigma_y = 0.4$, for $s = 1, 2$, and 3 pixels, respectively. The lines with plus symbols, circles, and triangles-down represent the Q values obtained using a smoothing procedure when $s = 1$, 2, and 3 pixels, respectively. Figure 4.28 illustrates the retrieved object phase distribution obtained at the maximum deformation when $s = 3$ pixels and when the smoothing procedure is applied. Note that the case shown in Figure 4.28 is the most unfavorable one. Clearly, the outliers were easily removed and the Q values were also increased.

Therefore, it is seen that the three-dimensional directional wavelet transform approach is a very efficient technique to recover transient object phase distribution in the analysis of TSPI signals when a significant number of set of adjacent nonmodulating pixels are present in the recorded data.

4.5 Concluding Remarks

Temporal speckle pattern interferometry has become a well-established technique being used in several fields of science and industry. Particularly, it seems that an important part of future works in TSPI will depend on the necessity of analyzing more complex dynamical problems with high accuracy. With the continuous progress in the development of imaging sensor technology and computers, a challenging task will be the development of more powerful numerical methods for phase recovery.

The numerical methods to evaluate phase distributions that are described in this chapter use the time dimension in the analysis of TSPI signals and therefore give significant advantages with respect to other traditional data processing techniques.

Furthermore, the approach adopted in this chapter and contained within the framework of time–frequency representations can take into account the treatment of transients and consequently improves the accuracy produced by temporal phase recovery approaches.

Several strategies based on the use of atom basis decompositions, adaptive data-driven techniques, and energetic descriptions are described in this chapter and it is expected that the readers can gain insight into the issues discussed here. Promising results were also shown when three-dimensional wavelet approaches are used, which can take into account the information contained in a neighborhood of pixels and consequently improve of the accuracy of the phase estimation.

References

1 Patil, A. and Rastogi, P.K. (2007) Moving ahead with phase. *Opt. Lasers Eng.*, **45**, 253–257.

2 Mohan, N.K. and Rastogi, P.K. (2009) Recent developments in interferometry for microsystems metrology. *Opt. Lasers Eng.*, **47**, 199–202.

3 Bruno, L. and Poggialini, A. (2007) Back to the future: from speckle to holography. *Opt. Lasers Eng.*, **45**, 538–549.

4 Rastogi, P.K. (ed.) (2001) *Digital Speckle Pattern Interferometry and Related Techniques*, John Wiley & Sons, Inc., Chichester.

5 Malacara, D. (ed.) (2007) *Optical Shop Testing*, John Wiley & Sons, Inc., New Jersey.

6 Goudemand, N. (2006) Theoretical investigations on dual-beam illumination electronic speckle pattern interferometry. *Appl. Opt.*, **45**, 5092–5106.

7 Federico, A. and Kaufmann, G.H. (2008) Phase recovery in temporal speckle pattern interferometry using the generalized S-transform. *Opt. Lett.*, **33**, 866–868.

8 Mallat, S. (1998) *A Wavelet Tour of Signal Processing*, Academic Press, San Diego.

9 Karaalioğlu, C. and Skarlatos, Y. (2003) Fourier transform method for measurement of thin film thickness by speckle interferometry. *Opt. Eng.*, **42**, 1694–1698.

10 Yagnik, J., Siva, G.S., Ramakrishnan, K.R., and Rao, L.K. (2005) 3D shape extraction of human face in presence of facial hair: a profilometric approach. Proceedings of TENCON 2005 IEEE Region 10, pp. 1–5.

11 Pérez-López, C., De la Torre-Ibarra, M.H., and Mendoza-Santoyo, F. (2006) Very high speed cw digital holographic interferometry. *Opt. Exp.*, **14**, 9709–9715.

12 Quan, C., Wang, S.H., and Tay, C.J. (2006) Nanoscale surface deformation inspection using FFT and phase-shifting combined interferometry. *Prec. Eng.*, **30**, 23–31.

13 Abdul-Rahman, H.S., Gdeisat, M.A., Burton, D.R., Lalor, M.J., Rilley, F., and Abid, A. (2008) Three-dimensional Fourier fringe analysis. *Opt. Lasers Eng.*, **46**, 446–455.

14 Guo, H. and Huang, P.S. (2009) Absolute phase technique for the Fourier transform method. *Opt. Eng.*, **48**, 043609-1–043609-7.

15 Chen, L. and Tay, C.J. (2006) Carrier phase component removal: a generalized least-squares approach. *J. Opt. Soc. Am. A*, **23**, 435–443.

16 Quan, C., Tay, C.J., and Chen, L.J. (2007) A study on carrier-removal techniques in fringe projection profilometry. *Opt. Lasers Eng.*, **39**, 1155–1161.

17 Kaufmann, G.H. and Galizzi, G.E. (2002) Phase measurement in temporal speckle pattern interferometry: comparison between the phase-shifting and the Fourier transform methods. *Appl. Opt.*, **41**, 7254–7263.

18 Kaufmann, G.H. (2003) Phase measurement in temporal speckle

pattern interferometry using the Fourier transform method with and without a temporal carrier. *Opt. Commun.*, **217**, 141–149.

19 Vander, R., Lipson, S.G., and Leizerson, I. (2003) Fourier fringe analysis with improved spatial resolution. *Appl. Opt.*, **42**, 6830–6837.

20 Spataru, A. (1970) *Theorie de la transmission de l'information – I: signaux et bruits*, Translation from Editura Technica, Bucharest, Masson.

21 Hahn, S.L. (1996) *Hilbert Transforms in Signal Processing*, Artech House, Norwood.

22 Boashash, B. (1992) Estimating and interpreting the instantaneous frequency of a signal – Part 1: fundamentals. *Proc. IEEE*, **80**, 520–538.

23 Bedrosian, E. (1963) A product theorem for Hilbert transforms. *Proc. IEEE*, **51**, 868–869.

24 Wang, Z. and Bovik, A.C. (2009) Mean squared error: love it or leave it? *IEEE Signal Process. Mag.*, **26**, 98–117.

25 Wang, Z. and Bovik, A.C. (2002) A universal quality index. *IEEE Signal Process. Lett.*, **9**, 81–84.

26 Ruiz, P.D., Huntley, J.M., Shen, Y., Russell Coggrave, C., and Kaufmann, G.H. (2001) Vibration-induced phase errors in high-speed phase-shifting speckle pattern interferometry. *Appl. Opt.*, **40**, 2117–2125.

27 Ruiz, P.D., Huntley, J.M., Shen, Y., Russell Coggrave, C., and Kaufmann, G.H. (2002) Effects of random vibration in high-speed phase-shifting speckle pattern interferometry. *Appl. Opt.*, **41**, 3941–3949.

28 Svanbro, A., Huntley, J.M., and Davila, A. (2003) Optimal re-referencing rate for in-plane dynamic speckle interferometry. *Appl. Opt.*, **42**, 251–258.

29 Davila, A., Huntley, J.M., Kaufmann, G.H., and Kerr, D. (2005) High-speed dynamic speckle interferometry: phase errors due to intensity, velocity, and speckle decorrelation. *Appl. Opt.*, **44**, 3954–3962.

30 Ruiz, P.D., Huntley, J.M., and Kaufmann, G.H. (2003) Adaptative phase-shifting algorithm for temporal phase evaluation. *J. Opt. Soc. Am. A*, **20**, 325–332.

31 Sciammarella, C.A. and Sciammarella, F.M. (2003) Heisenberg principle applied to the analysis of speckle interferometry fringes. *Opt. Lasers Eng.*, **40**, 573–588.

32 Daubechies, I. (1990) The wavelet transform, time–frequency localization and signal analysis. *IEEE Trans. Inf. Theory*, **36**, 961–1005.

33 Rioul, O. and Duhamel, P. (1992) Fast algorithms for discrete and continuous wavelet transforms. *IEEE Trans. Inf. Theory*, **38**, 569–586.

34 Press, W., Teukolsky, S., Vetterling, W., and Flannery, B. (1992) *Numerical Recipes in FORTRAN*, 2nd edn, Cambridge University Press, Cambridge, Chapter 13, pp. 537–608.

35 Delprat, N., Escudié, B., Guillemain, P., Kroland-Martinet, R., Tchamitchian, P., and Torrésani, B. (1992) Asymptotic wavelet and Gabor analysis: extraction of instantaneous frequencies. *IEEE Trans. Inf. Theory*, **38**, 644–664.

36 Gradshteyn, I.S. and Ryzhik, I.M. (2007) *Table of Integrals, Series, and Products*, Academic Press, London.

37 Liu, H., Cartwright, A.N., and Basaran, C. (2004) Moire interferogram phase extraction: a ridge detection algorithm for continuous wavelet transforms. *Appl. Opt.*, **43**, 850–857.

38 Abid, A.Z., Gdeisat, M.A., Burton, D.R., Lalor, M.J., and Lilley, F. (2007) Spatial fringe pattern analysis using the two-dimensional continuous wavelet transform employing a cost function. *Appl. Opt.*, **46**, 6120–6126.

39 Guillemain, P. and Kronland-Martinet, R. (1992) Characterization of acoustic signals through continuous linear time–frequency representations. *Proc. IEEE*, **84**, 561–585.

40 Cherbuliez, M. (2001) Wavelet analysis of interference patterns and signals: development of fast and efficient processing techniques. Thèse 2377, École Polytechnique Fédérale de Lausanne.

41 Cherbuliez, M. and Jacquot, P. (2001) Phase computation through wavelet analysis: yesterdays and nowadays, in *Proceedings in FRINGE 2001*

(eds W. Osten and W. Jüptner), Elsevier, Paris.

42 Truchetet, F. and Laligant, O. (2008) Review of industrial applications of wavelet and multiresolution-based signal and image processing. *J. Electron. Imaging*, **17**, 031102.

43 Colona de Lega, X. (1996) Continuous deformation measurement using dynamic phase-shifting and wavelet transforms, in *Proceedings in Applied Optics and Optoelectronics* (ed. K.T.V. Grattan), Institute of Physics.

44 Colona de Lega, X. (1997) Processing of non-stationary interference patterns: adapted phase-shifting algorithms and wavelet analysis. Application to dynamic deformation measurements by holographic and speckle interferometry. Thèse 1666, École Polytechnique Fédérale de Lausanne.

45 Tomassini, P., Giulietti, A., Gizzi, L.A., Galimberti, M., Giulietti, D., Borghesi, M., and Willi, O. (2001) Analyzing laser plasma interferograms with a continuous wavelet transform ridge extraction technique: the method. *Appl. Opt.*, **40**, 6561–6568.

46 Federico, A. and Kaufmann, G.H. (2002) Evaluation of the continuous wavelet transform method for the phase measurement of electronic speckle pattern interferometry fringes. *Opt. Eng.*, **41**, 3209–3216.

47 Zhong, J. and Weng, J. (2005) Phase retrieval of optical fringe patterns from the ridge of the wavelet transform. *Opt. Lett.*, **30**, 2560–2562.

48 Watkins, L.R. (2007) Phase recovery from fringe patterns using the continuous wavelet transform. *Opt. Lasers Eng.*, **45**, 298–303.

49 Tay, C.J., Quan, C., Sun, W., and He, X.Y. (2007) Demodulation of a single interferogram based on continuous wavelet transform and phase derivative. *Opt. Commun.*, **280**, 327–336.

50 Fu, Y., Tay, C.J., Quan, C., and Chen, L.J. (2004) Temporal wavelet analysis for deformation and velocity measurement in speckle interferometry. *Opt. Eng.*, **43**, 2780–2787.

51 Fu, Y., Tay, C.J., Quan, C., and Miao, H. (2005) Wavelet analysis of speckle patterns with a temporal carrier. *Appl. Opt.*, **44**, 959–965.

52 Tay, C.J. and Fu, Y. (2005) Determination of curvature and twist by digital shearography and wavelet transforms. *Opt. Lett.*, **30**, 2873–2875.

53 Quan, C., Fu, Y., and Miao, H. (2006) Wavelet analysis of digital shearing speckle patterns with a temporal carrier. *Opt. Commun.*, **260**, 97–104.

54 Post, D., Han, B., and Ifju, P. (1994) *High Sensitivity Moiré*, Springer-Verlag, New York.

55 Walker, C.A. (ed.) (2004) *Handbook of Moiré Measurement*, Institute of Physics, London.

56 Kemao, Q., Soon, S.H., and Asundi, A. (2003) Instantaneous frequency and its applications to strain extraction in moiré interferometry. *Appl. Opt.*, **42**, 6504–6513.

57 Tay, C.J., Quan, C., Fu, Y., and Huang, Y., (2004) Instantaneous velocity displacement and contour measurement by use of shadow moiré and temporal wavelet analysis. *Appl. Opt.*, **43**, 4164–4171.

58 Quan, C., Fu, Y., Tay, C.J., and Tan, J.M. (2005) Profiling of objects with height steps by wavelet analysis of shadow moiré fringes. *Appl. Opt.*, **44**, 3284–3290.

59 Li, H.J., Chen, H.J., Zhang, J., Xiong, C.Y., and Fang, J. (2007) Statistical searching of deformation phases on wavelet transform maps of fringe patterns. *Opt. Lasers Eng.*, **39**, 275–281.

60 Lilley, F., Lalor, M.J., and Burton, D.R. (2000) Robust fringe analysis system for human body shape measurement. *Opt. Eng.*, **39**, 187–195.

61 Miao, H., Quan, C., Tay, C.J., and Fu, Y. (2007) Analysis of phase distortion in phase-shifted fringe projection. *Opt. Lasers Eng.*, **45**, 318–325.

62 Quan, C., Fu, Y., and Tay, C.J. (2007) Vibration measurement by use of wavelet transform and temporal carrier technique. *Meas. Sci. Technol.*, **18**, 1718–1723.

63 Abid, A.Z., Gdeisat, M.A., Burton, D.R., and Lalor, M.J. (2007) Ridge extraction

algorithms for one-dimensional continuous wavelet transform: a comparison. *J. Phys. Conf. Ser.*, **76**, 012045.

64 Quan, C., Chen, W., and Tay, C.J. Phase-retrieval techniques in fringe-projection profilometry. *Opt. Lasers Eng.*, **48**, 235–243.

65 Huang, L., Kemao, Q., Pan, B., and Asundi, A.K. Comparison of Fourier transform, windowed Fourier transform and wavelet transform methods for phase extraction from a single fringe pattern in fringe projection profilometry. *Opt. Lasers Eng.*, **48**, 141–148.

66 Osten, W. (ed.) (2007) *Optical Inspection of Microsystems*, CRC Press, Boca Raton.

67 Li, M., Quan, C., and Tay, C.J. (2008) Continuous wavelet transform for micro-component profile measurement using vertical scanning interferometry. *Opt. Lasers Eng.*, **40**, 920–929.

68 Liebling, M., Blu, T., and Unser, M. (2003) Fresnelets: new multiresolution wavelet bases for digital holography. *IEEE Trans. Image Process.*, **12**, 29–43.

69 Afifi, M., Fassi-Fihri, A., Marjane, M., Nassim, K., Sidki, M., and Rachafi, S. (2002) Paul wavelet-based algorithm for optical phase distribution evaluation. *Opt. Commun.*, **211**, 47–51.

70 Barj, E.M., Afifi, M., Idrissi, A.A., Nassim, K., and Rachafi, S. (2006) Speckle correlation fringes denoising using stationary wavelet transform. Application in the wavelet phase evaluation technique. *Opt. Lasers Eng.*, **38**, 506–511.

71 Zhong, J. and Weng, J. (2004) Dilating Gabor transform for the fringe analysis of 3-D shape measurement. *Opt. Eng.*, **43**, 895–899.

72 Zhong, J. and Weng, J. (2004) Spatial carrier-fringe pattern analysis by means of wavelet transform: wavelet transform profilometry. *Appl. Opt.*, **43**, 4993–4998.

73 Bethge, J. and Steinmeyer, G. (2008) Numerical fringe pattern demodulation strategies in interferometry. *Rev. Sci. Instrum.*, **79**, 073102.

74 Gdeisat, M.A., Abid, A., Burton, D.R., Lalor, M.J., Lilley, F., Moore, C., and Qudeisat, M. (2009) Spatial and temporal carrier fringe pattern demodulation using the one-dimensional continuous wavelet transform: recent progress, challenges, and suggested developments. *Opt. Lasers Eng.*, **47**, 1348–1361.

75 Vandewalle, P., Kovačević, J., and Vetterli, M. (2009) Reproducible research in signal processing. *IEEE Signal Process. Mag.*, **26**, 37–47.

76 Chen, Y.-K., Chakrabarti, C., Bhattacharyya, S., and Bougard, B. (2009) Signal processing on platforms with multiple cores, part 1. *IEEE Signal Process. Mag.*, **26**, 24–25.

77 Gao, W., Kemao, Q., Wang, H., Lin, F., and Seah, H.S. Parallel computing for fringe pattern processing: a multicore CPU approach in Matlab environment. *Opt. Lasers Eng.*, **47**, 1286–1292.

78 Kepner, J. (2009) *Parallel Matlab for Multicore and Multinode Computers*, SIAM, Philadelphia.

79 Claasen, T.A.C.M. and Mecklenbräuker, W.F.G. (1980) The Wigner distribution – a tool for time–frequency signal analysis. Part I: continuous-time signals. *Philips J. Res.*, **35**, 217–250.

80 Cohen, L. (1989) Time–frequency distributions: a review. *Proc. IEEE*, **77**, 941–981.

81 Boudreaux-Bartels, G.F. (2000) Fourier Transforms, in *The Transforms and Applications Handbook*, (ed. A, Poularikas) 2nd edn, CRC Press, Florida.

82 O'Neill, J.C. and Williams, W.J. (1999) Shift covariant time–frequency distributions of discrete signals. *IEEE Trans. Signal Process.*, **47**, 133–146.

83 Jeon, J. and Williams, W.J. (1992) Alias-free generalized discrete-time time–frequency distributions. *IEEE Trans. Signal Process.*, **40**, 2557–2765.

84 Jeon, J., Cunningham, G.S., and Williams, W.J. (1995) The discrete-time phase derivative as a definition of discrete instantaneous frequency and its relation to discrete time–frequency distributions. *IEEE Trans. Signal Process.*, **43**, 341–344.

85 Pitton, J.W. and Atlas, C.E. (1995) Discrete-time implementation of the cone kernel time–frequency representation. *IEEE Trans. Signal Process.*, **43**, 1996–1998.

86 Richman, M.S., Parks, T.W., and Shenoy, R.G. (1995) Discrete-time, discrete-frequency time–frequency representations. *Proc. IEEE Int. Conf. Acoust. Speech Signal Process.*, **2**, 1029–1032.

87 Madjarova, V.D., Kadono, H., and Toyooka, S. (2003) Dynamic electronic speckle pattern interferometry (DESPI) phase analyses with temporal Hilbert transform. *Opt. Exp.*, **11**, 617–623.

88 Madjarova, V.D., Kadono, H., and Toyooka, S. (2006) Use of dynamic electronic speckle pattern interferometry with the Hilbert transform method to investigate thermal expansion of a joint material. *Appl. Opt.*, **45**, 7590–7596.

89 Huang, N.E., Shen, Z., Long, S.R., Wu, M.C., Shih, H.H., Zheng, Q., Yen, N.-C., Tung, C.C., and Liu, H.H. (1998) The empirical mode decomposition and the Hilbert spectrum for nonlinear and non-stationary time series analysis. *Proc. R. Soc. Lond. A*, **454**, 903–995.

90 Baldi, A., Equis, S., and Jacquot, P. (2007) Phase extraction in dynamic speckle interferometry by empirical mode decomposition, in *Proceedings of Experimental Analysis of Nano and Engineering Materials and Structures* (ed. E.E. Gdoutos), Springer, pp. 719–720.

91 Equis, S. and Jacquot, P. (2009) The empirical mode decomposition: a must-have tool in speckle interferometry? *Opt. Exp.*, **17**, 611–623.

92 Marengo Rodriguez, F.A., Federico, A., and Kaufmann, G.H. (2007) Phase measurement improvement in temporal speckle pattern interferometry using empirical mode decomposition. *Opt. Commun.*, **275**, 38–41.

93 Marengo Rodriguez, F.A., Federico, A., and Kaufmann, G.H. (2008) Hilbert transform analysis of a time series of speckle interferograms with a temporal carrier. *App. Opt.*, **47**, 1310–1316.

94 Equis, S. (2009) Phase extraction of non-stationary signals produced in dynamic interferometry involving speckle waves. Thèse 4514, École Polytechnique Fédérale de Lausanne.

95 Wu, Z. and Huang, N.E. (2004) A study of the characteristics of white noise using the empirical mode decomposition method. *Proc. R. Soc. Lond. A*, **460**, 1597–1611.

96 Huang, N.E., Wu, Z., Long, S.R., Arnold, K.C., Chen, X., and Blank, K. (2009) On instantaneous frequency. *Adv. Adapt. Data Anal.*, **1**, 177–229.

97 Stockwell, R.G. (1999) S-transform analysis of gravity wave activity from a small scale network of airglow imagers. PD thesis. Faculty of Graduate Studies, The University of Western Ontario London, Ontario.

98 Pinnegard, C.R. and Mansinha, L. (2003) The bi-Gaussian S-transform. *SIAM J. Sci. Comput.*, **24**, 1678–1692.

99 Pinnegard, C.R. and Mansinha, L. (2004) Time-local Fourier analysis with a scalable, phase-modulated analyzing function: the S-transform with a complex window. *Signal Process.*, **84**, 1167–1176.

100 Pinnegard, C.R. (2006) A new subclass of complex-valued S-transform windows. *Signal Process.*, **86**, 2051–2055.

101 Sejdić, E., Djurović, I. and Jiang, J. (2007) S-transform with frequency dependent Kaiser window. Proceedings of International Conference on Acoustic, Speech and Signal Processing, IEEE, pp. III 1165–III 1168.

102 Federico, A. and Kaufmann, G.H. (2003) Phase retrieval in digital speckle pattern interferometry by use of a smoothed space–frequency distribution. *Appl. Opt.*, **42**, 7066–7071.

103 Equis, S. and Jacquot, P. (2009) A new application of the Delaunay triangulation: the processing of speckle interferometry signals, in *Fringe 2009: International Workshop on Advanced Optical Metrology* (eds W. Osten and M. Kuzawinska), Springer.

104 Su, X., Chen, W., Qc, Z., and Chao, Y. (2001) Dynamic 3-D shape measurement method based on FTP. *Opt. Lasers Eng.*, **36**, 46–64.

105 Chen, L.-C., Ho, H.-W., and Nguyen, X.-L. (2010) Fourier transform profilometry (FTP) using an innovative band-pass filter for accurate 3-D surface reconstruction. *Opt. Lasers Eng.*, **48**, 182–190.

106 Quian, K., Fu, Y., Liu, Q., Seah, H.S., and Asundi, A. (2006) Generalized three-dimensional windowed Fourier transform for fringe analysis. *Opt. Lett.*, **31**, 2121–2123.

107 Kemao, Q. (2004) Windowed Fourier transform for fringe pattern analysis. *Appl. Opt.*, **43**, 2695–2702.

108 Kemao, Q. (2004) Addendum: Windowed Fourier transform for fringe pattern analysis. *Appl. Opt.*, **43**, 3472–3473,

109 Kemao, Q. (2007) Two-dimensional windowed Fourier transform for fringe pattern analysis: principles, applications and implementations. *Opt. Lasers Eng.*, **45**, 304–317.

110 Fu, Y., Groves, R.M., Pedrini, G., and Osten, W. (2007) Kinematic and deformation parameter measurement by spatiotemporal analysis of an interferogram sequence. *Appl. Opt.*, **46**, 8645–8655.

111 Antoine, J.P. and Murenzi, R. (1996) Two-dimensional directional wavelets and the scale-angle representation. *Signal Process.*, **52**, 259–281.

112 Combes, J.M., Grossman, A., and Tchamitchian, Ph. (eds) (1989) *Wavelets: Time–Frequency Methods and Phase Space*, Springer, Berlin.

113 Federico, A. and Kaufmann, G.H. (2009) Robust phase recovery in temporal speckle pattern interferometry using a 3D directional wavelet transform. *Opt. Lett.*, **34**, 2336–2338.

5
Optical Vortex Metrology

Wei Wang, Steen G. Hanson, and Mitsuo Takeda

5.1
Introduction

Speckle photography, whose basic idea is to compare images of speckle intensity patterns, one from a rough object in an initial state and a second from the same object after it is subjected to some form of displacement, has been studied extensively, and various techniques have been developed during the past decades [1–5]. Although they differ in specific technical details, these existing techniques are based exclusively on the cross-correlation function of intensity of the speckle field, while the underlying phase information has received much less attention. Recently, the so-called pseudophase information from a complex signal representation of a speckle field has been proposed, and the improved performance based on the pseudophase-only correlation has been demonstrated [6, 7]. Furthermore, because the pseudophase can be obtained without recourse to interferometry, this technique can be applied to speckle-like artificial markings, such as printed random dots and random textures of natural origin (e.g., biological tissues), even under incoherent natural illumination. This versatility of the pseudophase information opens up new possibilities in a wide range of applications beyond those known for laser speckle metrology. It is the goal of this chapter to elucidate some of these techniques and the associated benefits supported by experimental findings with particular emphasis on singularities in the pseudophase map.

5.2
Speckle and Optical Vortices

The negative-exponential probability density function for the intensity of a fully developed speckle pattern attains its maximum value at zero intensity. The occurrence of precisely zero intensity at a point in a speckle pattern is an event that can and indeed does occur [8]. These points, which are extremely general features in optical fields, are known by various terms encompassing these properties: nodal points, phase singularities, wave dislocations, and optical vortices; and a considerable literature exist on their properties [9–11].

Advances in Speckle Metrology and Related Techniques. Edited by Guillermo H. Kaufmann

ISBN: 978-3-527-40957-0

In general, when three or more waves interfere rather than vanish on fringes, the light is zero at points in two dimensions. At these positions where the intensity of the wave attains a zero value, the phase is undefined (i.e., singular). In optics, this is naturally the complex amplitude of a scalar optical field, whose modulus is the real amplitude, and argument is the phase. For this field, we write

$$\tilde{U} = \mathrm{Re}(\tilde{U}) + j\mathrm{Im}(\tilde{U}) = A\exp(j\theta), \tag{5.1}$$

where Re and Im are the real and imaginary parts, respectively, A is the amplitude, and θ is the phase. Just as the angle in a polar coordinate system is not defined at the origin, the phase is not defined when $\tilde{U} = 0$, and near such a point the whole 2π range of phases occurs. This means that there is a net change of phase in a circuit C enclosing the zero point, quantized in units of 2π:

$$q = \frac{1}{2\pi}\oint_C \nabla\theta \cdot \mathrm{d}\vec{r}. \tag{5.2}$$

The integer q (positive or negative) is called the topological charge of the singularity. The sign of q is called the sign of the singularity, positive if the phase increases in a right-handed sense. In a scalar optical field, the optical current, which is the Poynting vector of the electric–magnetic field, takes the form [12]

$$\vec{J} = \mathrm{Im}[\tilde{U}^* \nabla \tilde{U}] = A^2 \nabla\theta. \tag{5.3}$$

With $\vec{J}$ pointing in the direction of phase change $\nabla\theta$, the phase singularity is therefore the optical vortex of the optical current flow.

Phase singularities were recognized as a general phenomenon of wave physics in a seminal paper by Nye and Berry [9], although they have similarities to vortices and phase singularities elsewhere in science, such as in fluid dynamic, chemical, and biological systems. As a ubiquitous phenomenon in interference, optical vortices frequently occur in reflection from – or propagation through – random, turbulent media. In a random optical wave such as a speckle field, there are many vortices interspersed in between the bright speckles. A simple representation of such a speckle field is shown in Figure 5.1.

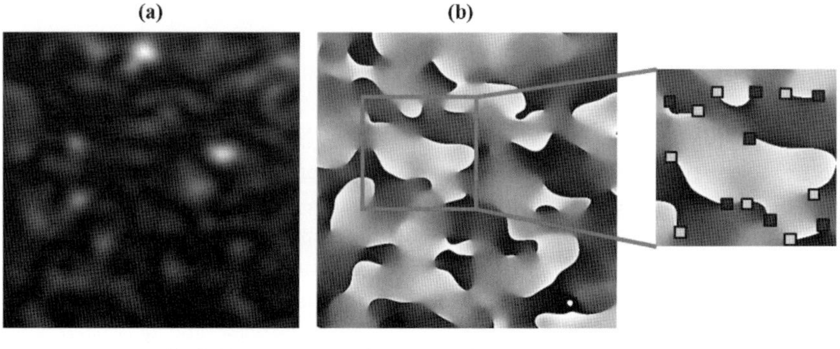

Figure 5.1 A speckle field. (a) Intensity. (b) Phase with inserted optical vortices (charge +1 (dark gray squares), −1 (light gray squares)).

However, due to the singularity nature in the phase map, the presence of optical vortices has caused considerable trouble in the measurement and interpretation of such scattered waves. The occurrence of phase singularities in random fields is often regarded as a nuisance, because different continuous paths of phase around a singularity lead to a difference of 2π upon unwrapping. The phase unwrapping problem is an important issue in optics in general, and especially important in optical interferometry. Several algorithms have been suggested in dealing with the singularities in general phase unwrapping problems [13–15].

5.3
Core Structure of Optical Vortices

A reason frequently given for studying phase singularities is that they organize and reduce the global structure of the optical field into a structure like a "skeleton" draped with the phase and intensity structure [11, 16–18]. Since vortices are zeros, they do not contain any information directly. The information about the rest of the field must lie in their spatial configuration and the local phase and intensity structure near the vortex point.

Figure 5.2 shows the amplitude contours and phase structure in the neighborhood of a phase singularity. The phase singularity is located at the center of the elliptical contours of the amplitude, which is the intersection of the zero crossings of the real and imaginary parts of optical field [19, 20]. The phase has a characteristic feature of a 2π helical structure. Usually, the change of the phase around the phase singularities is nonuniform as a function of the azimuthal angle. Since the phase singularity is a vortex of the optical current, it has its vorticity, defined as

$$\vec{\Omega} = \frac{1}{2}\nabla \times \vec{J} = \frac{1}{2}\nabla \tilde{U}^* \times \nabla \tilde{U}. \tag{5.4}$$

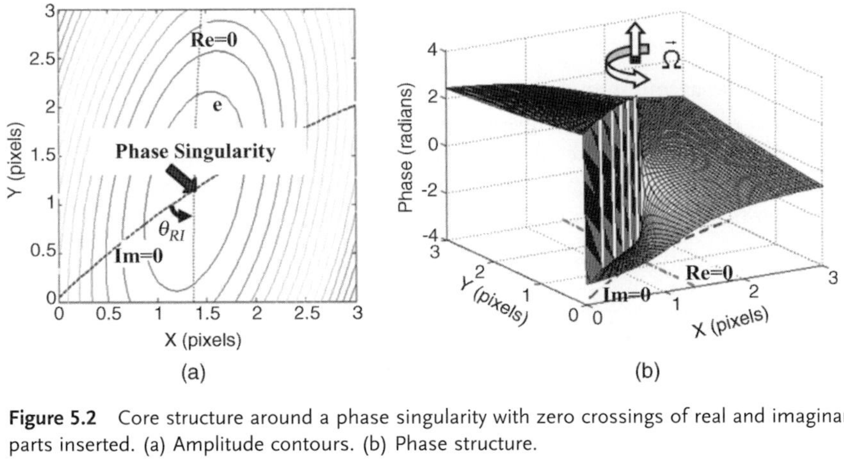

Figure 5.2 Core structure around a phase singularity with zero crossings of real and imaginary parts inserted. (a) Amplitude contours. (b) Phase structure.

In two-dimensional fields, $\vec{\Omega}$ is in the $\hat{z}$-direction, around which the optical current circulates in a right-handed sense [21].

The most general way of describing the local elliptic anisotropy of an optical vortex follows from the Stokes parameters used to describe elliptic polarization [22]. The elliptical intensity contours are entirely described by the complex gradient vector $\nabla\tilde{U}$. The ellipsoidal geometry can therefore be completely described by the following parameters, analogous to the Stokes parameters [23–25]:

$$S_0 = |\partial_x\tilde{U}|^2 + |\partial_y\tilde{U}|^2, \tag{5.5a}$$

$$S_1 = |\partial_x\tilde{U}|^2 - |\partial_y\tilde{U}|^2, \tag{5.5b}$$

$$S_2 = 2\,\mathrm{Re}(\partial_x\tilde{U}^*\partial_y\tilde{U}), \tag{5.5c}$$

$$S_3 = 2\,\mathrm{Im}(\partial_x\tilde{U}^*\partial_y\tilde{U}). \tag{5.5d}$$

These parameters describing the anisotropic ellipse for $\nabla\tilde{U}$ are mathematically analogous, but physically unrelated, to the Stokes parameters in polarization. For instance, the sign of S_3 gives the sign (topological charge) of the vortex (i.e., the sense of phase increase around the singularity), analogous to the handedness of polarization. Meanwhile, it follows from Equations 5.5a–5.5d that only three of them are independent since they are related by the identity valid for a fully polarized field:

$$S_0^2 = S_1^2 + S_2^2 + S_3^2. \tag{5.6}$$

Following the same procedure as used for polarization, we can represent the ellipsoidal parameters as in the following equations:

$$S_1 = S_0\cos(2\chi)\cos(2\varphi), \tag{5.7a}$$

$$S_2 = S_0\cos(2\chi)\sin(2\varphi), \tag{5.7b}$$

$$S_3 = S_0\sin(2\chi). \tag{5.7c}$$

Here, φ is the azimuth angle between the major semiaxis of the ellipse and the x-axis, and χ is the ellipticity angle, which is the arctangent of the ratio between the length of the two axes of the ellipse, as shown in Figure 5.3a. Equations 5.7a–5.7c indicate a simple geometrical representation for all the parameters S_i $(i = 1, 2, 3)$ as components in a three-dimensional sphere surface occupied in space by the unit vector:

$$\vec{S} = \begin{pmatrix} s_1 \\ s_2 \\ s_3 \end{pmatrix} = S_0^{-1}\begin{pmatrix} S_1 \\ S_2 \\ S_3 \end{pmatrix}. \tag{5.8}$$

Figure 5.3b shows the Poincaré sphere representation for the anisotropic core structure of the pseudophase singularity. Here, $\hat{s}_1$, $\hat{s}_2$, $\hat{s}_3$ may be regarded as the

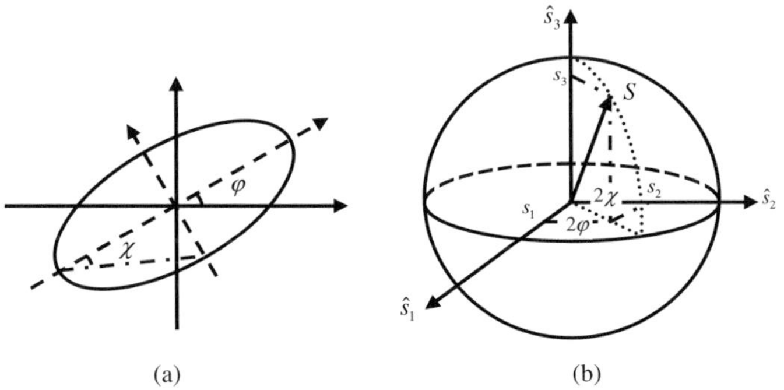

Figure 5.3 (a) Amplitude contour ellipse for describing the anisotropic core structure of a phase singularity. (b) Stokes-like parameters for amplitude contour ellipse and the Poincaré's sphere representation.

Cartesian coordinates of a point $\vec{S}$ on a surface of the sphere with unit radius, such that 2φ and 2χ are the spherical angular coordinates of this point. Thus, to every possible state of the anisotropy ellipse for the core structure of a phase singularity, there corresponds one point on the sphere surface, and vice versa.

5.4 Principle of Optical Vortex Metrology

In the field of optical interferometry, many efforts have been concentrated on how to avoid the nuisance from phase singularities because they hinder unique phase unwrapping of the measured phase map. Contrary to the general belief in optical metrology, a new technique, referred to as optical vortex metrology (OVM), which makes use of phase singularities, has been developed recently, that is, "if you can't fight them, join them!" The technique was based on the fact that phase singularities are well-defined geometrical points with unique core structures and spatial configuration, which serve as their unique fingerprints and endow the phase singularities with valuable information and capability as identifiable optimal markers [26–30].

5.4.1 Complex Signal Representation of a Speckle-Like Pattern

Before explaining optical vortex metrology, we first briefly review the two-dimensional complex signal representation of a speckle pattern. Usually, ordinary real-valued images do not *per se* have phase singularities, as they are not complex analytical functions. In engineering and physics, it is common practice to represent a real-valued signal by a complex-valued signal. Based on the real-valued image, one can construct a unique complex analytical field by deriving the imaginary part based on a Hilbert transform of the intensity, and subsequently merge the real – and imaginary – parts [31]. Unfortunately, this way of creating an analytical function is asymmetrical with respect to the coordinates, and thus less useful [6, 26]. To obtain a two-dimensional isotropic complex signal representation of a 2D speckled pattern, a rotationally symmetric transformation (here the Riesz transform or Laguerre–Gauss (LG) transform) has been applied for a complex signal representation and will better serve the purpose [32–34]. Having converted the incoherent image into an analytical field facilitates the finding of the phase singularities, which uniquely depicts the structure under investigation.

Given a real-valued speckle pattern $g(x, y)$, a complex signal $\tilde{g}(x, y)$ can be defined by an isotropic filter:

$$\tilde{g}(x, y) = |\tilde{g}(x, y)| \exp[j\theta(x, y)] = g(x, y) * \mathbb{LG}(x, y), \tag{5.9}$$

where $*$ denotes the convolution operation and $\mathbb{LG}(x, y)$ is a Laguerre–Gauss function in the spatial signal domain,

$$\begin{aligned} \mathbb{LG}(x, y) &= (j\pi^2\omega^4)(x + jy)\exp[-\pi^2\omega^2(x^2 + y^2)] \\ &= (j\pi^2\omega^4)[r\exp(-\pi^2 r^2\omega^2)\exp(j\alpha)], \end{aligned} \tag{5.10}$$

with $r = \sqrt{x^2 + y^2}$, $\alpha = \arctan(y/x)$ being the spatial polar coordinates. Alternatively, the complex signal $\tilde{g}(x, y)$ may also be calculated by means of the inverse Fourier transform of the spectrum $G(f_x, f_y)$ of $g(x, y)$ multiplied by a Laguerre–Gauss filter. That is,

$$\tilde{g}(x, y) = \mathbb{F}^{-1}\{\mathrm{LG}(f_x, f_y) \cdot G(f_x, f_y)\}. \tag{5.11}$$

Here, $\mathbb{F}^{-1}$ is the inverse Fourier transform and $\mathrm{LG}(f_x, f_y)$ is a Laguerre–Gauss filter in the frequency domain defined as follows:

$$\begin{aligned} \mathrm{LG}(f_x, f_y) &= (f_x + jf_y)\exp[-(f_x^2 + f_y^2)/\omega^2] \\ &= \varrho\exp(-\varrho^2/\omega^2)\exp(j\beta), \end{aligned} \tag{5.12}$$

where $\varrho = \sqrt{f_x^2 + f_y^2}$ and $\beta = \arctan(f_y/f_x)$ are the polar coordinates in the spatial frequency domain. Because the spiral phase in LG filter has its unique property of a signum function along any section through the origin, a spatial isotropy exists in the generated complex signal representation for a real-valued image. It is also readily seen from Equations 5.11 and 5.12 that the generated complex signal $\tilde{g}(x, y)$ can be referred to as a Laguerre–Gauss signal due to its derivation from a Laguerre–Gauss filter.

The phase associated with $\tilde{g}(x, y)$ is referred to as the pseudophase to distinguish it from the true phase of a laser speckle field. The complex signal representation has the advantage that it permits one to use the associated pseudophase information for general random patterns other than laser speckles, such as a printed or projected random dot pattern or the random texture found on a natural object. Therefore, such transform of an incoherent image into a complex signal has the versatility that expands applications beyond those known for laser speckle metrology.

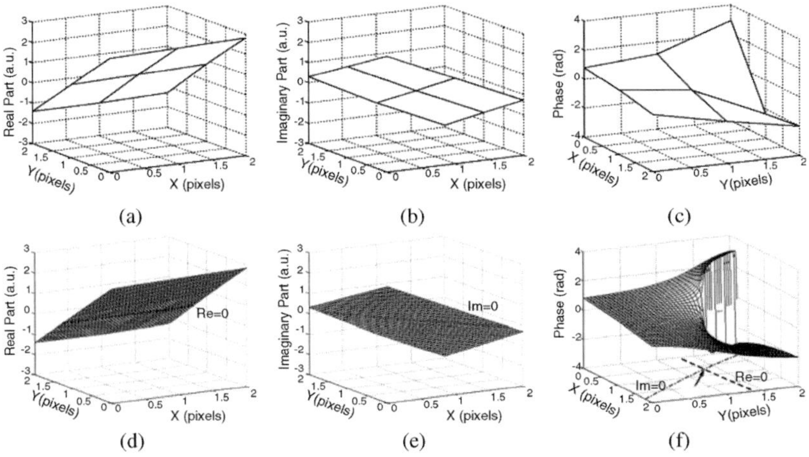

Figure 5.4 Real and imaginary parts of complex signal representation, and the corresponding phase structure around a phase singularity. (a–c) Before interpolation. (d–f) After interpolation.

5.4.2 Principle of Optical Vortex Metrology

As we have observed in Figure 5.2, the phase singularities are located at the intersections of the two zero-crossing lines in the plane of observation:

$$\mathrm{Re}[\tilde{g}(x,y)] = 0 \quad \text{and} \quad \mathrm{Im}[\tilde{g}(x,y)] = 0. \tag{5.13}$$

Figure 5.4a shows the real part of a complex analytic signal around a phase singularity, and Figure 5.4b is the corresponding imaginary part. The reconstructed phase is shown in Figure 5.4c, where a typical 2π helix structure around the phase singularity cannot be observed clearly. This means that the exact location of the phase singularity cannot be identified with subpixel resolution.

Note that in contrast to the discontinuous and complicated phase structure, the real and imaginary parts have an extremely simple structure consisting of a smooth monotonic surface [20]. It is this local monotonic behavior that has made possible the highly precise reconstruction of the detailed local structure of the complex analytic signal around the phase singularity by a two-dimensional interpolation of the real and imaginary parts. In the immediate vicinity of the phase singularity, the real and imaginary parts of the analytic signal can be expressed as

$$\mathrm{Re}[\tilde{g}(x,y)] = a_r x + b_r y + c_r, \quad \mathrm{Im}[\tilde{g}(x,y)] = a_i x + b_i y + c_i. \tag{5.14}$$

By the least-square method, the coefficients a_k, b_k, c_k $(k = r, i)$ can be obtained that make the planes best fit to the values of the complex analytic signal detected at the pixel grids surrounding the phase singularity. The interpolated surfaces for the real and imaginary parts are shown in Figure 5.4d and e, with the contour lines $\mathrm{Re}[\tilde{g}(x,y)] = 0$ and $\mathrm{Im}[\tilde{g}(x,y)] = 0$ inserted, respectively. From these interpolated real and imaginary parts, the detailed phase profile around an optical vortex can be obtained, as shown in

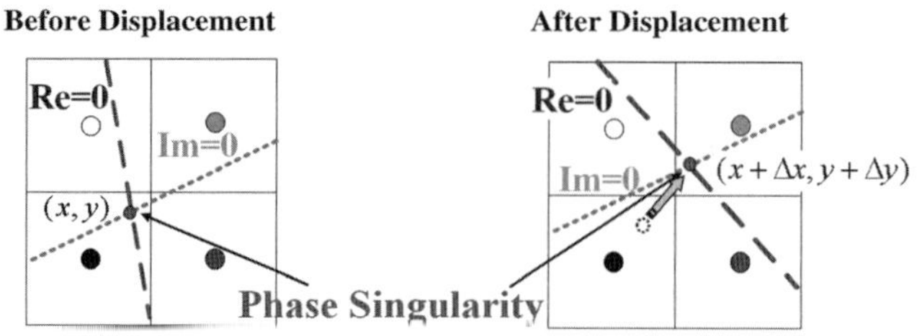

Figure 5.5 Schematic diagram of the principle of optical vortex metrology.

Figure 5.4f. The phase singularity, occurring at the intersection of the zero contour lines, is a point in the plane. From the coordinates (x, y) of the point of intersection, the location of the phase singularity with subpixel accuracy is given by

$$x = (c_i b_r - c_r b_i)/(a_r b_i - a_i b_r), \quad y = (a_i c_r - a_r c_i)/(a_r b_i - a_i b_r). \tag{5.15}$$

Just as a random speckle intensity patterns imprint marks on a coherently illuminated object surface, randomly distributed phase singularities in the pseudo-phase information associated with the speckle patterns imprint unique marks, that is, singularities, related to the object surface with positive and negative topological charges. When observed with focus on the object surface, the displacement of each phase singularity is directly related to the local displacement of the object surface. Thus, the displacement of an object can be estimated from the coordinate change for each phase singularity within the entire probing area, as shown in Figure 5.5.

5.5 Some Applications

Just as for the case of conventional speckle photography, where the intensity information of a speckle image has been made use of, optical vortex metrology requires no interferometer and can also be applied to speckle-like artificial markings, such as printed random dots and random textures of natural origin (e.g., biological tissues), even under incoherent natural illumination. Therefore, the versatility of the pseudo-phase of the complex signal representation of a speckle-like pattern provides new possibilities in a wider range of applications beyond those known for laser speckle metrology. It is the goal of this section to explore some of the application of OVM with particular emphasis on the benefits from using the information of phase singularities.

5.5.1 Nanometric Displacement Measurement

Among various applications of speckle photography, in-plane displacement measurement is one of the most fundamental and important ones. Therefore, we start our

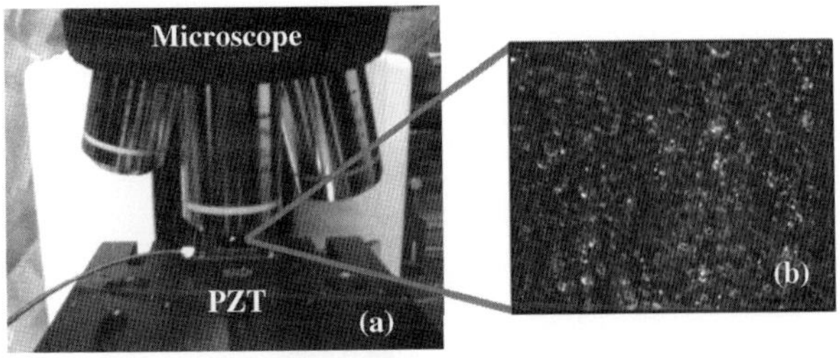

Figure 5.6 (a) Experimental setup for generation and recording of a speckle-like pattern with nanometric displacement. (b) Recorded speckle-like pattern.

introduction of optical vortex metrology from its application to nanometric displacement measurement.

Since a white light speckle pattern is known to be less prone to decorrelation [35], a speckle-like image has been generated by directly illuminating the surface of piezoelectric transducer (PZT) stage with a halogen lamp from a microscope, as shown in Figure 5.6; this also demonstrates that OVM can be applied to an object illuminated by incoherent light. The speckle-like pattern was imaged by the microscope (with a 20× objective lens and a 0.45× relay lens) onto an image sensor. Through introducing controllable microdisplacement with the PZT, the grayscale images for speckle-like patterns have been recorded by a CMOS camera with pixel size 5.2 μm × 5.2 μm. With this nominal magnification and the pixel separation of the CMOS image sensor, the unit pixel displacement corresponds to an object displacement of 578 nm.

From the recorded speckle-like pattern for each nanometric displacement introduced by the PZT, an isotropic complex signal has been generated by the aforementioned Laguerre–Gauss filtering and the pseudophase information has been retrieved. To adjust the average speckle size and control the density of phase singularities, a proper bandwidth of the Laguerre–Gauss filter ω in Equation 5.12 has been chosen carefully so that a single speckle includes about 40 pixels along the line traversing it. After determining the coordinates of all the phase singularities in the probe area, the given displacement by the proposed optical vortex metrology can be measured.

Figure 5.7a shows the histogram for the x-coordinate changes of the phase singularities for the displacement when 0.008 V was applied to the PZT, and Figure 5.7b is the histogram for the y-coordinate changes. From the locations of the maximum of the histograms, the displacements of the object in the two directions can be determined. Based on the histogram, the standard deviation σ can also be calculated, which serves as a reliability measure of the proposed technique.

Notice that as both histograms look like a symmetric Lorentzian form, the peak locations can be identified by simply calculating the mean values of coordinate differences in the x- and y-directions. In this measurement, the displacement along the x-direction has its mean value equal to 0.03973 pixels (23.0 nm) with a standard deviation $\sigma = 0.01518$ pixels (8.8 nm), which gives an estimation of the uncertainty for one-dimensional displacement based on merely using a single singularity.

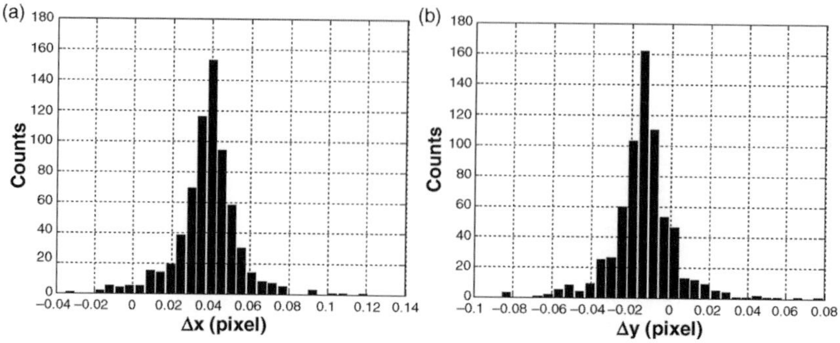

Figure 5.7 Histograms of coordinate changes of phase singularities for speckle pattern before and after displacement: (a) *x*-direction and (b) *y*-direction. Unit pixel corresponds to 578 nm.

Similarly, the mean value of Δy is equal to -0.01366 pixels (7.9 nm), with its corresponding standard deviation equal to 0.01516 pixels (8.8 nm). As expected, the generated complex Laguerre–Gauss signal is indeed isotropic, because the two displacement histograms for phase singularities have equal peak widths with identical standard deviations. From these two mean values, the object displacement distance can be obtained as $\Delta L = (\langle \Delta x \rangle^2 + \langle \Delta y \rangle^2)^{1/2} = 0.04201$ pixels (24.3 nm).

Within the linear region of the PZT, the microdisplacement can be increased by applying higher voltages. Figure 5.8 shows how the histograms of the displacements of the phase singularities change with the voltages applied to the PZT. As expected, the peak position varied with the applied voltages. The location of the histogram peak for an applied voltage of 0.008 V is observed at the correct position, but the peak height is less than that for 0.004 V. This phenomenon may arise from two different origins. One is that the speckles begin to change their shapes in addition to their pure lateral displacement; in other words, decorrelation occurs when the displacement is increased. This gives rise to creation and annihilation of the phase singularities, and

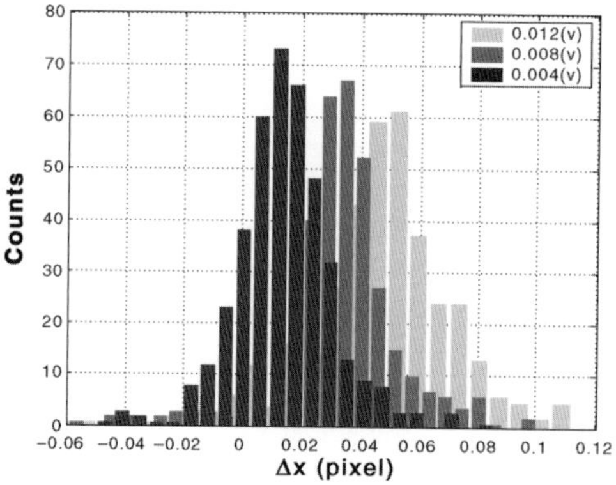

Figure 5.8 Variation of the peak positions and peak heights of the displacement histograms of phase singularities, with varying voltages applied to piezoelectric transducer (unit pixel corresponds to 578 nm).

the newly created or annihilated phase singularities are unable to find their counterparts for correct calculation of their coordinate difference. The other origin is the flow of the phase singularities across the boundary of the probe area. The object displacement causes some of the phase singularities to move into (or out of) the probe area across the boundary. This also introduces a decrease in counts of matching phase singularities in the histogram due to the lack of counterparts in the second pseudophase map. The same phenomenon can be observed for an applied voltage equal to 0.012 V. As is the case for the shape change of the cross-correlation function in the conventional correlation-based technique [36], the displacement histogram in the proposed optical vortex metrology becomes broader with a decrease of the peak height for increased applied voltage. In this measure, the standard deviations are 0.0183, 0.0297, and 0.0340 pixels for voltages equal to 0.004, 0.008, and 0.012 V, respectively. This indicates that the standard deviation is proportional to the square root of the displacement. This is understandable if one notes that the decrease of the histogram peak height of phase singularities is closely related to the decrease of the correlation peak in conventional correlation-based speckle metrology, because both effects are caused by speckle decorrelation.

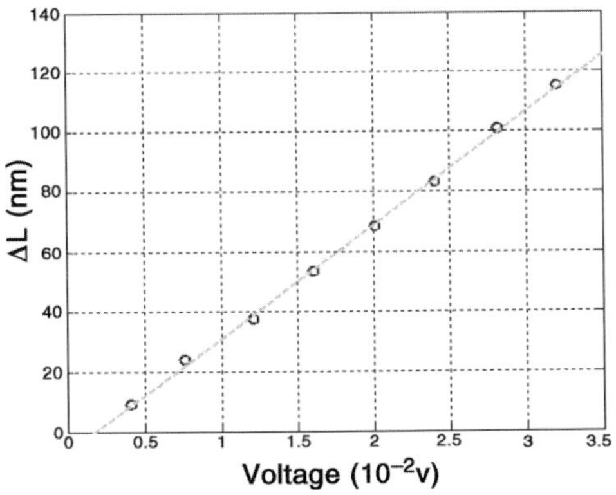

Figure 5.9 Relation between the applied voltage to PZT and the displacement detected from the peak position of the displacement histogram of the phase singularities.

Figure 5.9 shows the relation between the applied voltage to the PZT and the detected nanoscale displacement ΔL, where the dashed line gives its linear fitting with a standard deviation of 1.2 nm. With an increase of the applied voltage, the amount of measured displacement increases linearly. Therefore, Figure 5.9 serves as an experimental demonstration of the validity of the optical vortex metrology for microdisplacement measurement with nanoscale resolution.

5.5.2
Linear and Angular Encoder

In the previous section, a high-resolution measurement of nanometric displacement by OVM has been demonstrated, which makes use of the information about the locations of phase singularities before and after a displacement. To facilitate this, the corresponding phase singularities between the pre- and postdisplacement phase maps have to be identified. If the displacement is known to be small a priori, the searching process can be restricted only to the closest neighbor phase singularities of the same topological charge. However, when the displacement is large and/or nonuniform and no a priori information is given, it is difficult or maybe even impossible to uniquely identify the corresponding phase singularities.

To solve this problem, the use of additional information about the core structure of the phase singularity for identification becomes indispensable. As shown in Figure 5.2 for the core structure of the phase singularities, the change of the phase around a phase singularity is nonuniform, and the typical core structure around the phase singularities is strongly anisotropic. Besides, the phase singularity is located at the center of the elliptical contours of the amplitude, which is the intersection of the zero crossings of the real and imaginary parts of a Laguerre–Gauss signal $\tilde{g}$. Note that the eccentricity of the contour ellipse e and the zero-crossing angle θ_{RI} between the real and imaginary parts are invariant to the in-plane rigid-body motion of the object involving translation and rotation. Thus, these two geometric parameters can be adopted to describe the local properties of the phase singularities. In addition, each phase singularity has its own topological charge and vorticity, which can also be assumed to be invariant to an in-plane rigid-body displacement involving translation and rotation. Just as no fingers have exactly the same fingerprint pattern, no phase singularities have exactly the same local properties with identical eccentricity e, zero-crossing angle θ_{RI}, topological charge q, and vorticity $\vec{\Omega}$. It is this uniqueness of the core structure that enables the correct identification and the tracking of the complicated movements of phase singularities.

Based on their definitions, the geometrical and physical parameters characterizing the phase singularity can be expressed in terms of the four fitting coefficients:

$$e = \sqrt{1 - \frac{(a_r^2 + a_i^2 + b_r^2 + b_i^2) - \sqrt{(a_r^2 + a_i^2 - b_r^2 - b_i^2)^2 + 4(a_r b_r + a_i b_i)^2}}{(a_r^2 + a_i^2 + b_r^2 + b_i^2) + \sqrt{(a_r^2 + a_i^2 - b_r^2 - b_i^2)^2 + 4(a_r b_r + a_i b_i)^2}}}, \tag{5.16}$$

$$\theta_{RI} = \begin{cases} |\arctan[(a_r b_i - a_i b_r)/(a_r a_i + b_r b_i)]| & |\theta_{RI}| < \pi/2 \\ \pi - |\arctan[(a_r b_i - a_i b_r)/(a_r a_i + b_r b_i)]| & |\theta_{RI}| > \pi/2 \end{cases}, \tag{5.17}$$

$$\Omega = |a_r b_i - a_i b_r|, \tag{5.18}$$

$$q = \operatorname{sgn}(\vec{\Omega} \cdot \vec{e}_z) = \operatorname{sgn}(a_r b_i - a_i b_r). \tag{5.19}$$

Although the coefficients a_k, b_k, c_k $(k = r, i)$ may change with the displacement of the object, the parameters $(e, \theta_{RI}, \Omega, q)$ characterizing the local properties of the pseudophase singularities remain stable without significant changes because they are invariant to translation and rotation. When observed with focus on the object surface, the displacement of each phase singularity can be directly related to the local displacement of the object surface. After identifying corresponding phase singularities before and after displacement making use of their core structures as fingerprints, the local displacement of an object can be estimated by tracing the movements of the phase singularities.

To find the correctly matching phase singularities for the object after displacement, it is necessary to prescribe several conditions for the correct identification, which are expressed as follows:

$$q = q', \tag{5.20}$$

$$|\Delta e| = |e - e'| < \varepsilon_1, \tag{5.21}$$

$$|\Delta\Omega| = |(\Omega - \Omega')/(\Omega + \Omega')| < \varepsilon_2, \tag{5.22}$$

$$|\Delta\theta_{RI}| = |\theta_{RI} - \theta'_{RI}| < \varepsilon_3, \tag{5.23}$$

where primed parameters are related to the phase singularity after displacement. After selection of appropriate threshold values for $\varepsilon_1, \varepsilon_2$, and ε_3, most of the phase singularities with large differences in their core structures can be efficiently eliminated. Next, within a few survived candidates of phase singularities, the figure-of-merit for the best matching of phase singularities has been chosen as

$$E = (e - e')^2 + \left(\frac{\Omega - \Omega'}{\Omega + \Omega'}\right)^2 + \left[\frac{2}{\pi}(\theta_{RI} - \theta'_{RI})\right]^2. \tag{5.24}$$

After calculation of the merit function for each pair of phase singularities, the correct counterparts can be identified from the minimum value of E. Thus, the in-plane displacement of an object can be estimated from the coordinate change for each pseudophase singularity within the probing area.

Figure 5.10a shows the movements of phase singularities with a translational displacement, where the locations of phase singularities before and after displacement have been indicated by ○ and ■, respectively. During this identification process to find the correct counterparts for phase singularities, the threshold values in Equations 5.21–5.23 have been selected empirically as $\varepsilon_1 = 0.05$, $\varepsilon_2 = 0.1$, and $\varepsilon_3 = 0.174$ (rad), and the search process has been performed over the whole probe area, since no prior knowledge about the local displacement of the object have been assumed. As expected, most of the phase singularities have found their counterparts in the upper-right direction as indicated with a short straight line connecting the correct pair of phase singularities, and the coordinate differences between the corresponding phase singularities gave a good estimation for the local displacement of the object as far as only those phase singularities are used. However, it is obvious that some phase singularities failed to find their correct counterparts as indicated with long straight lines stretching in random directions, and thus became the source of errors for the proposed optical vortex metrology. Figure 5.10b is the histogram for the $\hat{x}$-coordinate changes of the phase singularities, and the histogram for the $\hat{y}$-coordinate changes is shown in Figure 5.10c. From the locations of the histogram

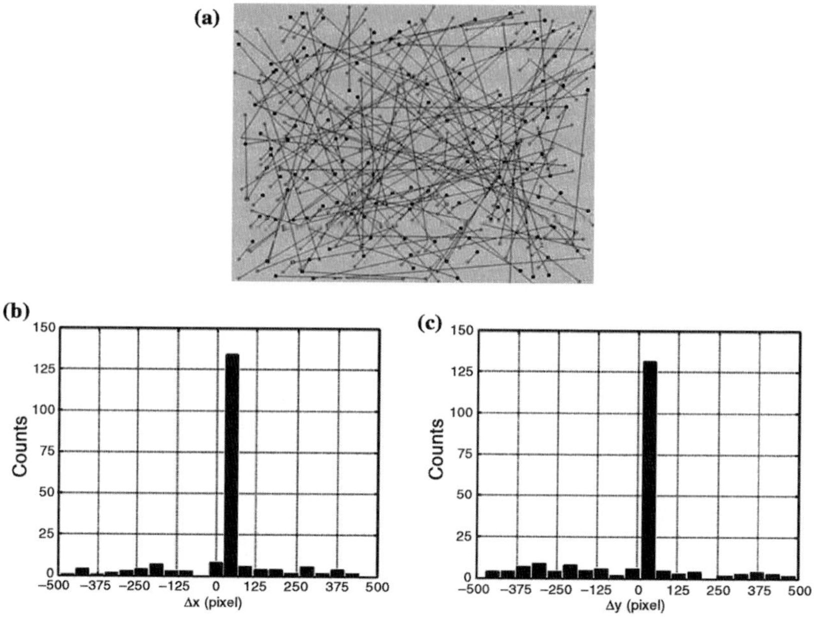

Figure 5.10 (a) Displacement of phase singularities; ○: before displacement, ■: after displacement. Histograms of coordinate changes of phase singularities: (b) $\hat{x}$-direction, (c) $\hat{y}$-direction.

peaks, an initial estimate of the displacements of the object can be made by simply calculating the mean values of coordinate differences in the two directions using only the samples belonging to these histogram peaks. In this measurement, the displacement along the x-direction has its mean value equal to 36.3 pixels (21.0 μm) with a standard deviation $\sigma = 17.9$ pixels (10.3 μm), which indicates a very large uncertainty in the estimation for one-dimensional displacement measurement based on a single singularity. Similarly, we get the mean value of Δy equal to 37.2 pixels (21.5 μm) and a corresponding standard deviation equal to 17.4 pixels (10.1 μm). Thus, the 2D displacements of the object can be roughly estimated in the initial step. Meanwhile, the standard deviations for Δe and $\Delta\theta_{RI}$ have also been calculated, which serve as stability measures for the core structures of the pseudophase singularities during the object movement. In this gauge, the standard deviation for Δe is 0.0148 and that of $\Delta\theta_{RI}$ is 0.066 (rad).

At both sides of the main histogram peaks in Figure 5.10, very long tails (side lobes) in both histograms can be observed, which indicate the region of errors in the displacement measurement arising from those phase singularities that failed to find their correct counterparts. Note that in contrast to the small proportion of phase singularities spreading out in the histogram, the most coordinate differences for the phase singularities remain concentrated in the histogram with very narrow bar width. It is this high concentration that makes possible the precise initial estimate of the displacement.

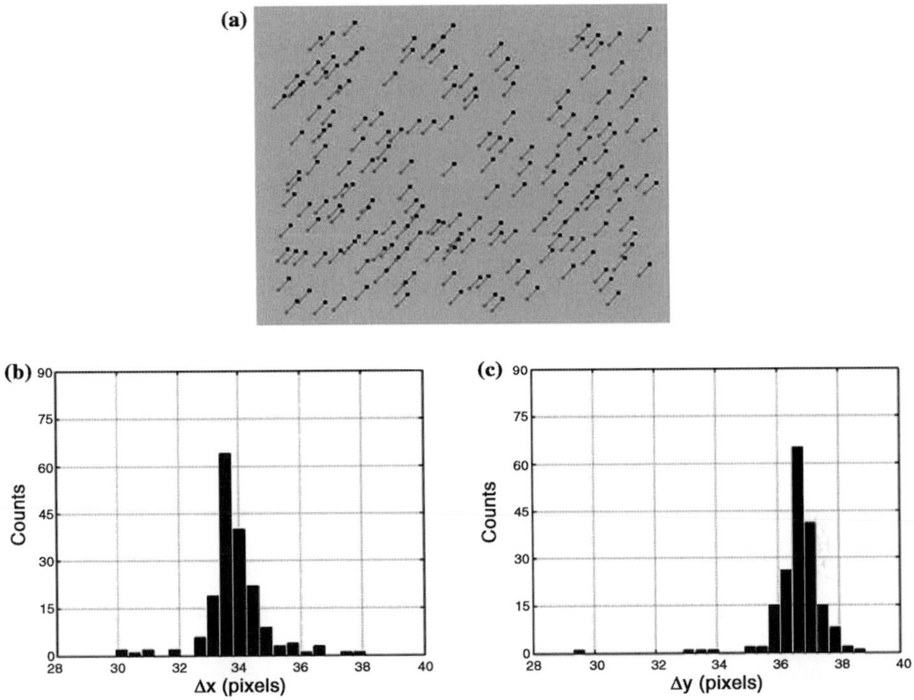

Figure 5.11 Results of phase singularities after refined identification. (a) Displacement of phase singularities. Histograms of coordinate changes of phase singularities: (b) $\hat{x}$-direction, (c) $\hat{y}$-direction. Please note the change of scale between this figure and the scale in Figure 5.10.

Based on this initial displacement estimation and the a priori knowledge of the translational rigid-body displacement, the next interrogation area can be restricted to a small window of 20×20 pixels around the initial estimate of the displacement given by the rounded pixel coordinates (36, 37). Figure 5.11a shows the results for the displacement of the phase singularities after the local search performed for the restricted area described above. The histograms for the coordinate changes of the phase singularities along the $\hat{x}$- and the $\hat{y}$-directions are shown in Figure 5.11b and c, respectively. As anticipated, all the phase singularities have been shifted with uniform displacement in the upper-right direction. From the location of the histogram peak, the displacement along the $\hat{x}$-direction has its mean value equal to 33.9 pixels (19.6 μm) with a standard deviation $\sigma = 0.7$ pixels (0.4 μm), which indicates a significant improvement as compared with the result derived from the first preliminary estimation. Similarly, the mean value of Δy is equal to 36.7 pixels (21.2 μm), and the corresponding standard deviation is 0.5 pixels (0.3 μm). In Figure 5.11a, the standard deviations for Δe and $\Delta\theta_{RI}$ become 0.0159 and 0.0429 (rad), respectively. From these two mean values, the object displacement distance can be obtained as $\Delta L = (\langle\Delta x\rangle^2 + \langle\Delta y\rangle^2)^{1/2} = 49.96$ pixels (28.9 μm), and the direction angle of the displacement is $\delta = \arctan(\langle\Delta y\rangle/\langle\Delta x\rangle) = 0.825$ rad.

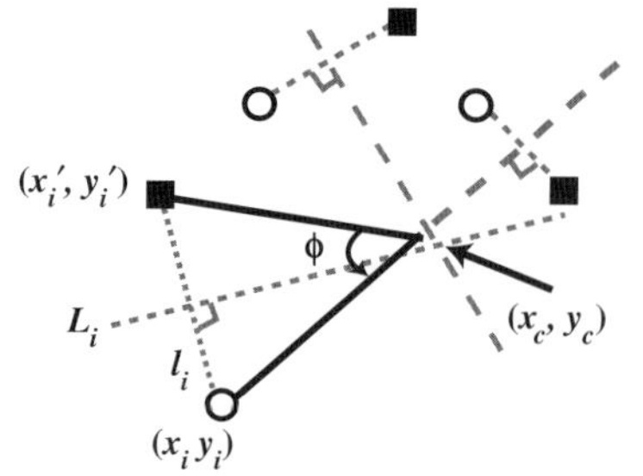

Figure 5.12 Schematic diagram for in-plane rotation measurement by optical vortex metrology.

In many industrial control systems with various kinds of rotating parts, an accurate measurement of angular displacement in dynamic environments is particularly important. Due to its inherent advantage of noncontact measurements, optical detections of angular displacement based on the intensity cross-correlation have been put forward mainly during the past decades [37, 38]. Here, the application of optical vortex metrology is extended to the case of rotational displacement measurement.

Figure 5.12 shows a schematic diagram for in-plane rotation measurement by using OVM. After the identification of the corresponding phase singularities in the pseudophase maps for the object before and after rotational displacement, the perpendicular bisector L_i between this pair of phase singularities can be expressed as

$$A_i x + B_i y + C_i = 0, \quad A_i = 2(x_i - x_i'), \quad B_i = 2(y_i - y_i'), \quad \text{and} \\ C_i = (x_i'^2 + y_i'^2 - x_i^2 - y_i^2), \tag{5.25}$$

where (x_i, y_i) and (x_i', y_i') are the coordinates of the phase singularity before and after a rotational displacement, respectively.

Ideally, the center of rotation of the object, to be found at the intersections of all the perpendicular bisectors connecting each pair of the corresponding phase singularities, should be a single point. From Equation 5.25 for each perpendicular bisector, the location of the center of rotation can be determined by applying the least-square fitting method giving

$$x_c = \frac{\left(\sum_i B_i^2\right)\left(-\sum_i A_i C_i\right) - \left(\sum_i A_i B_i\right)\left(-\sum_i B_i C_i\right)}{\left(\sum_i A_i^2\right)\left(\sum_i B_i^2\right) - \left(\sum_i A_i B_i\right)^2}, \tag{5.26}$$

$$y_c = \frac{\left(\sum_i A_i^2\right)\left(-\sum_i B_i C_i\right) - \left(\sum_i A_i B_i\right)\left(-\sum_i A_i C_i\right)}{\left(\sum_i A_i^2\right)\left(\sum_i B_i^2\right) - \left(\sum_i A_i B_i\right)^2}. \tag{5.27}$$

From the law of cosines, the rotation angle ϕ_i can be obtained for each pair of phase singularities from the coordinates of the three points: (x_i, y_i), (x_i', y_i'), and (x_c, y_c). Thus, the rotational displacement of an object can be estimated from the histogram of rotation angle for all phase singularities within the probing area.

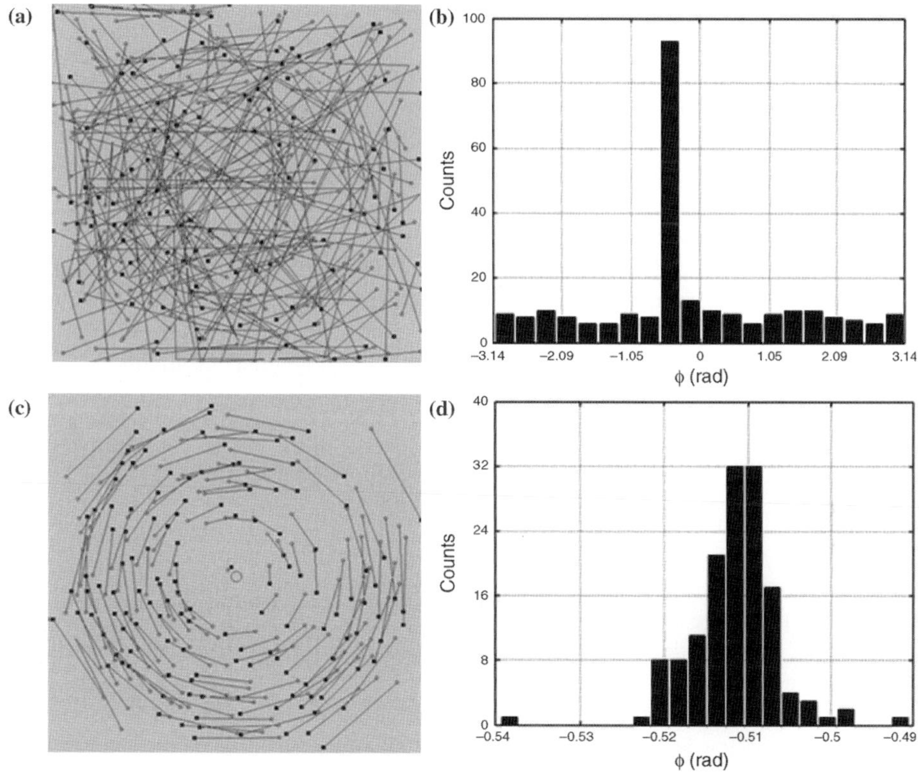

Figure 5.13 Results of optical vortex metrology applied for rotational displacement. *Upper row*: rough searching over whole probe area. *Lower row*: local searching for fine identification. (a and c) Displacement of phase singularities. (b and d) Histogram of rotation angle of phase singularities (note the difference in scaling of the horizontal axis).

By introduction of the rotational displacement with a turntable of the microscope, the rotation angle can be measured by using the OVM. After appropriate choice of the threshold values as before, the preliminary search is performed for the phase singularity identification within the entire probe area. Figure 5.13a shows the rotational displacements of phase singularities after this initial, unrestricted search. The histogram of rotation angles for the phase singularities is shown in Figure 5.13b. Just as what was the case for the translational displacement, most of the phase singularities have found their correct counterparts during this preliminary identification process. Based on the location of the maximum for the histogram, the rotation angle ϕ has its mean value equal to -0.527 rad with a standard deviation of $\sigma = 0.082$ rad. In addition, the coordinate of the rotation center has been obtained $(x_c, y_c) = (485.9, 484.7)$ in units of pixels from Equations 5.26 and 5.27. Based on these results derived from the initial rotation estimation, a location-dependent spatial initial shift has been derived for the object after displacement given by

$$x = x_c + (x_i - x_c)\cos\phi + (y_i - y_c)\sin\phi, \tag{5.28}$$

$$y = y_c - (x_i - x_c)\sin\phi + (y_i - y_c)\cos\phi. \tag{5.29}$$

Within a small search window of 20 × 20 pixels, the fine identification for phase singularities has been performed, and the results are shown in Figure 5.13c and d.

As expected with the rotation of the rigid-body specimen, all phase singularities have found their correct counterparts, and the local displacements in azimuth direction increase linearly with the radial distance from the center of rotation. From the location of the histogram peak, the rotation angle has its mean value equal to $\phi = -0.518$ rad with a standard deviation of $\sigma = 0.002$ rad, which also indicates a significant improvement as compared with the initial estimation. Figure 5.13 serves as an experimental demonstration of the validity of the proposed technique for speckle in-plane rotation measurement with 2 mrad angular resolution.

5.5.3 Fluid Mechanical Analysis

Velocity is one of the most important parameters in fluid mechanics due to its cardinal influence on transport phenomena. Velocity measurements by using speckle and speckle-related techniques have been explored extensively, and various techniques have been developed [39, 40]. Though they differ in specific technical details, these speckle-related methods have at least two common features, namely, to seed the flow with particles and to detect the velocity vectors through cross-correlation of the intensity distributions for the recorded images. The seeding particles moving with a fluid flow provide kinematic information for the recording plane, and the local fluid velocity is derived from the ratio between the measured spacing for matching tracers and the time between exposures. Although many attempts have been made to extend its applications during the past decades, the most serious problem for particle image velocimetry lies in the lack of an autonomous method for high-precision tracking of the seeding particles from the recorded speckle-like patterns. Due to the resulting velocity vectors being derived through a cross-correlation over small interrogation areas of the flow, the spatial resolution for particle image velocimetry has been restricted because of the fact that the resulting velocity distribution is a spatially averaged representation for the actual velocity field.

Since phase singularities are well-defined geometrical points with unique core structure acting as fingerprints, they can serve as optimal encoders for position marking. Experiments have been conducted to extend the application of optical vortex metrology to fluid mechanical investigations. In these experiments, tiny pieces of tea leaves were used as the microparticles with random shape floating on the water surface. The tea leaves moving with the flow are imaged by a lens onto the image sensor plane of a high-speed camera. The sequences of the water flow seeded by tea leaves were recorded at a frame rate of 125 frames per second.

Two examples of the recorded images of the floating tea leaves at different instants of time are shown in Figure 5.14a and b. The image in Figure 5.14b was recorded 8 ms after the recording of the image in Figure 5.14a. From the recorded image, the Laguerre–Gauss signal can be generated with a proper bandwidth of the LG filter in

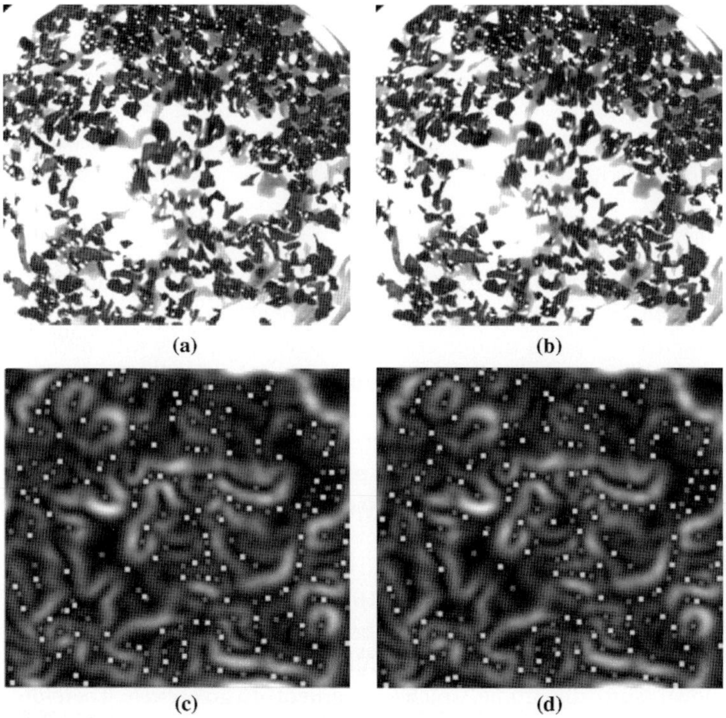

Figure 5.14 Recorded images (a and b) of the floating tea leaves on the water surface at different instants of time separated by 8 ms and the corresponding LG signals (c and d), with positive and negative pseudophase singularities indicated by dark gray and light gray squares, respectively.

Equation 5.12, so that the pseudophase singularities on the water surface are clearly separated, and each tea leaf corresponds to one pseudophase singularity. The corresponding amplitude distributions of the generated LG signals of the recorded images are shown in Figure 5.14c and d, where the locations of the positively and negatively signed pseudophase singularities are indicated by dark gray and light gray squares, respectively.

To uniquely identify the corresponding pseudophase singularities before and after displacement, the ellipticity of the contour ellipse, zero-crossing angle, and vorticity have been chosen like what was the case for the singularity fingerprints in the previous investigation. Though the performance of OVM has been demonstrated with large dynamic range and high spatial resolution, these three parameters are not the most mathematically natural measures. Noting that the local structure of a pseudophase singularity depends only on the gradient field of the LG signal, another fingerprint based on the aforementioned Stokes-like parameters has been chosen for the purpose of unique identification.

Based on Equations 5.5a–5.5d for the Stokes-like parameters and the fitting coefficients a_k, b_k, c_k $(k = r, i)$ for the real and imaginary part of LG signal, the parameters describing the anisotropy ellipse of each pseudophase singularity may be rewritten as

$$S_0 = a_r^2 + b_r^2 + a_i^2 + b_i^2, \tag{5.30a}$$

$$S_1 = a_r^2 + a_i^2 - b_r^2 - b_i^2, \tag{5.30b}$$

$$S_2 = 2(a_r b_r + a_i b_i), \tag{5.30c}$$

$$S_3 = 2(a_r b_i - a_i b_r). \tag{5.30d}$$

Figure 5.15 shows the obtained Stokes-like parameters for each pseudophase singularity displayed on the surface of a Poincaré sphere. As anticipated, each pseudophase singularity has its unique anisotropic core structures with different ellipticity and azimuth angles, and these ellipticities and azimuth angles have almost uniform distribution on the Poincaré sphere with a slight concentration at the two poles. It is this uniqueness of the anisotropy that enables a correct identification and the tracking of the complicated movement of pseudophase singularities through their fingerprints expressed by the Stokes-like parameter.

To find the correctly matching pseudophase singularities for the tea leaf after displacement, it is necessary to introduce a similarity measure that gives a criterion for correct identification. As shown in Figure 5.16, the distance along a geodesic line can be chosen as the figure-of-merit for the best matching of the pseudophase singularities. The geodesic line is the shortest path between two points on the sphere surface, also known as an orthodrome, which is a segment of a great circle. Therefore, the merit function based on the distance along the geodesic line is given by

$$\left|\Delta\vec{S}\right| = \left|\arccos(\vec{S}\cdot\vec{S}')\right| \le \varepsilon, \tag{5.31}$$

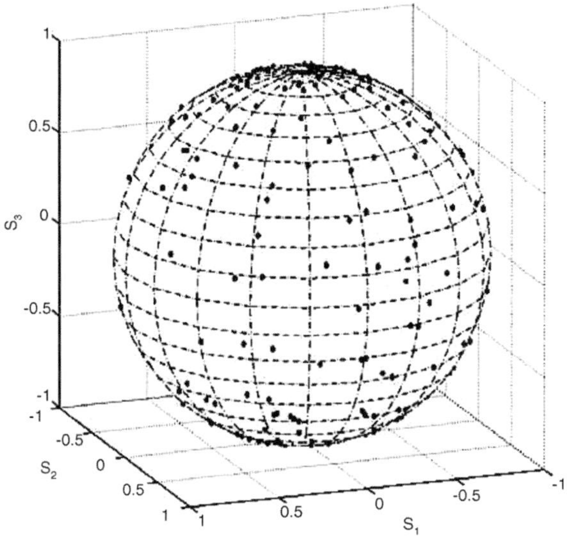

Figure 5.15 Distribution of anisotropic pseudophase singularities on the Poincaré sphere.

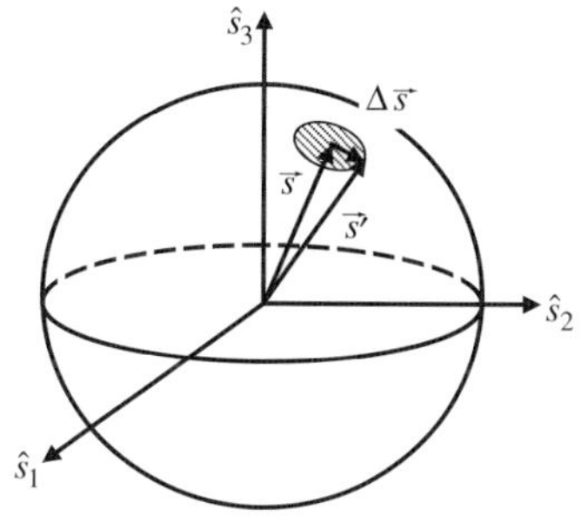

Figure 5.16 Orthodrome on Poincaré sphere as the merit function of the best matching for pseudophase singularities.

where primed parameters are related to the pseudophase singularities after displacement. After an appropriate selection of the threshold value ε and some calculation of the merit function for each pair of pseudophase singularities, the correct counterparts can be identified from the minimum value of $\left|\Delta\vec{S}\right|$. Thus, the in-plane displacement of a particle can be estimated from the coordinate change $(\Delta x, \Delta y)$ of the corresponding pseudophase singularity attached to the tea leaf, and the linear velocity of the measured flow at a given position can be estimated from the rate of change for the coordinate of each singularity between two sequential exposures. Meanwhile, when observed with focus on the seeded particle surface, the change of azimuth angle φ can be directly related to the rotation or spin of each pseudophase singularity associated with the tea leaf, and the angular velocity and the applied torque of the particle spin can be estimated through the change of the azimuth angle $\Delta\varphi$.

In the experiment, every tea leaf has experienced a different amount of movement during the recording period, and the pseudophase singularities have maintained their spatial structures even after being translated with respect to each other by an amount corresponding to its complicated movement. Figure 5.17 shows the in-plane movements of phase singularities, where the locations of phase singularities before and after displacement are indicated by ○ and ■, respectively. During the identification process in order to match the pseudophase singularities, the threshold value in Equation 5.31 was selected as $\varepsilon = 0.1$, and the search was performed in the neighborhood area within 30×30 square pixels; this is due to the recording speed of the high-speed camera being much faster than the fluid velocity so that the phase singularities do not move more than 30 pixels between successive frames. After identification of the corresponding pseudophase singularities for each pair of consecutive images making use of their Poincaré sphere representation of the core structures as fingerprints, the complicated movement of the pseudophase singularities can be traced. Figure 5.18 shows the trajectories for the selected four tea leaves for 99 frames of recorded images, where the arrows indicate their movement directions at different instants of time. As expected when blowing onto the water surface, the central flow had a large velocity pushing the tea leaves aside, so that these

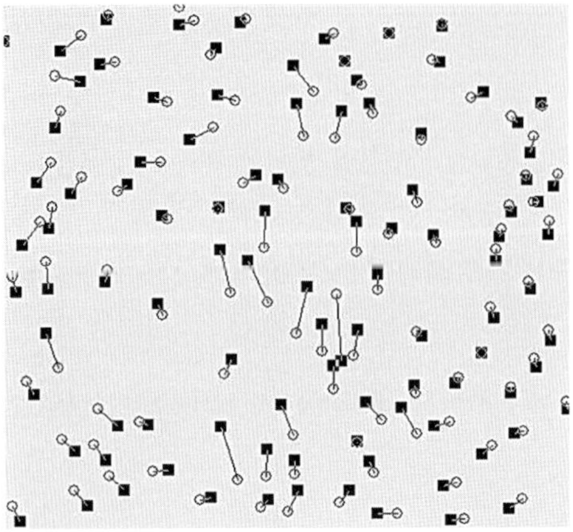

Figure 5.17 Displacement of pseudophase singularities; singularities before and after displacement are indicated by open circles (○) and filled squares (■), respectively.

floating particles at the left and right part of the water have counterclockwise and clockwise movements, respectively, being repelled by the wall of the circular container. It can be also observed that some phase singularities changed the directions of their movement abruptly due to collision between leaves, as shown by trajectory C in Figure 5.18.

From the coordinate changes $(\Delta x, \Delta y)$ for each registered pseudophase singularity, the horizontal and vertical components of the velocity can be estimated from $v_x = \Delta x/\Delta t$ and $v_y = \Delta y/\Delta t$, where the sampling time interval is $\Delta t = 8$ ms in this example. Figure 5.19a shows the velocity history for tea leaf A in Figure 5.18, where the dashed–dotted line, dashed line, and the solid line indicate v_x, v_y, and $v = (v_x^2 + v_y^2)^{1/2}$, respectively. As anticipated, the tea leaf velocity has a fluctuating structure due to the advection with the water. At the beginning of the measurement, the vertical

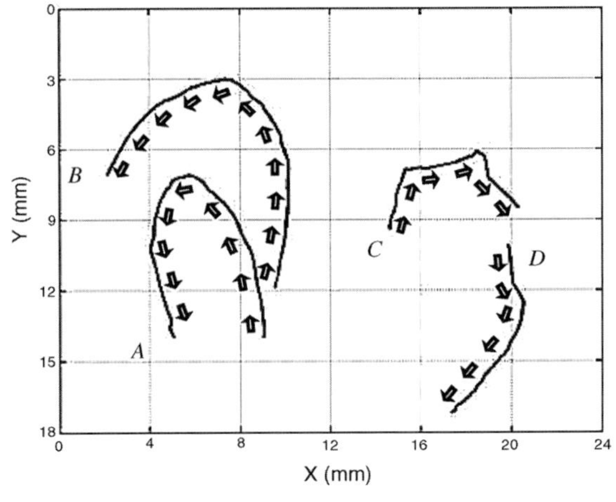

Figure 5.18 Trajectories of different tea leaves on the water surface.

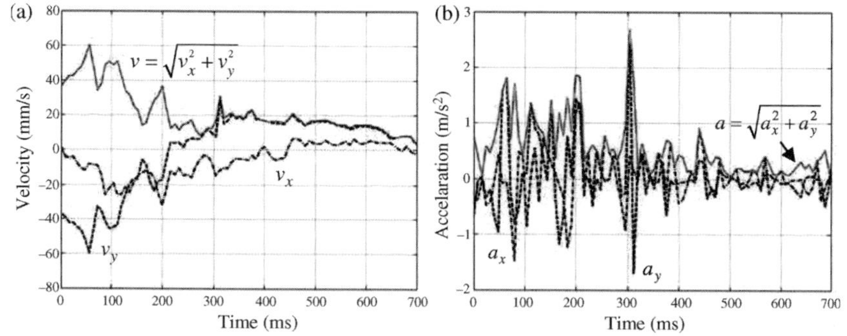

Figure 5.19 Measurement results of the velocity (a) and acceleration (b) for the tea leaf A in Figure 5.18.

component of the velocity v_y played a dominant role with a relatively larger value than the horizontal part of the velocity. As time goes by, the two velocity components decrease, and the tea leaf finally ceases its movement due to fluid damping. After calculation of the finite difference for the horizontal and vertical velocity components, the temporal variation of the acceleration can be obtained as shown in Figure 5.19b. Under the reasonable assumption that the mass of the particle is constant during the recording process, Figure 5.19b provides a local force diagrams for the tea leaf A at the different instants of time. From the solid curve for the acceleration sum, we can also observe the local acceleration peaks stemming from the collision force that the tea leaf received from other tea leaves during their random movements.

Meanwhile, the spinning angular displacement $\Delta\varphi$ around the center of the phase singularity can be obtained from the change of azimuth angle for each registered pseudophase singularity, and the spin angular velocity can be estimated from $\Omega = \Delta\varphi/\Delta t$, as shown in Figure 5.20a. After the calculation of the finite difference for the spin angular velocity, the history of the spin angular acceleration has been obtained in Figure 5.20b, which is proportional to the applied torque on the tea leaf A because of the constant moment of inertia. Therefore, Figures 5.19 and 5.20

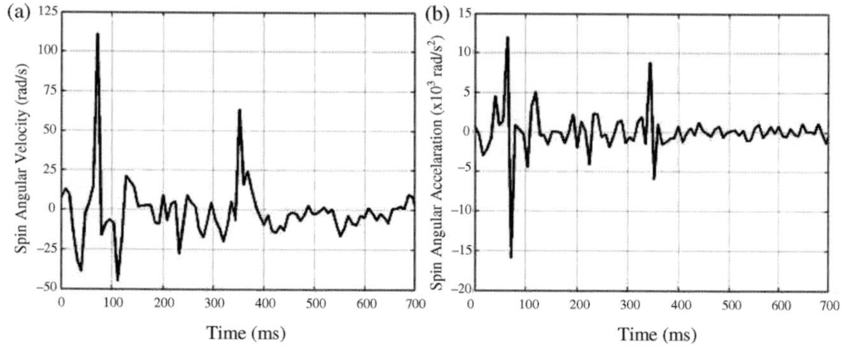

Figure 5.20 Measurement results of the spin angular velocity (a) and spin angular acceleration (b) for the tea leaf A in Figure 5.18.

experimentally demonstrate the validity of the proposed optical vortex metrology for fluid mechanics measurement based on the anisotropic core structure of pseudophase singularities.

5.5.4
Biological Kinematic Analysis

In modern design of engineering systems, the transfer of technology from natural life forms to synthetic constructs is often advantageous because evolutionary pressure typically forces natural systems to become highly optimized and efficient [41]. To imitate mechanisms found in nature successfully, *in vivo* measurement of dynamic structures plays a key role in the field of both mechanics and, to a lesser extent, biology. This objective has previously been pursued with partial success by establishing fixed points on the object's surface as markers to be followed later. Needless to say, the rather arbitrarily placing of these points will inevitably influence the measurement result. Furthermore, these methods will usually not possess the desired accuracy. Meanwhile, dynamic measurements by using speckle techniques have been explored. Although many attempts have been made to extend the applications of speckle metrology, the most difficult issue for its biological applications lies in the fact that the structures themselves are flexible and thus are not easily followed by standard cross-correlation techniques. Moreover, the entire structure might undergo rotation and maybe even scaling during movement. Finally, it is usually of importance to track the dynamical behavior of parts of the sample individually. Therefore, the need for establishing an autonomous method for tracking the dynamics of biological specimen with a high accuracy becomes obvious with an increase of interests in bionics.

On the other hand, though the validity of optical vortex metrology has been demonstrated by using the anisotropic core structures of phase singularities as fingerprints for identification, the performance was affected or undermined due to the instability of phase singularities stemming from the decorrelation in the speckle-like pattern. To solve this problem, the spatial structure of a group of pseudophase singularities has been made use of for tracking purpose in addition to the information about the location and the core structures of the individual phase singularities.

As shown in Figure 5.21, a cluster of pseudophase singularities are connected together to form a specific figure or picture. This spatial structure can serve as a constellation of pseudophase singularities with a fixed configuration that uniquely characterizes the mutual position between the individual phase singularities. Similar to what is the case for stars in a constellation in the sky with different size, brightness, and other characteristics, each optical vortex in a constellation of pseudophase singularities has its own anisotropic core structure. After identification of the corresponding phase singularities of the speckle pattern after displacement through their core structures, a cluster of pseudophase singularities can be tracked, and therefore a kinematic analysis based on the spatial structure in the constellation of singularities can be conducted. Just as a stellar constellation will experience translation, rotation, and scaling during different seasons of a year or as viewed from

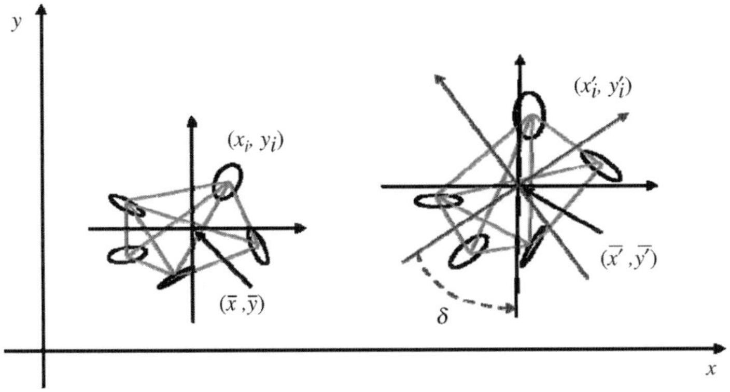

Figure 5.21 Schematic diagram for the constellation of pseudophase singularities. *Left*: before movement. *Right*: after movement.

different places on the earth, the constellation of pseudophase singularities will have similar movements and still maintain its particular configuration. It is this constancy and stability of the configuration in a constellation of phase singularities that has made possible the decorrelation–robust tracing with high reliability even for a complicated movement of the specimen.

Let (x_i, y_i), (x'_i, y'_i) be the coordinates of the ith pseudophase singularity in one constellation before and after movement, respectively. Making use of the theory of elasticity [42], it is possible to decompose a complicated movement into a sum of translation, rotation, and deformation (scaling). That is,

$$\begin{pmatrix} x'_i - \bar{x}' \\ y'_i - \bar{y}' \end{pmatrix} = \beta \begin{pmatrix} \cos\delta & \sin\delta \\ -\sin\delta & \cos\delta \end{pmatrix} \begin{pmatrix} x_i - \bar{x} \\ y_i - \bar{y} \end{pmatrix}, \tag{5.32}$$

where β is a scale factor and δ is a rotation angle. Here, $(\bar{x}, \bar{y})$ and $(\bar{x}', \bar{y}')$ can be considered as the coordinates of the gravity centers before and after movement given by

$$\bar{x} = \sum\nolimits_i x_i/N \quad \text{and} \quad \bar{y} = \sum\nolimits_i y_i/N, \tag{5.33a}$$

$$\bar{x}' = \sum\nolimits_i x'_i/N \quad \text{and} \quad \bar{y}' = \sum\nolimits_i y'_i/N, \tag{5.33b}$$

with N being the total number of pseudophase singularities in this constellation. Therefore, the figure-of-merit for best matching of the phase singularities constellation has been chosen as

$$E = \sum\nolimits_i \{[x'_i - \bar{x}' - C(x_i - \bar{x}) - S(y_i - \bar{y})]^2 + [y'_i - \bar{y}' + S(x_i - \bar{x}) - C(y_i - \bar{y})]^2\}, \tag{5.34}$$

where $C = \beta\cos\delta$ and $S = \beta\sin\delta$. On the basis of the least-square fitting routine, the condition for E to be a minimum is that $\partial E/\partial C = 0$ and $\partial E/\partial S = 0$. After straightforward algebra, C and S can be rewritten as

$$C = \sum_i (\mathbf{x}'_i \mathbf{x}_i + \mathbf{y}'_i \mathbf{y}_i) / \sum_i (\mathbf{x}_i^2 + \mathbf{y}_i^2), \tag{5.35}$$

$$S = \sum_i (\mathbf{x}'_i \mathbf{y}_i - \mathbf{x}_i \mathbf{y}'_i) / \sum_i (\mathbf{x}_i^2 + \mathbf{y}_i^2), \tag{5.36}$$

where $\mathbf{x}'_i = x'_i - \bar{x}'$, $\mathbf{x}_i = x_i - \bar{x}$, $\mathbf{y}'_i = y'_i - \bar{y}'$, and $\mathbf{y}_i = y_i - \bar{y}$. Thus, the kinematic analysis of an object under a complicated movement can be conducted with the information of translation $(\Delta x, \Delta y)$, deformation β, and rotation angle δ by tracing the dynamics of the constellation of the pseudophase singularities. These are

$$\beta = \sqrt{S^2 + C^2}, \tag{5.37}$$

$$\delta = \arctan(S/C), \tag{5.38}$$

$$\Delta x = \bar{x}' - \bar{x}, \tag{5.39}$$

$$\Delta y = \bar{y}' - \bar{y}. \tag{5.40}$$

In the experiments, the biological sample is a swimming fugu fish with a speckle-like intensity pattern on its body surface. Two examples of the recorded temporally separated images of the swimming fugu are represented in Figure 5.22a and b. The image in

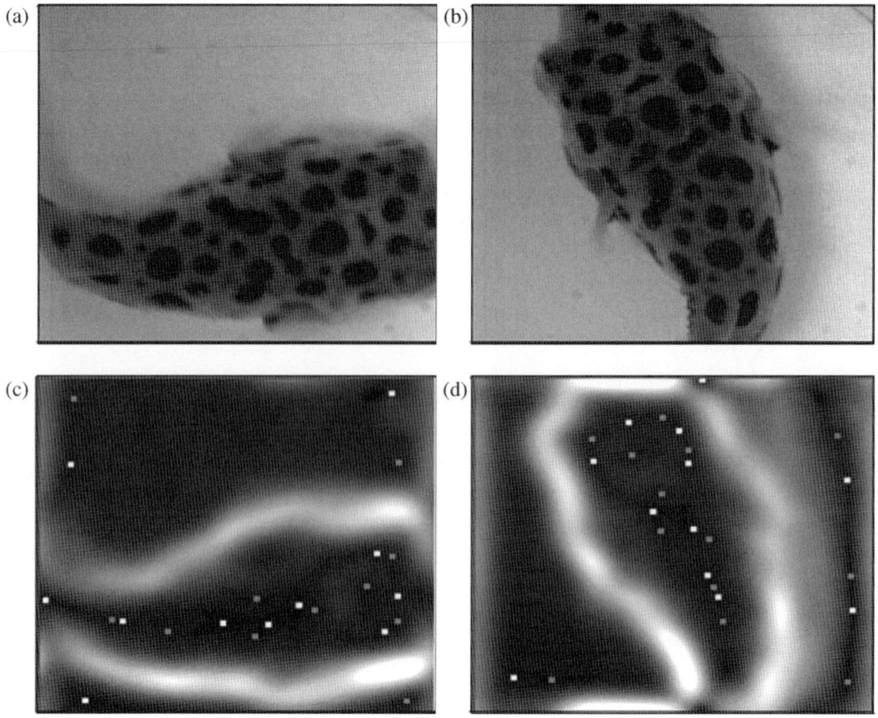

Figure 5.22 Recorded images for the swimming fugu at different instants of time (a,b) and the generated Laguerre–Gauss signals with pseudophase singularities inserted (c,d).

Figure 5.22b is delayed by 3.3 s from that in Figure 5.22a. The corresponding amplitude distributions of the generated complex signals for the recorded images are shown in Figure 5.22c and d, where the locations for the positive and negative pseudophase singularities have been indicated by dark gray and light gray squares, respectively. As expected, a big constellation of pseudophase singularities can be observed on the surface of the fugu body through its complex signal representation, and the constellation boundary is readily drawn up due to the edge enhancement effect introduced by the Laguerre–Gauss filter [34]. Meanwhile, it is easily observed from Figure 5.22c and d that the constellation configuration of pseudophase singularities from the LG signal is in all respects similar to the network of optical vortices and polarization singularities in scalar and vector speckle field [16, 18], because the generated complex signal representation shares the same statistical properties as the laser speckle field [7].

In all these figures, it can easily be observed that the fugu has experienced a large rotation during the recording period, and the pseudophase singularities show structures similarly transformed with respect to each other by an amount that corresponds to the complicated movements of the fugu. Thus, most of the pseudophase singularities can find their correct counterparts, and constitute a stable constellation with specific spatial configuration. The coordinate differences between the corresponding phase singularities in the constellation gave a good estimation for

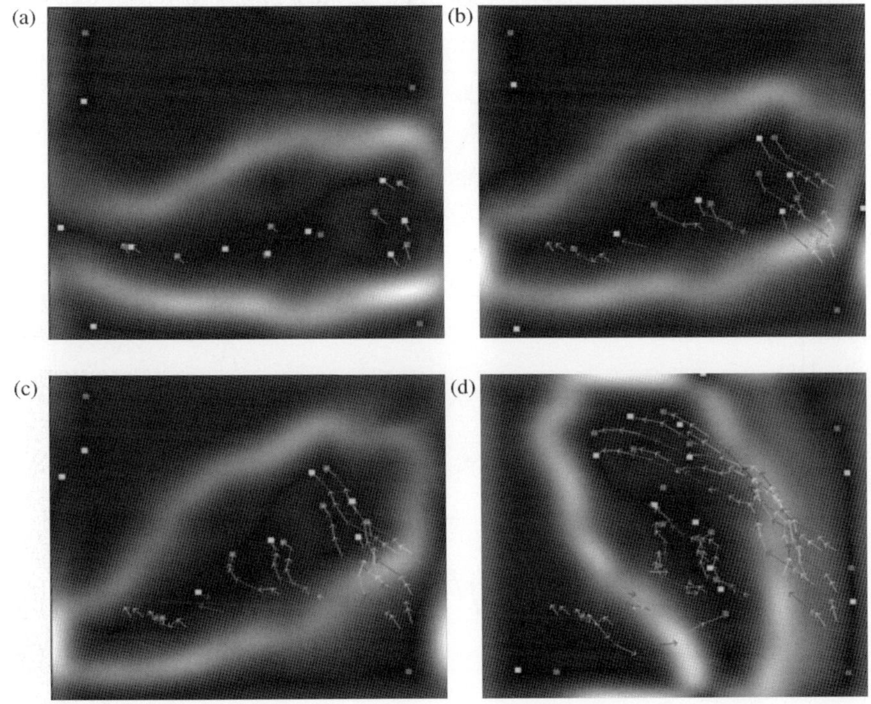

Figure 5.23 The trajectory of the pseudophase singularities in the fugu's body at different instants of time: (a) $t = 0.067$ s, (b) $t = 0.699$ s, (c) $t = 1.665$ s, (d) $t = 3.33$ s.

the swimming fugu, and could be used for biological kinematic analysis as long as we only use matching phase singularities.

After identifying corresponding pseudophase singularities for each pair of consecutive images making use of their core structures as fingerprints, the movement of the swimming fugu can be traced through its trajectory, as shown in Figure 5.23, where the arrows indicate their movement directions at different instants of time. As expected with the rotation of fugu's swimming, the trajectory exhibits an arch shape and body parts far from the gravity center of fugu show a larger arc length. With the coordinate information for each phase singularity in the constellation, the *in vivo* measurement can be conducted and the instantaneous information can be obtained for this swimming fugu about its translation, rotation, and scaling based on Equations 5.37–5.40. Meanwhile, the fugu's dynamic information about the linear and angular velocities can also be obtained from the calculated displacement in translation $(\Delta x, \Delta y)$ and rotation angle δ for every time interval of exposures with $\Delta t = 33.3$ ms.

Figure 5.24 shows the movement decomposition of a swimming fugu based on the constellation of phase singularities. As shown in Figure 5.24a, the linear velocity for translation has a typical periodical structure with an average frequency around 5 Hz, where the solid and dashed line stand for v_x and $v_y (v_x = \Delta x/\Delta t,\ v_y = \Delta y/\Delta t)$, respectively. This rhythmic swimming behavior was produced by the cyclical rotation of fugu's fin to provide the internal force by the pushing of water, which resulted in

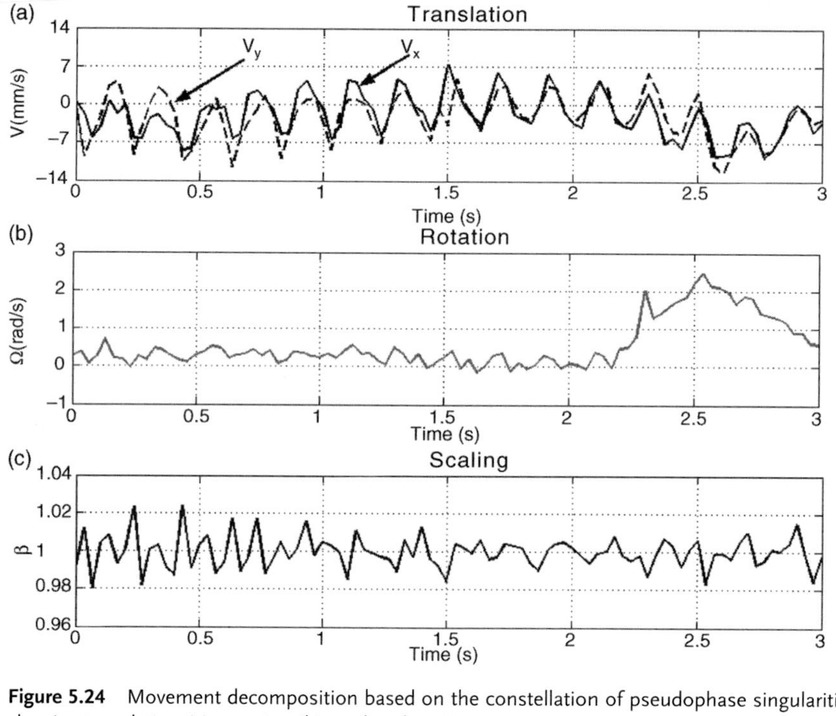

Figure 5.24 Movement decomposition based on the constellation of pseudophase singularities showing translation (a), rotation (b), and scaling (c).

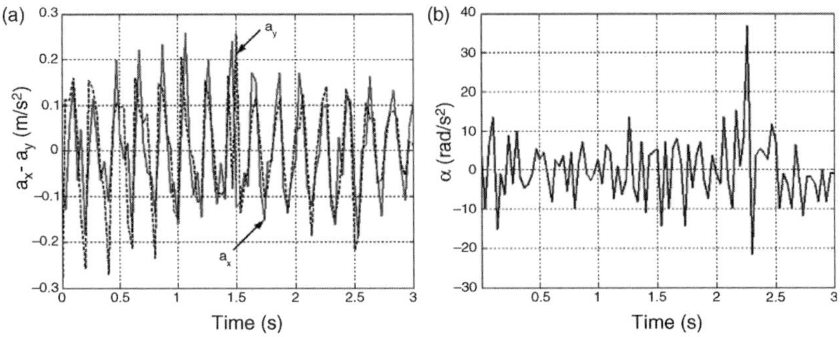

Figure 5.25 History of the sum force (a) and torque (b) applied to the fugu.

the forward-and-back movements. As expected, it is evident from Figure 5.24b that this swimming fugu has a large angular velocity ($\Omega = \Delta\delta/\Delta t$) at time t equal to 2.6 s. Figure 5.24c shows the scale change, which may be attributed to a slight depth change in the water.

After calculating the finite difference for the horizontal and vertical velocity components and the angular velocity, we obtained the variation of the linear and angular accelerations, as shown in Figure 5.25. Under the reasonable assumption that the mass and the moment of inertia for the fugu are constant during the recording process, Figure 5.25a and b provides the history for sum force and torque applied to the fugu at different instants of time, because the linear and angular accelerations are proportional to the force F and torque τ, that is, $F_x \propto a_x = \Delta v_x/\Delta t$, $F_y \propto a_y = \Delta v_y/\Delta t$, and $\tau \propto \alpha = \Delta\Omega/\Delta t$, respectively. As anticipated for the periodical linear velocity, the fugu, which can be regarded as a driven harmonic oscillator, had experienced a sinusoidal-like force to produce an oscillation. From the torque diagram shown in Figure 5.25b, the local torque peak corresponding to a quick change of angular velocity around $t = 2.3$ s can be observed, which stems from the clockwise bending of fugu's tail to give a large rotation.

5.6 Conclusion

In modern optics, the study of phase singularities or optical vortices has become a key area since the seminal work of Nye and Berry in the 1970s. Due to the fact that optical vortex possesses orbital angular momentum stemming from its unique helical phase with an azimuthal component of the optical energy flow, the phase singularities have come to attract new attentions in the fields of application of photon manipulation. In holographic interferometry, for example, scientists are interested in phase singularities because they hinder unique phase unwrapping of the measured phase map. Therefore, a major effort in optical metrology has been concentrated on how to get rid of the effect of phase singularities. In this chapter, we have introduced another approach to the study of phase singularities. Rather than regarding the phase singularities as obstacles in phase unwrapping, we explored the new possibilities of making use of phase singularities in optical metrology.

The optical vortex metrology and conventional speckle metrology share the same philosophy of utilizing the physical phenomena that were initially regarded as obstacles in optical metrology. In early times, speckles were regarded as a nuisance degrading image quality, but now they are well appreciated as a very useful vehicle for optical metrology. Learning from history, we believe that phase singularities can prove very useful in optical metrology although they are commonly regarded as obstacles in connection with phase unwrapping. Just as the speckle is ubiquitous in any experiment involving scattered laser light due to the easy availability of coherent source, the generation of optical vortices is inevitable. Therefore, the proposed optical vortex metrology can serve as an initial trial in the application of phase singularities to optical metrology, while their exploitation for other applications is still in its infancy. Maybe the next decade will see the advanced application of phase singularities to optical metrology.

Acknowledgments

We are indebted to M. R. Dennis and Yu Qiao for their helpful discussion and Tomoaki Yokozeki, Reika Ishijima, Akihiro Matsuda, and Nobuo Ishii for their help in experimental preparation. This work was supported by Grant-in-Aid of Japan Society for the Promotion of Science B (21360028).

References

1 Erf, R.K. (ed.) (1978) *Speckle Metrology*, Academic Press, New York.

2 Francon, M. (1979) *Laser Speckle and Applications in Optics*, Academic Press, New York.

3 Ennos, A.E. (1984) Speckle interferometry, in *Laser Speckle and Related Phenomena*, 2nd edn (ed J.C. Dainty), Springer-Verlag.

4 Sirohi, R.S. (ed.) (1993) *Speckle Metrology*, Marcel Dekker, New York.

5 Fomin, N.A. (1998) *Speckle Photography for Fluid Mechanics Measurements*, Springer-Verlag, Berlin.

6 Wang, W., Ishii, N., Hanson, S.G., Miyamoto, Y., and Takeda, M. (2005) Pseudophase information from the complex analytic signal of speckle fields and its application. Part I: micro-displacement observation based on phase-only correlation in the signal domain. *Appl. Opt.*, **44**, 4909–4915.

7 Wang, W., Ishii, N., Hanson, S.G., Miyamoto, Y., and Takeda, M. (2005) Pseudophase information from the complex analytic signal of speckle fields and its applications. Part II: statistical properties of the analytic signal of a white-light speckle pattern applied to the micro-displacement measurement. *Appl. Opt.*, **44**, 4916–4921.

8 Goodman, J.W. (2006) *Speckle Phenomena in Optics: Theory and Application*, Roberts & Company Publishers, Colorado.

9 Nye, J.F. and Berry, M.V. (1974) Dislocations in wave trains. *Proc. R. Soc. Lond. A*, **366**, 165–190.

10 Soskin, M.S. and Vasnetsov, M.V. (2001) Singular optics, in *Progress in Optics* (ed. E. Wolf), Elsevier, Amsterdam.

11 Dennis, M.R., O'Holleran, K., and Padgett, M.J. (2009) Singular optics: optical vortices and polarization singularities, in *Progress in Optics* (ed. E. Wolf), Elsevier, Amsterdam.

12 Jackson, J.D. (1998) *Classical Electrodynamics*, 2nd edn, John Wiley & Sons, Inc., New York.

13 Robinson, D.W. (1993) Phase unwrapping methods, in *Interferogram Analysis* (eds D.W. Robinson and G.T. Reid), Institute of Physics Pub, pp. 194–229.

14 Takeda, M. (1996) Recent progress in phase unwrapping techniques, in *Optical Inspection and Measurements*, vol. 2782 (ed. C. Gorecki), S.P.I.E., Bellingham, WA, pp. 334–343.

15 Ghiglia, D.C. and Pritt, M.D. (1998) *Two-Dimensional Phase Unwrapping*, John Wiley & Sons, Inc., New York.

16 Nye, J.F. (1999) *Natural Focusing and Fine Structure of Light*, Institute of Physics Pub.

17 Berry, M.V. and Dennis, M.R. (2001) Knotted and linked phase singularities in monochromatic waves. *Proc. R. Soc. Lond. A*, **457**, 141–155.

18 Soskin, M., Denisenko, V., and Egorov, R. (2004) Topological networks of paraxial ellipse speckle fields. *J. Opt. A Pure Appl. Opt.*, **6**, S281–S287.

19 Berry, M.V. and Dennis, M.R. (2000) Phase singularities in isotropic random waves. *Proc. R. Soc. Lond. A*, **456**, 2059–2079.

20 Wang, W., Hanson, S.G., Miyamoto, Y., and Takeda, M. (2005) Experimental investigation of local properties and statistics of optical vortices in random wave fields. *Phys. Rev. Lett.*, **94**, 103902-1–103902-4.

21 Berry, M.V. (1998) Much ado about nothing: optical dislocation lines (phase singularities, zeros, vortices . . .), in *Proceedings of International Conference on Singular Optics*, vol. 3487 (ed. M.S. Soskin), SPIE, Bellingham, WA, pp. 1–5.

22 Born, M. and Wolf, E. (1999) *Principles of Optics*, Cambridge University Press.

23 Dennis, M.R. (2004) Local structure of wave dislocation lines: twist and twirl. *J. Opt. A Pure Appl. Opt.*, **6**, S202–S208.

24 Egorov, Y.A., Fadeyeva, T.A., and Volyar, A.V. (2004) The fine structure of singular beams in crystals: colours and polarization. *J. Opt. A Pure Appl. Opt.*, **6**, S217–S228.

25 Roux, F.S. (2004) Coupling of noncanonical optical vortices. *J. Opt. Soc. Am. B*, **21**, 664–670.

26 Wang, W., Ishii, N., Hanson, S.G., Miyamoto, Y., and Takeda, M. (2005) Phase singularities in analytic signal of white-light speckle pattern with application to micro-displacement measurement. *Opt. Commun.*, **248**, 59–68.

27 Wang, W., Yokozeki, T., Ishijima, R., Wada, A., Hanson, S.G., Miyamoto, Y., and Takeda, M. (2006) Optical vortex metrology for nanometric speckle displacement measurement. *Opt. Express*, **14**, 120–127.

28 Wang, W., Yokozeki, T., Ishijima, R., Hanson, S.G., and Takeda, M. (2006) Optical vortex metrology based on the core structures of phase singularities in Laguerre–Gauss transform of a speckle pattern. *Opt. Express*, **14**, 10195–10206.

29 Wang, W., Dennis, M.R., Ishijima, R., Yokozeki, T., Matsuda, A., Hanson, S.G., and Takeda, M. (2007) Poincaré sphere representation for the anisotropy of phase singularities and its applications to optical vortex metrology for fluid mechanical analysis. *Opt. Express*, **15**, 11008–11019.

30 Qiao, Y., Wang, W., Minematsu, N., Liu, J., Takeda, M., and Tang, X. (2009) A theory of phase singularities for image representation and its applications to object tracking and image matching. *IEEE Trans. Image Process.*, **18**, 2153–2166.

31 Goodman, J.W. (2000) *Statistical Optics*, Wiley–Interscience, New York.

32 Riesz, M. (1927) Sur les fonctions conjuguees. *Math. Zeitschrift*, **27**, 218–244.

33 Larkin, K.G., Bone, D.J., and Oldfield, M.A. (2001) Natural demodulation of two-dimensional fringe pattern. I. General background of the spiral phase quadrature transform. *J. Opt. Soc. Am. A*, **18**, 1862–1870.

34 Guo, C., Han, Y., Xu, J., and Ding, J. (2006) Radial Hilbert transform with Laguerre–Gaussian spatial filters. *Opt. Lett.*, **31**, 1394–1396.

35 Asundi, A. and North, H. (1998) White-light speckle method: current trends. *Opt. Laser Eng.*, **29**, 159–169.

36 Beck, M.S. and Plaskowski, A. (1987) *Cross Correlation Flowmeters: Their Design and Application*, Adam Hilger, Bristol.

37 Dai, X., Sasaki, O., Greivenkamp, J.E., and Suzuki, T. (1995) Measurement of small rotation angles by a parallel interference pattern. *Appl. Opt.*, **34**, 6380–6388.

38 Rose, B., Imam, H., Hanson, S.G., Yura, H.T., and Hansen, R.S. (1998) Laser speckle angular displacement sensor: theoretical and experimental study. *Appl. Opt.*, **37**, 2119–2129.

39 Fomin, N.A. (1998) *Speckle Photography for Fluid Mechanics Measurements*, Springer-Verlag, Berlin.

40 Raffel, M., Willert, C., and Kompenhans, J. (2002) *Particle Image Velocimetry*, Springer-Verlag, Berlin.

41 Benyus, J.M. (1998) *Biomimicry: Innovation Inspired by Nature*, William Morrow & Co., New York.

42 Landau, L.D. and Lifshitz, E.M. (1984) *Theory of Elasticity*, Butterworth-Heinemann, Oxford.

6
Speckle Coding for Optical and Digital Data Security Applications

Arvind Kumar, Madan Singh, and Kehar Singh

6.1
Introduction

Laser speckles have been successfully applied to a large number of applications such as in engineering measurements, optical information processing, biomedical engineering, and so on. Engineering measurements by speckle techniques are important because they are noncontact and noninvasive. Detailed information on speckles and their numerous applications can be found in Dainty [1, 2], Erf [3], Francon [4], Rastogi [5, 6], Sirohi [7, 8], and Goodman [9, 10], to name a few sources.

Laser speckles also play an important role in optical information processing that has been developed and established as an important area of optics. It has shown great potential for useful manipulation of data and images for their storage, security, and transfer applications. Speckles play an important role in optical encryption that is one of the techniques used for safe data and image transfer. Encryption can be defined as the transformation of data/image that is as close to impossible as possible to read without the appropriate knowledge referred to as key to the encrypted information. People with the span of time and requirement have investigated and applied different methods for secure storage and secret communication. Toward this end, traditional methods have relied on the principle of cryptography [11]. Thus, cryptography as the science of securing information is one of the tools that ensures security, integrity, and authentication of the electronic data.

In the present era, which is often called the "information age," data security is of paramount importance in many application areas such as communication, e-Commerce, computer network, software and hardware, banknotes anticounterfeits, intellectual property and copyright protection, personal identification, authentication, and verification, to name a few. Data security has become of utmost concern following a sharp rise in the rate at which the information is being disseminated. An area where security is of utmost concern is the personal identification where the demand is for fast and reliable identification of the people through the verification of their documents. The identification of passports, passwords, credit cards, and personal identification number is often required. As these proofs of identity have

Advances in Speckle Metrology and Related Techniques. Edited by Guillermo H. Kaufmann

ISBN: 978-3-527-40957-0

become increasingly necessary, they have also increased the opportunity for deception. In a similar manner, the information in the form of images and communication signals faces threat of interception on the way. The property documents, wealth materials, money transfer through credit cards, data/information about the security forces, or secret documents face the problem of manipulation/fraud.

Due to the high profit/gain involved in the sale of counterfeit/duplicate products, a technology war is going on between the developers of data security system and counterfeiters. The optical information fraud has become increasingly common due to the rapid technological advancements in computers, charge-coupled devices (CCDs), printers, scanners, copiers, and image processing software. Nowadays, it has become possible to produce authentic-looking counterfeits of portraits, logos, symbols, money bills, and other complex patterns. As a result, the counterfeit goods such as computer chips and machine tools are now arriving in the market in greater numbers. Even the anticounterfeiting techniques that represent a high state of the art have become more vulnerable, and anyone can counterfeit a product that can pass casual inspection by the average person.

Due to the increasing threat of terrorism worldwide, security and defense are receiving increased attention. Inspection requirements in public places, ports of entry, and transportation hubs around the globe have increased. So the optical images, patterns, and information to be sent from one place to another need a coding device. There is a need to keep the record in such a manner [11] that it ensures the following:

- **Confidentiality**: ensures that information either transmitted or stored is accessible to an authorized person.
- **Authentication**: ensures that origin of the information is correctly identified with an assurance that the identity is not false.
- **Access control**: requires that access to information resources may be limited to authorized persons only.

As is well known, the security of cryptosystems is enhanced by using more powerful algorithm and large key length to encode data, which in turn requires more computing time and power. When large amount of data are to be encoded, these requirements may turn out to be the system's bottleneck. To make the work of counterfeiters more difficult, many techniques are being explored and developed. Scientists have explored various ways of using commercially available optical systems attractive for data security applications. First, optics offers many degrees of freedom to encode data more securely:

- Information can be hidden in any of several dimensions, such as the phase, wavelength, spatial frequency, polarization, or orbital angular momentum of the light.
- Optical signals are two dimensional (2D), whereas electrical signals are one dimensional. Moreover, most of the electronic processors are limited to serial processing. However, optical systems can process and transmit large amount of information in parallel.

- Each pixel of an optical system can represent a complex value. Modulating the amplitude and phase of light using a spatial light modulator (SLM) may represent a complex number.
- To break an optical security system, a hacker needs to employ sophisticated and expensive optical devices, systems, and techniques, which are less accessible and flexible compared to the digital techniques.

All these properties may substantially increase the number of possibilities the would-be code breaker must consider when faced with these new optical encryption techniques.

In an encryption system, we wish to encode information in such a fashion that only application of correct key will reveal the original information. Modern encryption systems are basically based upon a combination of optical and electronic systems, and therefore called "hybrid systems.". Such a system uses CCDs, spatial light modulators, computers, and optical setup for encryption and decryption. Digital techniques of encryption are falling short of expectations due to the fact that when the encryption key chain length becomes longer, the speed goes down. On the other hand, optical methods process information in a parallel way, and are therefore inherently faster. Over the past few years, some researchers have explored various optical/optoelectronic methods for data encryption due to its above-mentioned advantages. Scientists have explored various ways of using commercially available optical systems attractive for data security applications. In addition to the already mentioned advantages, the optical systems also have the following advantages.

- Better security is possible because of the larger key length.
- Due to the use of photorefractive and photopolymers as recording materials, desired capacity of the memory is possible.
- It is advantageous to implement a system to securing the data with biometrics like fingerprints and facial features.

Since 1995, the time when Refregier and Javidi [12] investigated a technique of optical encryption by using two random phase masks (RPMs), one in the input plane and another in the Fourier plane of a 4-f imaging setup, the scientists all over the world have done considerable work on the optical and digital encryption. Two books edited by Javidi [13, 14], review articles by Matoba and Javidi [15], Unnikrishnan and Singh [16, 17], Singh and Nishchal [18], Singh *et al.* [19], Hennelly and Sheridan [20, 21], Alfalou and Brosseau [22], Matoba *et al.* [23], and chapters by Tajahuerce *et al.* [24, 25], Chalasinska–Macukow and Kotynski [26], Rosen [27], Kim *et al.* [28], Kishk and Javidi [29], Marvel [30], Matoba and Javidi [31], Naughton and Javidi [32], Nomura and Javidi [33, 34], Nomura *et al.* [35], Tanida and Sasaki [36], Rosen and Javidi [37], Nomura [38], Choi *et al.* [39], Shortt *et al.* [40], Frauel *et al.* [41], Sinzinger *et al.* [42], Cuadrado-Laborde [43], and Gluckstad and Palima [44] have been published on the subject. Apart from these sources, there are several national and international conference proceedings containing a large number of relevant papers.

In the present chapter, an attempt has been made to describe certain speckle coding techniques for optical and digital data security applications, with an emphasis on the work carried out by the authors' group. A review is first given of various optical encryption and decryption methods. The topics covered in the literature are summarized below, though the subject classification is an extremely difficult task due to cross-linkages between the various topics involved.

6.2 Double Random Fourier Plane Encoding

In a simplified technique, it is sufficient to use a single random pattern at the Fourier transform (FT) plane of a 4-f optical system to encrypt the data. But to make the optical data more secure, double random patterns may be used. In the double random phase encoding (DRPE) method (Figure 6.1), two statistically independent random phase masks are used for encryption, one in the input plane and another in the Fourier plane of a 4-f imaging setup. In this way, the system encrypts a primary image to a white noise by using these two RPMs [12, 13, 45–48]. The RPMs serve as the keys to the encrypted data. Decryption can be done by using either the conjugate of the Fourier plane mask or the phase conjugation of the signal. If the phase conjugation technique is used to decrypt the image, one needs to use the same RPM as used during encryption to decode the data successfully. Figure 6.2 shows a view of the images taken in successive steps of encryption and decryption processes. The following steps are involved in the process.

We denote the input image to be encoded by a function $f(x)$. $R_1(x)$ and $R_2(u)$ are two statistically independent RPMs, with their phases being uniformly distributed over $[0,2\pi]$.

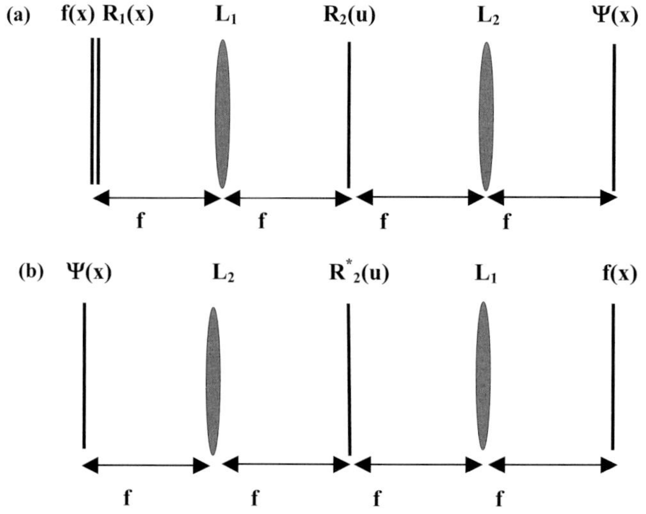

Figure 6.1 Optical setup for encryption and decryption. (a) Encryption. (b) Decryption. $R_1(x)$ and $R_2(u)$, RPMs; L_1, L_2, lenses; $f(x)$, input image; $\Psi(x)$, encrypted image; $f(x)$, decrypted image.

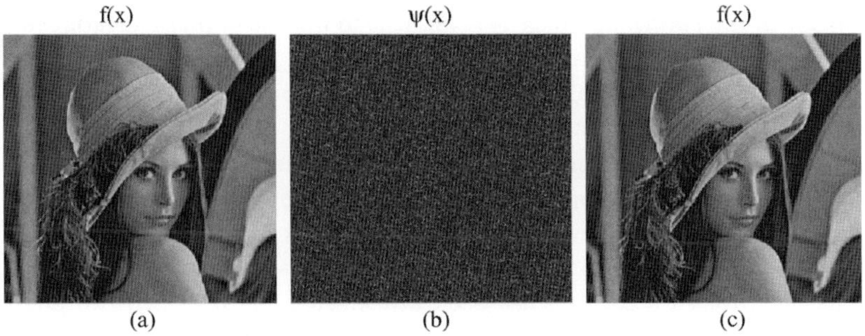

Figure 6.2 A view of images taken at different stages of encryption and decryption. (a) Input image $f(x)$. (b) Encrypted image $\Psi(x)$. (c) Decrypted image $f(x)$.

Here x denotes the spatial and u denotes the Fourier domain coordinates. For simplicity, the analysis has been presented in one dimension. An extension to 2D is straightforward. First, the input image $f(x)$ is multiplied by the first RPM $R_1(x)$ denoted as $\exp[i2\pi R_1(x)]$ to convert the input image into white noise but not encrypted. This complex function $f(x)\exp[i2\pi R_1(x)]$ is filtered through the second RPM $R_2(u)$ denoted as $\exp[i2\pi R_2(u)]$, changing it into the stationary white noise. The RPMs $\exp[i2\pi R_1(x)]$ and $\exp[i2\pi R_2(u)]$ are placed respectively in the input and the Fourier planes. The encrypted image obtained in the output plane is given by

$$\Psi(x) = \{f(x)\exp[i2\pi R_1(x)]\}^* h(x), \qquad (6.1)$$

where $h(x)$ is the impulse response of the phase-only filter $\exp[i2\pi\ R_2(u)]$ and * denotes the convolution operation. It is possible to show that $\psi(x)$ is a stationary white noise [35].

The decryption can be done in two ways. In one way, the decoding is done by multiplying the FT of $\psi(x)$ with $\exp[-i2\pi R_2(u)]$ and by performing a subsequent inverse Fourier transform (IFT) to get $f(x)\exp[i2\pi\ R_1(x)]$ that is multiplied by $\exp[-i2\pi R_1(x)]$ to obtain the decoded image $f(x)$ as

$$f(x) = \mathrm{IFT}\{\mathrm{FT}[\Psi(x)] \times \mathrm{conj}(R_2(u))\} \times \exp[-i2\pi R_1(x)]. \qquad (6.2)$$

The successful encryption is possible only with the right key $\exp[-i2\pi R_2(u)]$. Recording the result by using a CCD camera gives the decrypted image. In such type of recording, we do not need to multiply with $\exp[-i2\pi R_1(x)]$ as a CCD camera is an intensity detector.

Another way would be to start with the conjugate of the encrypted image, obtain one FT, then multiply with the same RPM kept at the Fourier plane, and then obtain IFT to get the decrypted image as

$$f(x) = \mathrm{IFT}\{\mathrm{FT}[\mathrm{conj}\{\Psi(x)\}] \times R_2(u)\}. \qquad (6.3)$$

Hence, in this case, the conjugate of the RPM is not required, but the conjugate of the encrypted image is required. Phase conjugation technique has been used to generate the conjugate of the encrypted image [49–58]. Due to the use of phase conjugation, the phase introduced during the encryption is canceled if the same RPMs are used at the respective locations. Also, the phase conjugation corrects other phase distortions, such as defects of the optical components.

Investigations on the DRPE and its variants have continued at a fast pace. There is a vast body of literature on the subject, and it is not possible to present a detailed review. In the following, we review very briefly some of the work.

Javidi *et al.* [59] have presented results in case of binarized encrypted images. A cascaded iterative FT algorithm has been investigated [60] in which two phase masks are designed and located in the input and the Fourier domain of a 4-f correlator. Nomura *et al.* [61] have proposed a design of an input phase mask that is iteratively designed so that the extent of the Fourier spectrum of the product of an input image and the input phase mask corresponds to the space bandwidth of the optical system. A method has been proposed [62] to simultaneously transmit DRPE key and an encrypted image by making use of the fact that an acceptable decryption result can be obtained when only partial data of the encrypted image is taken in the decryption. The authors have analyzed the security of amplitude encoding for the DRPE. A system discussed by Frauel *et al.* [63] is found to be resistant to brute force attacks, but vulnerable to chosen- and known-plaintext attacks.

Barrera *et al.* [64] have implemented a method using computer-generated toroidal zone plates. A numerical analysis of the DRPE technique has been carried out [65] to determine how, in the case of amplitude and phase encoding, the two decryption keys affect the output grayscale image when they are in error. Amaya *et al.* [66] investigated a multichanneled puzzle-like encryption method to increase the data security in transmission. Each decomposed part of the input object is encrypted separately in a 4-f DRPE architecture by setting the optical parameters in a determined status, each parameter set defining a channel. Situ *et al.* [67] have investigated the collision in the DRPE. Collision is a situation that occurs when two or more distinct inputs in a security system produce identical outputs.

DRPE has also been used for encryption by using Jigsaw transform [68], fully phase encryption [64], and matrix-added or matrix-multiplied input images [69]. Kumar *et al.* [70] have introduced RPMs in front of lenses in a 4-f setup in the conventional DRPE system to protect the system from the impulse attack. The random function is obtained at the Fourier plane in this technique that makes the impulse attack ineffective. A scheme has been investigated [71] in which a modified Fourier transform (MFT) is obtained by the randomization of the lens phase function and provides immunity against the impulse function attack. Barrera and Torroba [72] have explored a way to perform an undercover operation where one channel (amplitude) is used to depict decoy information to confuse intruders, while the other (phase) operates with the true information. They have used Gerchberg–Saxton algorithm to transform the amplitude and phase-encrypted information into pure phase data.

Monaghan *et al.* [73] have examined the effect of quantization on the decryption process due to the discrete values and quantized levels, which an SLM can physically display. Investigations have also been made [74] on the effect of introducing error into the decryption phase keys in the form of a pseudorandom distribution of incorrect pixels values. By quantifying the resulting error in the attempted decryption, for increasing amount of error in the decrypting phase keys, the authors investigated the effects of reducing the phase range. The DRPE technique has also been used for color objects, as would be discussed in Section 6.3.2.

6.2.1 Influence of Coded Image Perturbations, Noise Robustness, and SNR

In the DRPE, if the RPMs are not band limited, the optical reconstruction of the encrypted images is degraded by speckle noise. This aspect has been studied by Neto and Sheng [75]. Javidi *et al.* [76] have tested an optical system for security verification in the presence of different types of input noises and distortions such as scratches, fingerprint marks, and bending of the phase masks. The results show that encrypting using RPMs of limited number of pixels can provide well-encrypted images. It has also been shown [47] that the decoding operation of a DRPE system conserves noise, and the energy of perturbation in the decoded image is equal to that of the perturbation in the coded image. The distortions considered include different types of noise, loss of encrypted data, and the binarization of the encrypted image [59]. A more detailed analysis of the robustness to perturbation of the encoded image has also been carried out [77].

The robustness to noise has been compared with that of an XOR encryption method by Javidi *et al.* [78]. It has been shown that the DRPE method produces a lower mean square error (MSE) compared to the XOR encryption method. If the input image is phase encoded, the security is increased compared to that in the amplitude encryption. The fully phase encryption performs better than the amplitude-based encryption in the presence of additive noise with respect to the MSE [79]. Javidi *et al.* [80] have proposed a thresholding method of decryption that reduces MSE for the fully phase and amplitude-based encryption processes. The role of phase keys in the DRPE has been discussed by Monaghan *et al.* [65] who also carried out an error analysis.

6.3 Variants of the DRPE and Various Other Encryption Techniques

6.3.1 Fresnel and Fractional Fourier Transform Domain Encoding

Degrees of freedom used in the encoding can be increased by using three-dimensional (3D) positions of the RPMs in the DRPE method. When the RPMs are located at the Fresnel planes, the technique is called "Fresnel domain encryption" [81–84]. It is also interesting to utilize the gap between the space and frequency domains in optics for information processing. A continuum of domains referred to as "fractional Fourier domains" exists between the space and frequency domains. Several authors have described the systems based on the FRT, using fractional orders as additional degrees of freedom [16, 17, 20, 21, 50, 51, 85–90].

6.3.2 Color Image Encoding and Digital Simulation/Virtual Optics-Based Techniques

Color image encryption has been done by utilizing Fresnel, Fourier, fractional Fourier, and some other techniques of encoding [91–93]. Digital techniques-based image encryption systems use computers for encryption and decryption, and are known as "virtual optics systems". The security of computer based cryptosystems is enhanced by using more powerful algorithms and large key length to encode data. Therefore, the security and imperceptibility levels are dramatically increased [66, 94–99].

6.3.3 Phase Retrieval- and Polarization-Based Techniques

Use of phase retrieval techniques has opened a new way in security applications by designing phase masks on the basis of iterative phase retrieval algorithms [60, 100, 101]. In polarization encoding method, a grayscale image pattern is bonded to a polarization encoding mask, which consists of randomly oriented linear polarizers rotated at various angles varying between 0 and π. Mogenson and Gluckstad [102] have demonstrated a system based on the use of polarization-sensitive phase elements. A system for encrypting a binary image was demonstrated by Unnikrishnan *et al.* [103]. Barrera *et al.* [55] have presented results of the feasibility of using multiplexing, employing polarized light. Methods based on polarization encoding using speckle pattern correlation have also been described [104, 105].

6.3.4 Interference and Joint Transform Correlator Architecture-Based Techniques

Optical decryption can subsequently be implemented in the common path interferometer by use of single phase or if desired a combined phase/amplitude key. In a technique described by Zhang Yan and Wang [106], the image is encoded into two phase masks based on the optical interference. If two coherent light beams are modulated by the two phase-only masks and then are combined by a half mirror, the two beams interfere with each other at the output plane resulting the generation of the image. Wang Bo and Zhang [107] have presented a technique in which double image hiding is done based on interference. Han and Zhang [108] have presented optical image encryption based on two beam interference. Joint transform correlator (JTC) architecture has also been extensively used for optical security and validation [66, 109–111].

6.3.5 Fully Phase Encryption Techniques and Encrypted Holographic Memory

In the ideal case, transparent details of a phase object leave the intensity of the passing light unchanged, and an intensity detector cannot detect the information content of a phase-encoded image. If the encryption is performed using the input image as the phase image in a DRPE method, the security is enhanced compared to the encryption by using the amplitude image [51, 52, 74, 79, 112, 113]. Various encryption techniques have also been combined with numerous holographic storage methods for secure holographic memories [114–118].

6.4 Attacks against Random Encoding

Vulnerability of various encryption systems has now become a subject of increased activity. Carnicer *et al.* [119] have shown how methods based on DRPE keys are vulnerable to an organized attack of the chosen-ciphertext type. A method has been proposed [120] of chosen-plaintext attack on lensless DRPE in the Fresnel domain, by which an opponent can access two encryption keys with the help of the impulse functions as chosen plaintexts. In the approach by Peng *et al.* [121], an opponent can access both random keys with the help of the phase retrieval technique. A simulated annealing heuristic algorithm was used by Gopinathan *et al.* [122] to estimate the key, using a known-plaintext–ciphertext pair, which decrypts the ciphertext with arbitrary low error. Monaghan *et al.* [123] have checked the security of encryption and verification techniques with significant output images by known-plaintext attacks. Frauel *et al.* [63] have proposed several attacks against the DRP encryption scheme. Situ *et al.* [124] have investigated security of the encryption and verification techniques with significant output image by a known-plaintext attack. Naughton *et al.* [125] have considered all attacks proposed to date for DRPE.

6.5 Speckle Coding for Optical and Digital Data Security

As mentioned in Section 6.1, the speckles have been found as excellent information carriers. These have been used as a coding tool for storing data. Recently, the phase masks made by using speckle patterns have been used for data security applications [54, 55, 57, 126, 127]. Sun *et al.* [128] have used lateral shifting of a ground glass as a tool for holographic encryption and multiplexing. The effects of 3D shifting selectivity of volume hologram based on random phase encoding with a ground glass have also been investigated [116].

The shifting selectivity, being a key in phase encoding, can be used for holographic multiplexing and encryption. Barrera *et al.* [54] described a holographic memory system in which encrypted image of multiple images is recorded by successive different lateral shifting of RPM. In this process the successive images are encrypted and position encoded by speckle pattern arising from the RPM in-plane shifting between exposures. Each time the random pure phase mask kept at the FT plane is moved, a new speckle pattern is generated and recorded in the crystal volume. The situation is similar to introducing a general speckle pattern decorrelation, which occurs when the mask displacement reaches the average speckle size. The pattern that represents the encrypted information is essentially a speckle pattern. Polarized light has also been used for multiplexing encrypted data [55]. When storing each encrypted image, a polarization change is introduced in the system. This introduces decorrelation on the speckle pattern inside the storing medium. This approach has been used for multiple image encryption.

Barrera *et al.* [56] have investigated an architecture for multiple image encryption by using pupil aperture in the optical system. In this architecture, multiple image encryption is achieved by changing the pupil aperture mask in the recording lens between the successive pair of exposures to encrypt different images. So, in addition to the classical speckle phase mask, the geometrical parameter characterizing the apertures is introduced to increase the system security. In this scheme, random phase masks are located in the input and in the Fresnel region of the first lens, respectively. The crystal is kept in the imaging plane of second random phase mask. An additional third phase mask is attached with the pupil aperture of second lens and used as a key for a particular position of the pupil aperture. A pupil aperture system consisting of a pupil aperture along with the third phase mask serves as a tool to record multiple encrypted images. This technique increases the security of the image encryption without any degradation of its noise robustness with addition of degree of freedom provided by the aperture as an additional key. Matoba *et al.* [126] have used a phase mask made with speckle patterns for multiplexing by in-plane shifting the phase mask.

Sandwich specklegrams and holograms have earlier been used in nondestructive testing [129]. Due to their properties, the sandwich specklegrams may be used in optical security architectures. A sandwich phase diffuser is made by using two random phase diffusers sandwiching them together. Due to double diffraction, the new single sandwich diffuser (SD) offers enhanced security to the system. Singh and Kumar [57] have used a sandwich speckle phase mask for encryption. Fourier plane encoding with sandwich diffuser does the encryption. Experimental method has been described to overcome the practical difficulty for the generation of the key in case of using a sandwich diffuser. However, the alignment of the two diffusers constituting a single sandwich diffuser is quite difficult and cumbersome. A technique has been presented by us [57] to overcome this problem.

6.6 Encryption Using a Sandwich Phase Mask Made of Normal Speckle Patterns

We describe the encryption and decryption of a 2D image by using a sandwich diffuser, where Fourier plane encoding with sandwich diffuser results in the encryption. The encrypted image is holographically recorded in a PR $BaTiO_3$ crystal, and a phase conjugate of the encrypted image is then used for decryption.

6.6.1 Theoretical Analysis

Figure 6.3 shows the schematic diagram of a 4-f system used for encryption and decryption. Let (x, y) and (u, v) denote respectively the coordinates in the object

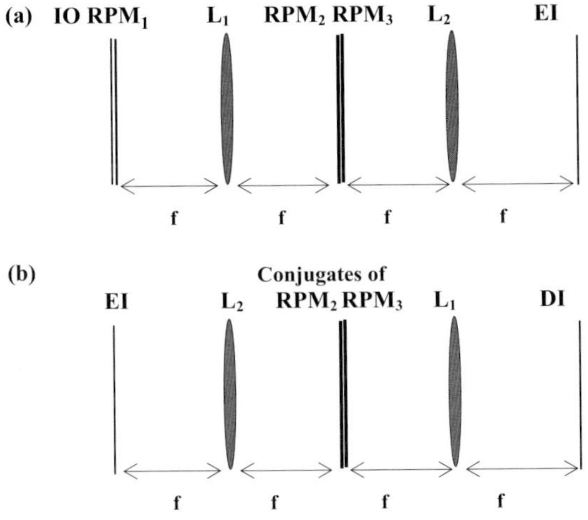

Figure 6.3 Schematic diagram, (a) encryption, and (b) decryption setup. RPM_1, RPM_2, and RPM_3, random phase masks; L_1, L_2, lenses; IO, input image; EI, encrypted image; DI, decrypted image.

and FT. Let $R_2(x, y)$ and $R_3(x, y)$ denote the inverse Fourier transforms of $R_2(u, v)$ and $R_3(u, v)$, respectively. The functions $R_2(u, v)$ and $R_3(u, v)$ are chosen to be phase functions denoted as $\exp[i\phi_2(u, v)]$ and $\exp[i\phi_3(u, v)]$, respectively, uniformly distributed in the interval $[0, 2\pi]$. Note that $R_2(x, y)$ and $R_3(x, y)$ are the impulse responses of the phase-only transfer functions $R_2(u, v)$ and $R_3(u, v)$ and thus provide stationary white noise.

The real-valued function $f(x, y)$ denotes the primary 2D image to be encrypted and $\Psi(x, y)$ denotes the encrypted image. The encryption of the input image $f(x, y)$ is done in two steps. First, $f(x, y)$ is multiplied by an RPM $R_1(x, y)$ at the input plane. This product is then convolved with $h(x, y)$, which is the impulse response of the phase-only transfer function. Here, $h(x, y)$ may be defined as impulse response of $R_2(u, v)$ and $R_3(u, v)$ combined. Alternatively the FT of the modified input image, that is, the multiplication of $f(x, y)$ and $R_1(x, y)$, is filtered by a combination of RPMs $R_2(u, v)$ and $R_3(u, v)$ in the Fourier plane.

These RPMs $R_1(x, y)$, $R_2(u, v)$, and $R_3(u, v)$ are chosen to be statistically independent. The image modified due to these RPMs is given by

$$\Psi(x, y) = \mathrm{FT}\{\mathrm{FT}[f(x, y) \times R_1(x, y)] \times [R_2(u, v) \times R_3(u, v)]\}. \tag{6.4}$$

It can be shown that the encrypted function in wide sense is a stationary white noise. The encrypted image is recorded in a holographic recording material such as a PR crystal.

Decryption can be carried out as follows:

A phase conjugate of the encrypted image is generated, Fourier transformed, and then multiplied by the RPMs $R_2(u, v)$ and $R_3(u, v)$. The resulting function is inverse Fourier transformed, thus giving the decrypted image. Because the detector at the recording plane is an intensity detector, this nullifies the phase introduced by $R_1(x, y)$:

$$f(x, y) = \text{IFT}\{\text{FT}[\text{Conj}\{\Psi(x, y)\}] \times [R_2(u, v) \times R_3(u, v)]\}. \tag{6.5}$$

If the sandwich RPM $\{R_2(u, v) \times R_3(u, v)\}$ is not used or used in a shifted position, it is almost impossible to retrieve the original data. The shift of the sandwiched RPM either in *X*- or *Y*-direction would be tolerable depending upon the correlation length of the RPM. If $R_2(u, v)$ is shifted by a few microns in the *X*-direction with respect to $R_3(u, v)$, it becomes impossible to retrieve the image.

6.6.2
Description of the Experimental Work

In this section, a method is described for making the phase diffusers by using speckle patterns as well as the gelatin granules. The phase diffusers are then combined to constitute a sandwich diffuser to be used in optical encryption and decryption. A technique is also described for making the alignment of two constituent phase diffusers along with the experiment results on encryption and decryption.

6.6.2.1 Preparation of Speckle Phase Masks

The RPMs R_1, R_2, and R_3 can be made in many ways: one method is by exposing a silver halide holographic plate with fully developed speckles generated by illuminating a strong diffuser with a coherent beam of light. Figure 6.4 shows three different normal speckle patterns recorded on three holographic plates separately. Using free-space geometry, the speckle size can be controlled either by controlling the illuminated area of the strong diffuser or by changing the distance between the diffuser and the recording plate.

The plates so exposed were processed (developed, fixed, and bleached) to get random speckle phase masks for use as random phase diffusers. RPMs R_1, R_2, and R_3 may contain different speckle patterns, or R_2 and R_3 may be similar depending upon the choice. In our case, we make R_2 and R_3 as identical phase diffusers by recording the same speckle pattern on the two plates separately.

Another method is by using gelatine granules. The gelatine granules are dissolved in hot water. The paste so formed is spread on to an optically flat transparent glass plate, and then a ground glass is pressed against the plate to impress its surface irregularities on the gelatine layer. The phase diffusers so made work well as the random phase diffusers. The structures of the random phase patterns obtained with a

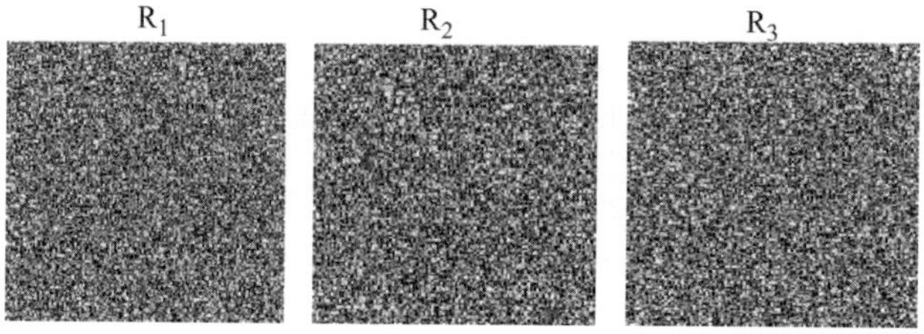

Figure 6.4 Structure of the random speckle patterns R_1, R_2, and R_3, as seen before bleaching the silver halide photographic plate.

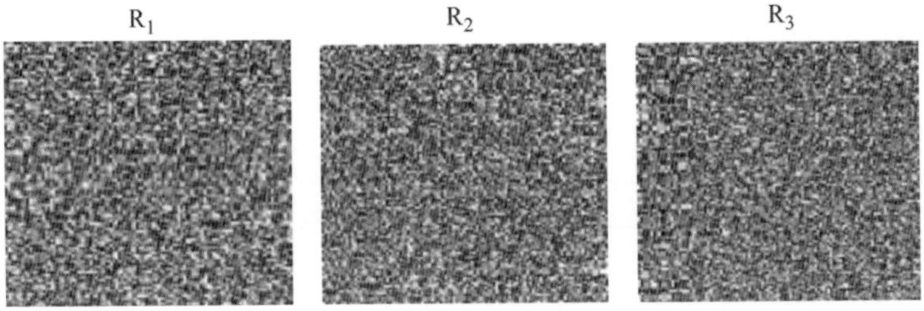

Figure 6.5 Structure of the random phase patterns as seen through phase contrast technique. R_1, R_2, and R_3 are fabricated with a thin film of gelatin on glass plates (grain size approx. 35 μm).

thin film of gelatin and seen through phase contrast technique are shown in Figure 6.5.

6.6.2.2 Making a Sandwich Phase Mask

Before their use in the encryption setup, the RPMs R_2 and R_3 are sandwiched together after giving a known in-plane shift to one of them. If the small marked portion of each constituent phase masks R_2 and R_3 contains the identical speckle patterns, the sandwich diffuser will generate a set of Young's fringes. At this position, the two constituent phase diffusers may be sandwiched together. Figure 6.6 shows the fringes generated respectively for 40 and 60 μm resultant in-plane shifts of one of the constituent diffusers of a sandwich diffuser in the X–Y plane.

6.6.2.3 Technique for Easy Alignment of the Constituent Speckle Phase Masks

As mentioned earlier, the alignment of two constituent phase diffusers of the sandwich diffuser is a difficult task. To overcome this difficulty, we make two identical phase diffusers R_2 and R_3 by exposing two holographic plates with identical speckle patterns. Another way is to expose same marked small areas of each recording plate with the identical speckle patterns. The rest of the areas may contain different speckle patterns. The small identical areas of these two plates containing the identical speckle patterns may not be used for coding, whereas the rest can be used for coding the information. If one of the constituent diffusers is kept at an in-plane shifted position with respect to other, Young's fringes are generated at the Fourier plane of a

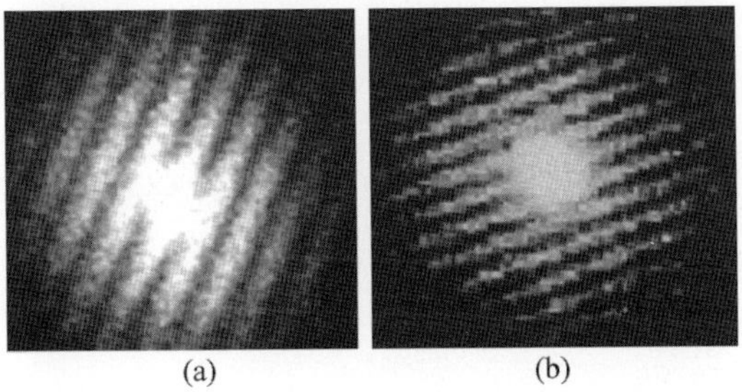

Figure 6.6 Fringes generated by a sandwich diffuser for (a) 40 μm and (b) 60 μm in-plane shift of R_2 with respect to R_3.

sandwich diffuser after illuminating with the coherent light. After the recording, if these two diffusers making a sandwich diffuser are separated, it becomes impossible to decrypt the image. For decryption, these two diffusers should be brought in the same position and orientation as they were during the recording, and then sandwiched together. In this condition, the sandwich diffuser generates the Young's fringes with the same orientation and width as it generated before separating the two constituent diffusers. However, in this case, the sandwich diffuser will generate the fringes with same orientation and width at two positions, that is, for $+d$ and $-d$ shifts. This ambiguity can be removed by making a third identical phase diffuser containing the same small marked area, and either of the two diffusers can be shifted in orthogonal directions. If either of the two identical diffusers making a sandwich diffuser is shifted in its plane with a different value as taken before recording, the sandwich diffuser will generate fringes of different width when illuminated with a parallel beam of light, and will act as a wrong key.

6.6.2.4 Experimental Results

The setup used is shown in Figure 6.7. A diode-pumped Nd:YAG laser (90 mW, Coherent Model DPSS Mini YAG, $\lambda = 532$ nm) is used as the coherent light source. A spatially filtered and expanded beam of light is incident on a given pattern at the object plane. In a 4-f geometry, lenses L_1 and L_2 are of focal length 150 mm each, and the PR crystal used is a C-cut $BaTiO_3$ ($10 \times 10 \times 5\ mm^3$). To record the encrypted image, we closed the shutter used in the readout beam so that the readout beam would not erase the recorded hologram.

The angle between the two beams was kept at 20° for an optimum SNR. After recording the encrypted image, we separated the two diffusers of the sandwich diffuser. To readout the encrypted image, one has to align the two diffusers R_2 and R_3

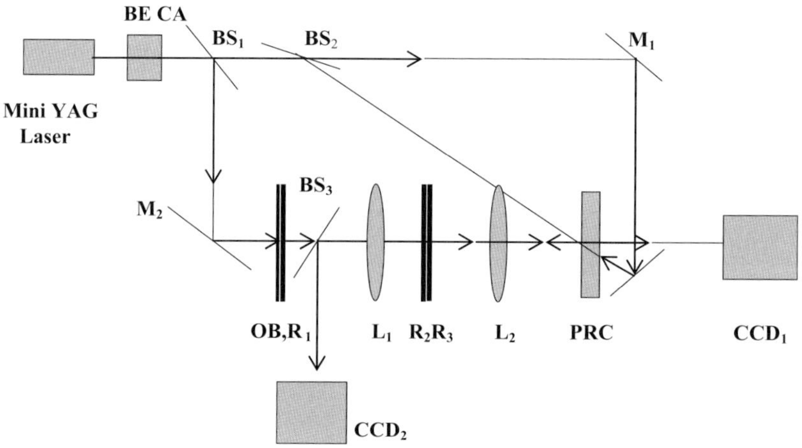

Figure 6.7 Experimental setup. BECA, beam expander collimator assembly; BSs, beam splitters; R_1R_2, sandwich diffuser; OB, input object; PRC, photorefractive crystal; CCDs, charge-coupled devices; Ls, lenses; Ms, mirrors; Rs, random phase masks.

constituting the sandwich in the same position and orientation in which they were during the recording. We can record multiple images using angular multiplexing in the same crystal. A CCD can capture the decoded amplitude image. For the first experiment, the RPMs were prepared by using gelatin granules. The image to be encrypted used for the study is shown in Figure 6.8a. The encrypted image formed at the PR crystal plane is imaged onto CCD_1 and is shown in Figure 6.8b. Figure 6.8c shows the decrypted image as collected by CCD_2. Figure 6.8d shows the decrypted image after using wrong key (phase mask at the displaced position).

For the second experiment, we prepared the RPMs R_1, R_2, and R_3 by exposing holographic plates with fully developed speckle patterns and then by processing the plates. The sandwich phase mask is made by using R_2 and R_3 with R_2 shifted in-plane by a value of 60 μm with respect to R_3. The image to be encrypted is shown in Figure 6.9a. The encrypted image formed at the PR crystal plane is imaged onto CCD_1 (Figure 6.9b). Figure 6.9c shows the decrypted amplitude image obtained with the right key, whereas Figure 6.9d shows the decrypted image if either of the two identical diffusers making a sandwich diffuser is shifted in-plane with a value different from that taken before recording. In this situation, the sandwich diffuser will generate fringes of different width or will not form fringes (if the shift is large enough) when illuminated with a parallel beam of light and will act as wrong key. In the conventional amplitude-based encryption techniques, of the two independent RPMs, only the Fourier domain RPM constitutes the key. However, in our demonstration for decryption, after generating conjugate of the encrypted image, both RPMs R_2 and R_3 constitute the key and have to be sandwiched together after matching with the previous positions as during the encryption. This is an essential requirement to retrieve the image. Hence, both the random codes $R_2(u, v)$ and $R_3(u, v)$ constitute the key for successful retrieval of the original data.

6.6.2.5 Computer Simulation

The simulation study was carried out on a MATLAB platform. The values of physical parameters are not the same as those used in the experiment. Amplitude data, a binary image, and a grayscale image of size (256 × 256) pixels (Figures 6.10a and 6.11a) were chosen for the study. The original image is multiplied by an RPM and is Fourier transformed. The Fourier transformed image is multiplied by the sandwich RPM. This

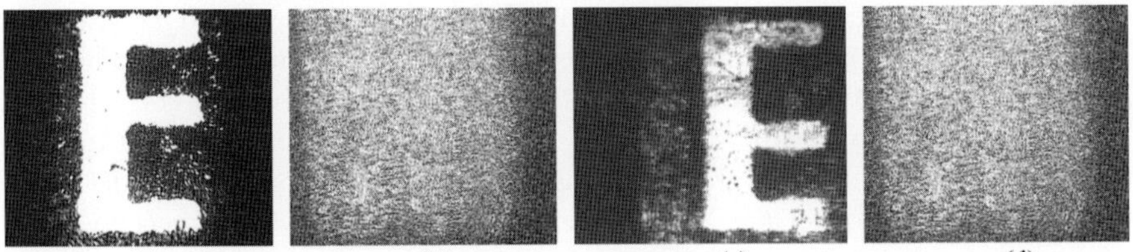

(a) (b) (c) (d)

Figure 6.8 Experimental results when the random diffusers R_1, R_2, and R_3 are made by using gelatine film. (a) Amplitude image to be encrypted. (b) Encrypted image. (c) Decrypted image with both phase masks at the initial positions (correct key). (d) Decrypted image when both the phase masks are displaced by 50 μm from their initial positions (wrong key).

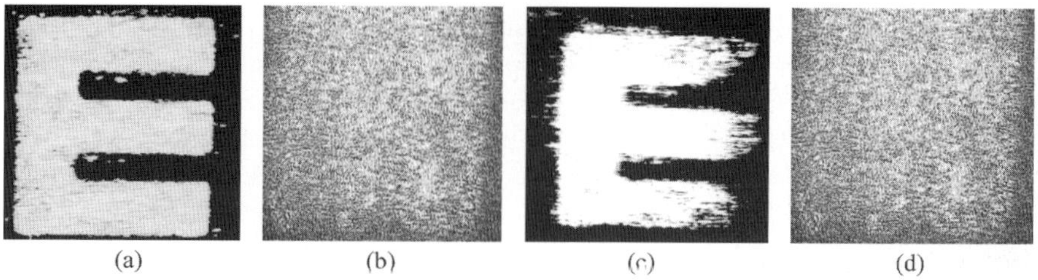

Figure 6.9 Experimental results when the random diffusers R_1, R_2, and R_3 are made photographically. (a) Original amplitude image to be encrypted. (b) Encrypted image. (c) Decrypted image with both identical diffusers R_2 and R_3 at the initial positions, that is, generating fringe with same orientation and width (correct key). (d) Decrypted image when both identical diffusers are displaced by 50 μm from their initial positions (wrong key).

resultant expression is now inverse Fourier transformed to get the encrypted images (Figures 6.10b and 6.11b) separately in case of a binary and a grayscale image. The random phase codes are generated from a random number generator in the MATLAB, the range of variation of phase encoding is $[0, 2\pi]$. For decryption, the encrypted image is Fourier transformed and multiplied by the conjugate of the sandwiched RPM, used in Fourier plane during encryption. The resultant image is now Fourier transformed to obtain the decrypted image. Since the detector can record the intensity distribution only, further multiplication of the conjugate of $R_1(x, y)$ is not required. The recovered image with correct key is shown in Figures 6.10c and 6.11c in case of a binary and a grayscale image. Figures 6.10d and 6.11d show the decrypted images in case of a binary and a grayscale image when a wrong key is used to recover the original image. In this program, three RPMs have been employed in the conventional DRPE random encoding algorithm. The use of an additional RPM as a sandwiched RPM enhances the security of the system. The first RPM R_1 is placed in close contact with the original image and other two RPMs (R_2, R_3) are in the Fourier plane.

To evaluate the reliability of the proposed algorithm, MSE was calculated by using the MSE criterion

$$\mathrm{MSE} = E\frac{1}{N \times M}\left[\sum_{x=0}^{N-1}\sum_{y=0}^{M-1} |\Psi(x, y)| - |f(x, y)|^2\right], \tag{6.6}$$

where $N \times M$ is the size of the image in numbers of pixels, $\Psi(x, y)$ and $f(x, y)$ are the decrypted and the primary images, respectively, and $E\{\cdot\}$ is the expected value. The MSE between the decrypted image and the primary image when the correct key is used are nearly zero ($\sim 10^{-27}$ and $\sim 10^{-32}$) and with the wrong key are 0.165 and 1256.8 in case of binary and grayscale images, respectively. When one of the RPMs R_3 is shifted by one pixel (i.e., 10 μm in units of distance) with respect to R_2 in X-direction, the decrypted images are obtained as shown in Figures 6.10e and 6.11e. The MSE calculated are 0.1649 and 1256.6 in case of binary and grayscale images, respectively. After shifting the RPM R_3 by one pixel (i.e., 10 μm in units of distance) in both X- and Y-directions, the decrypted images obtained are as shown in Figures 6.10f and 6.11f. The MSE calculated are 0.1647 and 1266.6 in case of binary

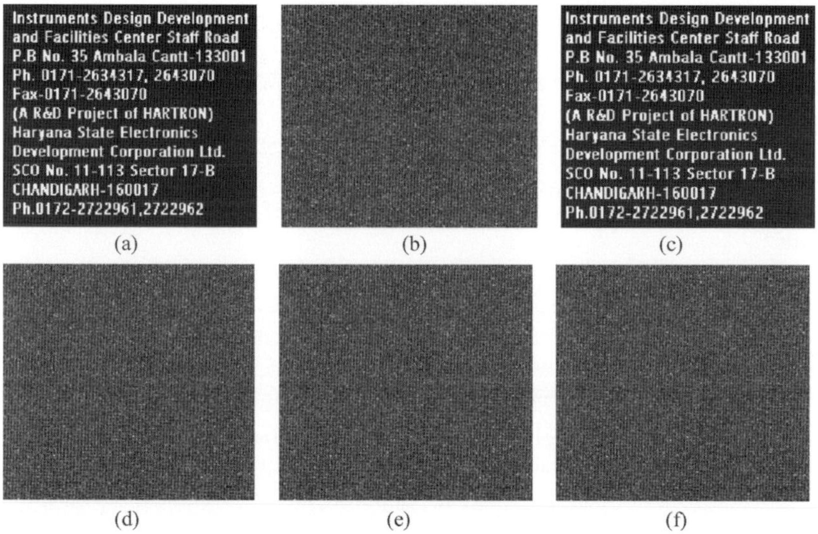

Figure 6.10 Simulation results of amplitude-based binary image. (a) Input image. (b) Encrypted image. (c) Decrypted image with correct key. (d) Decrypted image with wrong key. (e) Decrypted image after shifting the RPM R_2 by one pixel (10 μm) in X-direction (wrong key). (f) Decrypted image after shifting the RPM R_2 by one pixel (10 μm) in both X- and Y-directions (wrong key).

Figure 6.11 Simulation results of amplitude-based grayscale image. (a) Input image. (b) Encrypted image. (c) Decrypted image with correct key. (d) Decrypted with wrong key. (e) Decrypted image after shifting the RPM R_2 by one pixel (10 μm) in X-direction (wrong key). (f) Decrypted image after shifting the RPM R_2 by one pixel (10 μm) each in both X- and Y-directions (wrong key).

and grayscale images, respectively. These results clearly indicate that the encryption algorithm is very much sensitive to shift, that is, the decrypted image cannot be recovered even if the R_2 and R_3 are shifted by a single pixel dimension.

6.7 Optical Encryption Using a Sandwich Phase Mask Made of Elongated Speckle Patterns

To enhance the security of the data, DRPE architecture can be modified by replacing the amplitude information of the object with the phase-encoded information at the input plane. In this section, we describe the results of investigation [131] of an encryption technique using phase image for encryption and a sandwich phase mask made by using elongated speckle patterns as the key. The random phase masks made by using elongated speckles facilitate an easy alignment of the separated RPMs as compared to the case of encryption using a sandwich phase mask made with normal speckle patterns.

The elongated speckle patterns have earlier been investigated in the context of the effect of aberrations on the speckle shape [130]. Singh and Kumar [57] have investigated the encryption and decryption of a two-dimensional image by using a sandwich diffuser in the Fourier plane of the first lens of a 4-f imaging system. Singh *et al.* [131] have investigated an encryption system that uses in the Fourier plane a sandwich phase mask made with two elongated speckle patterns.

6.7.1 Preparation of the Elongated Speckle Phase Mask

To make the phase mask set by using elongated speckles, a strong diffuser is illuminated with a slit-shaped region. Here, the slit-shaped region may be 1 cm in length or more, while the width may be 25 μm or more. The length of the slit-shaped illuminated region decides the width of the elongated speckles, whereas the width of the slit decides the length of the speckles. After exposing the photographic plates with elongated speckles, the plates are processed to work as phase diffusers. However, in this section, we present the results of encryption and decryption by using a sandwich phase mask made with elongated speckles and gratings generated by a computer using MATLAB.

6.7.2 Description of the Method

For theoretical analysis, we have again used Figure 6.3. First, the input function $f(x, y)$ is changed into a phase function. A sandwich RPM $R(u, v)$ is made by using $R_2(u, v)$ and $R_3(u, v)$. $R_2(u, v)$ and $R_3(u, v)$ are made by using elongated speckle patterns. For encryption and decryption, same procedure as described in Section 6.6.1 is followed.

6.7.3
Computer Simulation Results

The simulation study was carried out on a MATLAB platform. A binary and a grayscale image of size 256 × 256 pixels have been chosen for study. One RPM at the input plane and two RPMs at the Fourier plane have been employed in the conventional DRPE algorithm. The first RPM_1 is placed in close contact with the original image and other two RPMs are placed in the Fourier plane. The original primary (binary) image (Figure 6.12d) is multiplied by the RPM_1 (Figure 6.12a), is Fourier transformed, and then multiplied by a sandwich of two RPMs (RPM_2 and RPM_3), as shown in Figure 6.12b and c. For decryption, the encrypted image is Fourier transformed and multiplied by the conjugate of the sandwich RPM used in Fourier plane during encryption.

The resultant image is further Fourier transformed to obtain the decrypted image. Since the detector can record the intensity only, further multiplication of the conjugate of $R_1(x, y)$ is not required. The recovered image with the correct key is shown in Figure 6.12f. The decrypted image recovered with a wrong key is shown in Figure 6.12g. The MSE calculated between the decrypted image and the primary image with a correct key is negligible ($\sim 10^{-27}$). The use of additional RPM as a sandwiched RPM enhances the security of the system. The same process is repeated for the grayscale image (Figure 6.12h) and the encrypted image is shown in Figure 6.12i. The decrypted images recovered are shown in Figure 6.12j and k with the correct and wrong keys, respectively. In this case also, the MSE between the decrypted image and the primary image with the correct key has been found to be negligible ($\sim 10^{-31}$). It may be noted that the encryption in case of binary image is not complete (Figure 6.12e), whereas in case of grayscale image, the image is completely encrypted (Figure 6.12i).

To make the system more secure in case of binary images, one of the diffusers (RPM_3) has been rotated by 90° to make the elongation of speckles in the horizontal direction. The pixel matrices of RPM_2 and RPM_3 have sizes of 8(32 × 256) and 4 (64 × 256), as shown in Figure 6.13a and b, respectively, to be used for making a sandwich. The encrypted image is shown in Figure 6.13c that is obtained by using the first RPM as shown in Figure 6.12a and the sandwich made by using elongated speckles as shown in Figure 6.13a and b. The decrypted images are shown in Figure 6.13d and e with the correct and wrong keys, respectively. In this case, the key RPM_2 is chosen as wrong key by displacing it in its own plane. The MSE calculated between the decrypted image and the primary image for the correct key is again negligible (10^{-27}).

To further strengthen the feasibility of our techniques, we have presented results of the encryption and decryption processes for the pixel matrices of size 8(32 × 256) and as shown in Figure 6.13f and g of the phase masks RPM_2 and RPM_3, respectively. The mask RPM_3 has been rotated to get the elongation of speckles in horizontal direction. The same process is repeated to get the encrypted image as shown in Figure 6.13h. The decrypted images are shown in Figure 6.13i and j with the correct and wrong

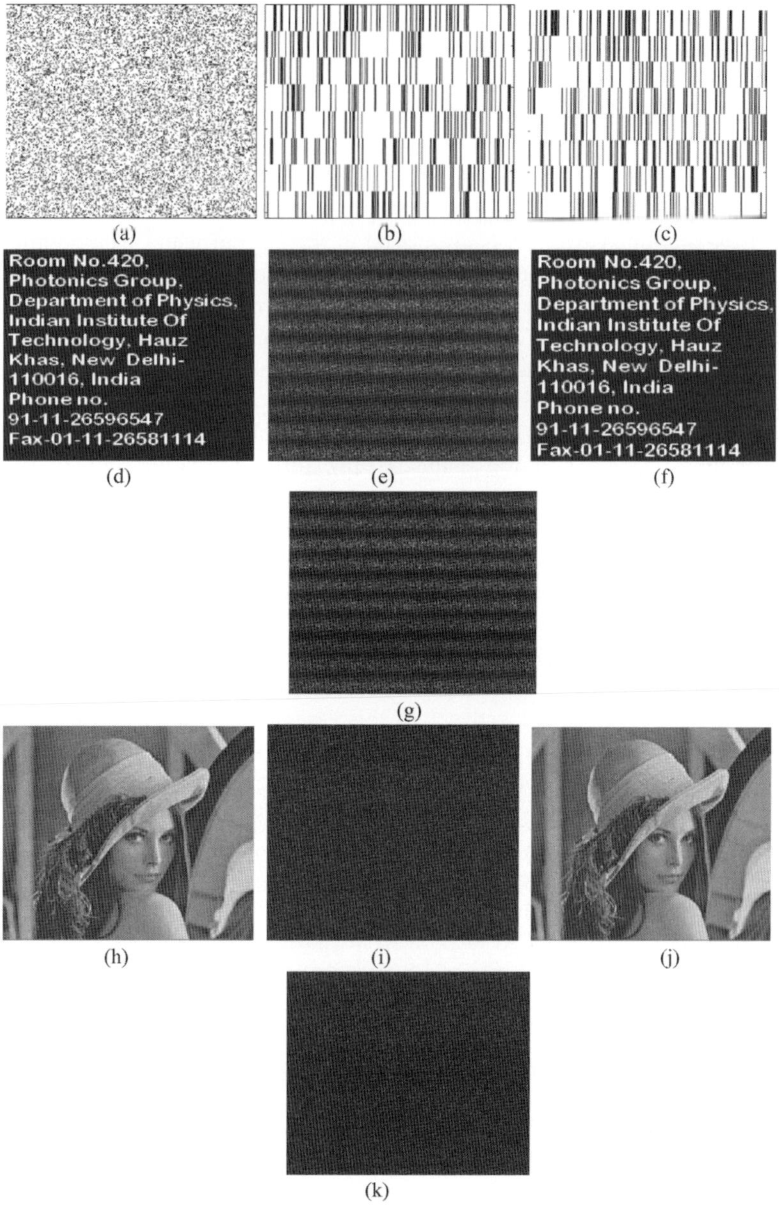

Figure 6.12 Simulation results with binary and grayscale images with elongated sandwich random mask generated by $8(32 \times 256)$ pixels matrix. (a) Random mask RPM_1. (b) Random mask RPM_2. (c) Random mask RPM_3. (d) Original binary image. (e) Encrypted image. (f) Decrypted image with the correct key. (g) Decrypted image with a wrong key. (h) Original grayscale image. (i) Encrypted image. (j) Decrypted image with the correct key. (k) Decrypted image with a wrong key (RPM_1, RPM_2, and RPM_3 are shown here as appeared before changing them into the phase masks).

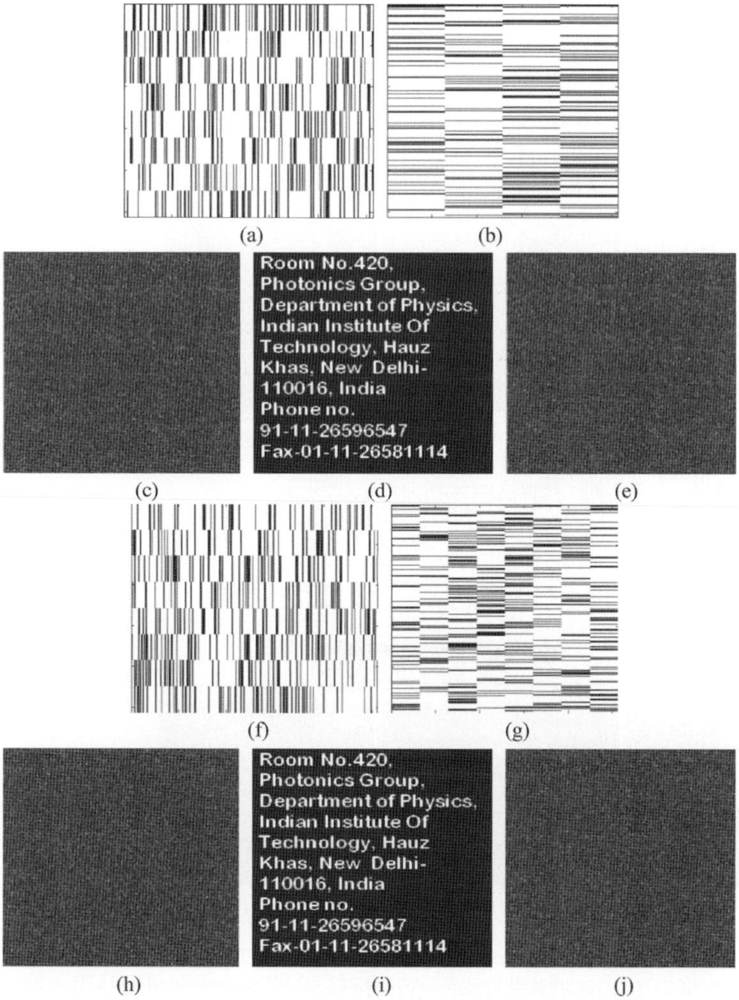

Figure 6.13 Binary image simulation results with sandwich RPM using elongated speckles and generated with different matrices. (a) Random mask RPM_2 generated by $8(32 \times 256)$ pixels matrix with vertical elongation of speckles. (b) RPM_3 generated by $4(64 \times 256)$ pixels matrix with horizontal elongation. (c) Encrypted image. (d) Encrypted image with the correct key. (e) Decrypted image with a wrong key. (f) RPM_2 generated by $8(32 \times 256)$ pixels matrix with vertical elongation. (g) RPM_3 generated by $8(32 \times 256)$ pixels matrix with horizontal elongation. (h) Encrypted image. (i) Decrypted image with the correct key. (j) Decrypted image with a wrong key (RPM_1, RPM_2, and RPM_3 are shown here as appearing before changing them into the phase masks).

keys, respectively. The MSE between the decrypted image and the input image when correct key is used is again negligible ($\sim 10^{-27}$).

We have also studied the encryption and decryption processes by replacing the first RPM by an RPM with elongated speckles. All the three RPMs have been generated by

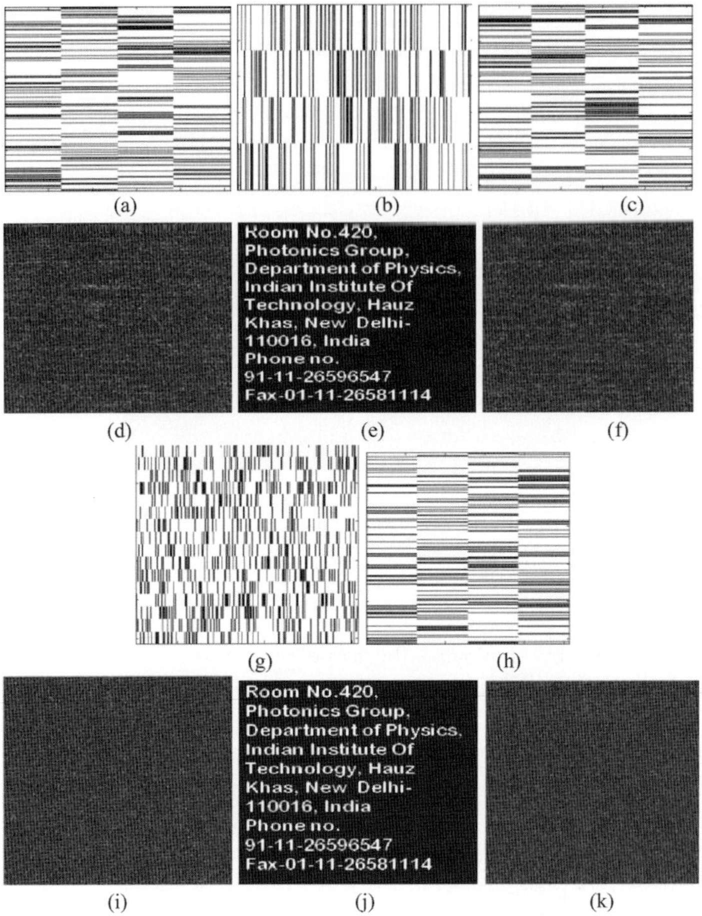

Figure 6.14 Binary image simulation results with sandwich random mask generated by different matrices. (a) Random mask RPM_1 with horizontal elongation of speckles generated by 4 (64×256) pixels matrix. (b) RPM_2 generated by $4(64 \times 256)$ pixels matrix with vertical elongation. (c) RPM_3 generated by $4(64 \times 256)$ pixels matrix with horizontal elongation. (d) Encrypted image. (e) Decrypted image with the correct key. (f) Encrypted image with a wrong key. (g) RPM_1 generated by $16(16 \times 256)$ pixels matrix with vertical elongation. (h) RPM_2 generated by $4(64 \times 256)$ pixels matrix with horizontal elongation. (i) Encrypted image. (j) Decrypted image with the correct key. (k) Decrypted image with correct key (RPM_1, RPM_2, and RPM_3 are shown here as appearing before changing them into the phase masks).

the pixel matrix of size $4(64 \times 256)$. RPM_1 rotated by 90° (Figure 6.14a) contains the elongation of speckles in horizontal direction. RPM_2 and RPM_3 contain the elongation of speckles in vertical and horizontal directions, respectively (Figure 6.14b and c). The same binary input image is multiplied by the RPM_1 placed in close contact with the input image. The encrypted image (Figure 6.14d) is obtained by repeating the procedure described earlier. The decrypted images with the correct key (Figure 6.14e) and with a wrong key (Figure 6.14f) are shown. We have supplemented our investigation by presenting results using two sandwich diffusers. First phase diffuser was made with elongated speckles as shown in Figure 6.14b and c and the second phase diffuser was made with two elongated speckle pattern random phase diffusers as shown in Figure 6.14g and h. The first sandwich phase diffuser is used in contact with the primary image and the second is kept at the Fourier plane. Both the RPMs of second sandwich phase diffuser have been prepared by considering pixel matrices of size $16(16 \times 256)$ and $4(64 \times 256)$ having elongation of speckles in the vertical and horizontal directions, respectively. The encrypted image is shown in Figure 6.14i, and the decrypted images are shown in Figure 6.14j and k with the correct and wrong keys, respectively.

In an another study, the sandwich phase mask $R(u, v)$ is constituted by using an RPM_2 made by using a normal speckle pattern as shown in Figure 6.12a and a sinusoidal phase grating (SPG). These patterns are generated by using the MATLAB® platform. A grayscale and a binary scale image of size 256×256 pixels have been chosen for the study. First the RPM_1 is made by a speckle pattern and used at the input plane, and then the second RPM, a sandwich phase mask, is made by using RPM_2 and an SPG denoted as $R_2(u, v)$ and $R_3(u, v)$, respectively. This sandwich phase mask is kept at the FT plane and acts as the key in the conventional DRPE algorithm.

The encryption is done by employing the same procedures as described in Section 6.6.1 by replacing one of the constituent phase masks $R_3(u, v)$ with the crossed grating (Figures 6.15b and 6.16b) with the grayscale image (Figure 6.15a) and

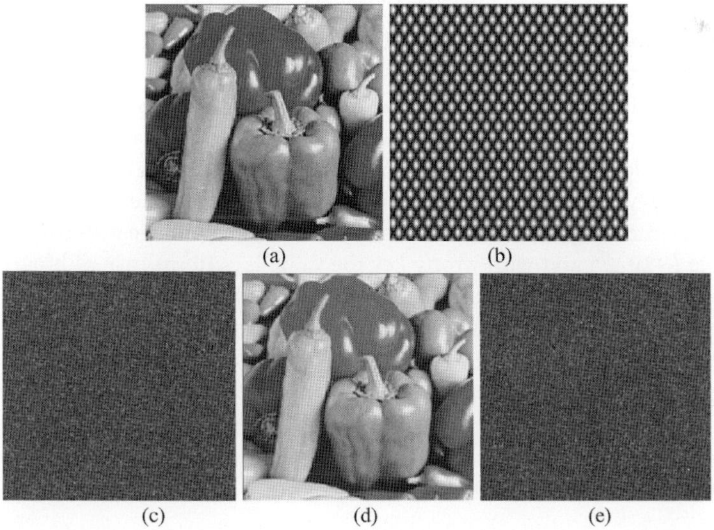

Figure 6.15 Simulation results of amplitude-based grayscale image with two-dimensional grating. (a) Input image. (b) Cross-grating. (c) Encrypted image. (d) Decrypted image with correct key. (e) Decrypted image with wrong key (without grating).

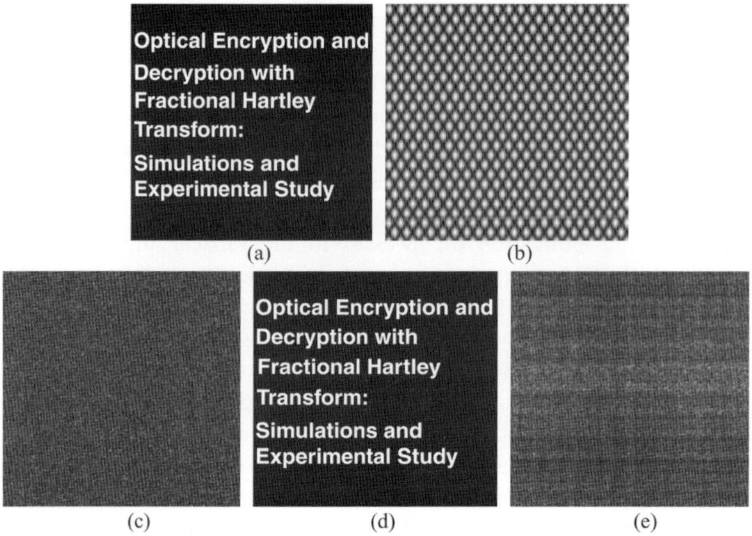

Figure 6.16 Simulation results of amplitude-based binary image with two-dimensional grating. (a) Input image. (b) Cross-grating. (c) Encrypted image. (d) Decrypted image with correct key. (e) Decrypted image with wrong key (without grating).

a binary image (Figure 6.16a). The encrypted images are shown in Figures 6.15c and 6.16c in case of grayscale and binary images, respectively.

After the encryption, we remove the phase grating. This act can be realized by using an SLM working in phase mode. This makes the system free from attack by using a delta function. For decryption, we have to display the same grating on the SLM without which it is not possible to get the desired image. The recovered image with the correct key is shown in Figure 6.15d in case of grayscale image and Figure 6.16d in case of binary image. Figures 6.15e and 6.16e are obtained with a wrong key (without grating) in case of grayscale and binary images, respectively. Similar results are observed on decryption with sandwich diffuser made using different SPGs. To evaluate the reliability of the proposed algorithm, MSE was calculated. The MSE calculated between the decrypted image and the primary (binary) image with correct key is negligible ($\sim 10^{-27}$) and with wrong key the value is $\sim 10^3$ in case of binary image. Also in case of grayscale image, the MSE between the decrypted image and the primary image with the correct key has been found to be negligible ($\sim 10^{-31}$) and with wrong key the value is 10^4.

6.8 Speckles for Multiplexing in Encryption and Decryption

Securing data in high-capacity storage systems in order to meet the users' requirements is one of the challenging tasks. There has been a continuous effort in developing multiplexing techniques for data storage and data security [114].

Techniques have been used in holography to store multiple images in a single crystal without cross talk by using angular, peristrophic (rotational), and wavelength multiplexing. Matoba and Javidi [115] have demonstrated encrypted optical storage based on DRPE by using angular multiplexing in a PR material. Denz *et al.* [132] have investigated several encryption techniques for holographic data storage, using orthogonal phase-code multiplexing and random phase encoding. Their system is capable of storing page-oriented data based on the selectivity of orthogonal phase codes in a PR $LiNbO_3$ crystal. In order to encrypt data, random phase multiplexing is added to the system. A ground glass is frequently used for phase-coded multiplexing of holograms.

Phase image-based content-addressable holographic data storage with security using random phase in the Fresnel domain has been described [133]. Situ and Zhang [134] investigated a technique of wavelength multiplexing into a DRPE system to achieve multiple image encryption. Islam and Alam [135] have proposed a technique for optical encryption and multiplexing of binary characters and images used for personal identification information.

Situ and Zhang [136] have investigated position multiplexing for multiple image encryption. Chang *et al.* [137] have concluded that under certain conditions there are few ways to decrypt the encrypted PR $LiNbO_3$ crystal holographic storage system without the proper knowledge of RPM. A method for color image encryption has been proposed [84] by use of wavelength multiplexing based on lensless Fresnel transform holography. An image is separated into three channels: red, green, and blue (R, G, and B), and each channel is independently encrypted. Gopinathan *et al.* [138] described an optical system that encodes an input signal to a polarization state, using an SLM. They also described multiplexing using two SLMs. As already discussed in Section 6.5, Barrera *et al.* [54–56] have carried out extensive work on multiplexing techniques in which speckle patterns play an important role. Shi *et al.* [139] have described a technique for multiple image hiding in the Fresnel domain to avoid the cross talk among the multiple images.

Sun *et al.* [140] have presented a study on the rotation selectivity of a volume hologram with random phase encoding. A technique has been described [141] by utilizing spread-space spread-spectrum of multiplexing images in which each image may be demultiplexed only if a set of random encryption keys is known. The technique utilizes the ability of the DRPE method to spread a signal's energy in both space and frequency domains in a controlled manner. The multiplex images are independently encrypted with different phase keys and then superimposed by recording subsequently in the same recording material. Inoue *et al.* [142] have proposed a simple and compact rewritable holographic memory system using a fieldstone of Ulexite that imposes random patterns on the reference beam to record plural images with the random-reference multiplexing scheme. The operations for writing and reading holograms are carried out by simply rotating the fieldstone in one direction. A system employing quadrature multiplexing has been suggested [143], in which pupil aperture masks kept in close contact with the second lens have been used to record multiple images by changing these between the exposures. Barrera *et al.* [144] have developed an undercover multiplexing technique to give additional protection for optical information encryption. To achieve multiplexing, they use the aperture size of the pupil in the optical system, as it governs the speckle size.

Hwang *et al.* [145] have presented multiple image encryption and multiplexing using a modified Gechberg–Saxton algorithm and phase modulation in the Fresnel transform domain. It has been shown [146] that multiple secure data recording under a wavelength multiplexing technique is possible in a JTC arrangement. This analysis reveals that the wavelength is a valid parameter to conduct image multiplexing encoding with the JTC architecture. An encryption undercover-selectable multiplexing technique has also been proposed by Barrera *et al.* [147] who have analyzed and compared the different alternatives and degrees of complexity this combination of mask brings to enhance the security of the system. Alfalou *et al.* [148] have presented a technique of multiplexing and simultaneous encoding of target images with the enhancement of encryption level of a DRPE system. In their scheme, first security level consists in using several independent image keys (randomly and structurally) in the multiplexing algorithm. At the second level, a standard DRP is included. The use of both levels can check the vandalism attempt done on transmitted encrypted images.

Xiao *et al.* [149] have presented a technique for multiple image encryption based on a cascaded phase retrieval algorithm (CPRA). The system consists of several stages of a standard 4-f correlator, in which the keys are not only the phase mask pairs produced by CPRA but also the phase distribution of the output plane of the front stage. The system can resist known-plaintext attack to some extent. Barrera and Torroba [150] investigated a combination of a free propagation scheme and a classical 4-f encrypting architecture to design a technique to simultaneously encrypt multiple input objects. A technique for noise-free recovery of optodigital encrypted and multiplexed images has been presented by Henao *et al.* [151]. We have presented in the following section the results of multiplexing in optical encryption of 2D images using an aperture system [152] and rotation [153] of one or both the constituent diffusers.

6.9 Multiplexing in Encryption Using Apertures in the FT Plane

In this section, we present the results of multiplexing in optical encryption of 2D images using an aperture system in the Fourier plane of a DRPE system. An aperture system having apertures of different shapes and sizes is made with the help of "paint brush" software. Successive recording of encrypted images is carried out by taking the input images one by one.

6.9.1 Methodology

Figure 6.17 shows the schematic of the 4-f system used for multiplexing. The encryption of the input image is carried out in two steps. First, the object pattern

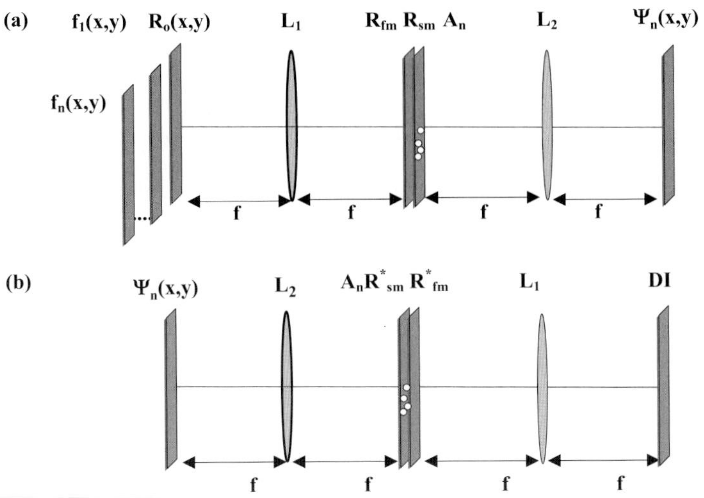

Figure 6.17 Schematic diagram of (a) multiplexing in encryption and (b) decryption setup. $R_0(x, y)$, R_{fm}, R_{sm}, random phase masks; R^*_{fm}, R^*_{sm}: conjugate of the random phase masks; L_1, L_2, lenses; A_n, apertures; $f_1(x, y)$, $f_2(x, y)$, . . ., $f_n(x, y)$, input images; $\Psi_n(x, y)$, multiplexed encrypted images; DI, decrypted image.

$f(x, y)$ is multiplied by an RPM $R_0(x, y)$ at the input plane. This product is then convolved with $R_c(x, y)$ and an aperture placed in contact with the sandwich diffuser made of diffusers $R_{fm}(u, v)$ and $R_{sm}(u, v)$. Here $R_c(x, y)$ may be defined as the impulse response of $R_{fm}(u, v)$ and $R_{sm}(u, v)$ combined. The FT of the modified input image, that is, the product of $f(x, y)$ and $R_0(x, y)$, is passed through the sandwich diffuser. The resultant information then passes through corresponding aperture of an aperture system kept in contact with the sandwich diffuser in the Fourier plane.

The random phase functions $R_0(x, y)$, $R_{fm}(u, v)$, and $R_{sm}(u, v)$ are chosen to be statistically independent. The modified encrypted image due to the first input image $f_1(x, y)$ is given by

$$\Psi_1(x, y) = \mathrm{FT}\{\mathrm{FT}[f_1(x, y) R_0(x, y)][R_{f_1}(u, v) R_{s_1}(u, v) A_1]\}\cdot \tag{6.7}$$

Similarly, the encrypted image of the second input image $f_2(x, y)$ with second sandwich diffuser and second aperture is given by

$$\Psi_2(x, y) = \mathrm{FT}\{\mathrm{FT}[f_2(x, y) R_0(x, y)][R_{f_2}(u, v) R_{s_2}(u, v) A_2]\}. \tag{6.8}$$

The n different input images can be recorded one by one, keeping the apertures at different locations in the Fourier plane. The encrypted image of the nth input image with mth sandwich diffuser and mth aperture is written as

$$\Psi_n(x, y) = \mathrm{FT}\{\mathrm{FT}[f_n(x, y) R_0(x, y)][R_{fm}(u, v) R_{sm}(u, v) A_m]\}. \tag{6.9}$$

The resultant encrypted image of N input images and M apertures may be expressed by

$$\psi_N(x,y) = \sum_{m=1}^{M}\sum_{n=1}^{N}[\mathrm{FT}\{\mathrm{FT}[f_n(x,y)R_0(x,y)][R_{\mathrm{fm}}(u,v)R_{\mathrm{sm}}(u,v)A_{\mathrm{m}}]\}]. \qquad (6.10)$$

After multiplexing, the constituent diffusers $R_{\mathrm{fm}}(u, v)$ and $R_{\mathrm{sm}}(u, v)$ in each recording may be separated, thus acting as $2M$ number of keys. Here, the number of apertures also acts as key, thus creating a total of $3M$ number of keys. In our case, m and n are kept equal, thus the number of keys is $3N$.

For decryption, it should be noted that the constituent diffusers making a sandwich diffuser should be placed at the same location and orientation that they occupied during the encryption, and the proper conjugate should be made. The decrypted image may be expressed as

$$f_n(x,y) = \mathrm{IFT}\{\mathrm{FT}[\Psi_N(x,y)][A_m \mathrm{Conj}\ [R_{fm}(u,v)\ R_{sm}(u,v)]\}. \qquad (6.11)$$

For $m = n$, we get the nth decrypted image with the right key. For $m \neq n$, we get the nth decrypted image $f_n(x, y)$ with wrong keys, that is, no image is obtained and we get only the noise. To get the decrypted images, the aperture system and the two diffusers $R_{\mathrm{fm}}(u, v)$ and $R_{\mathrm{sm}}(u, v)$ constituting the sandwich diffuser have to be aligned and brought back to the same position and orientation that they occupied during the recording.

For the optical implementation, multiple encrypted images can be stored in a single recording medium such as a PR $BaTiO_3$ crystal by using angular multiplexing. Changing the input object and the corresponding aperture position between the successive recordings facilitates multiplexing of the encrypted images. An intensity detector, that is, a CCD, can capture the decoded amplitude image that can be displayed onto an SLM for decryption process. The recorded multiplexed encrypted image hologram is read with the conjugate of the reference beam to obtain the decrypted images by phase conjugation.

The combination of other multiplexing techniques used with the aperture system increases the capacity of conventional multiplexing used for data storage.

6.9.2 Computer Simulation

The simulation study was carried out on a MATLAB-7.0 platform. The investigation has been supported by the calculation of MSE and signal-to-noise ratio (SNR) between the original and decrypted images. For the calculation of MSE, we have used the following expression:

$$\mathrm{MSE} = \frac{1}{P \times Q}\left[\sum_{x=0}^{P-1}\sum_{y=0}^{Q-1} |\Psi_s(x,y)| - |f_n(x,y)|^2\right]. \qquad (6.12)$$

For the calculation of SNR, the expression is

$$\mathrm{SNR} = \frac{\sum_{x=0}^{P-1}\sum_{y=0}^{Q-1} |\psi_s(x,y)|^2}{\sum_{x=0}^{P-1}\sum_{y=0}^{Q-1} ||\psi_s(x,y)| - |f_n(x,y)||^2}, \tag{6.13}$$

where P × Q is the size of the image in numbers of pixels. $\Psi_s(x, y)$ and $f_n(x, y)$ are respectively the decrypted and primary images.

6.9.3 Effect of Aperture Size on the Encryption and Decryption

Binary and grayscale images of size 256 × 256 pixels (Figures 6.18a and 6.19a) have been chosen for the study. We have studied the effect of changing the size of the circular apertures (Figures 6.18b–e and 6.19b–e) in encryption and decryption. The results showing the effect of changing the size, shape, and orientation/position of the apertures on the decrypted image have been presented. Apertures of different sizes (Figures 6.18b–e and 6.19b–e) are kept in contact with the sandwich diffuser at the Fourier plane and the recording is done for the binary and grayscale objects. The encrypted images obtained are shown in Figures 6.18f–j and 6.19f–j in case of binary and grayscale images. Figures 6.18f and 6.19f are the encrypted images obtained with full open aperture with binary and grayscale images, respectively. Figures 6.18g–j and 6.19g–j are the encrypted images recorded with the apertures as shown in Figures 6.18b–e and 6.19b–e, respectively. The resultant decrypted images are shown in Figures 6.18k–o and 6.19k–o. The MSE values as calculated for Figures 6.18 and 6.19 are 4.2×10^{-32}, 0.041, 0.083, 0.125, 0.223, and 7.30×10^{-27}, 0.058, 0.174, 0.262, 0.291 for binary and grayscale images by using full apertures and 80%, 40%, 10%, 5% of the full apertures, respectively. From Figures 6.18 and 6.19 and the MSE values, it is very much evident that the decrease in the aperture size results in a decrease of the resolution of the decrypted images and for a very small size aperture the decrypted image is transformed into a speckle pattern. So a compromise has to be made between the size of the aperture used and the resolution of the image required.

6.9.4 Effect of Increasing the Number and Size of the Apertures

A binary image of size 256 × 256 pixels (Figure 6.20a) has been chosen for the study to show the effect of increasing the number and size of circular apertures in performing encryption and decryption. Following the procedure described in the previous section, the encrypted images obtained with various apertures (Figure 6.20b–f) are shown in Figure 6.20g–k, respectively. The recovered images with the corresponding apertures are shown in Figure 6.20l–p. The MSE values for binary images corresponding to Figure 6.20 are 0.105, 0.102, 0.097, 0.084, and 0.071

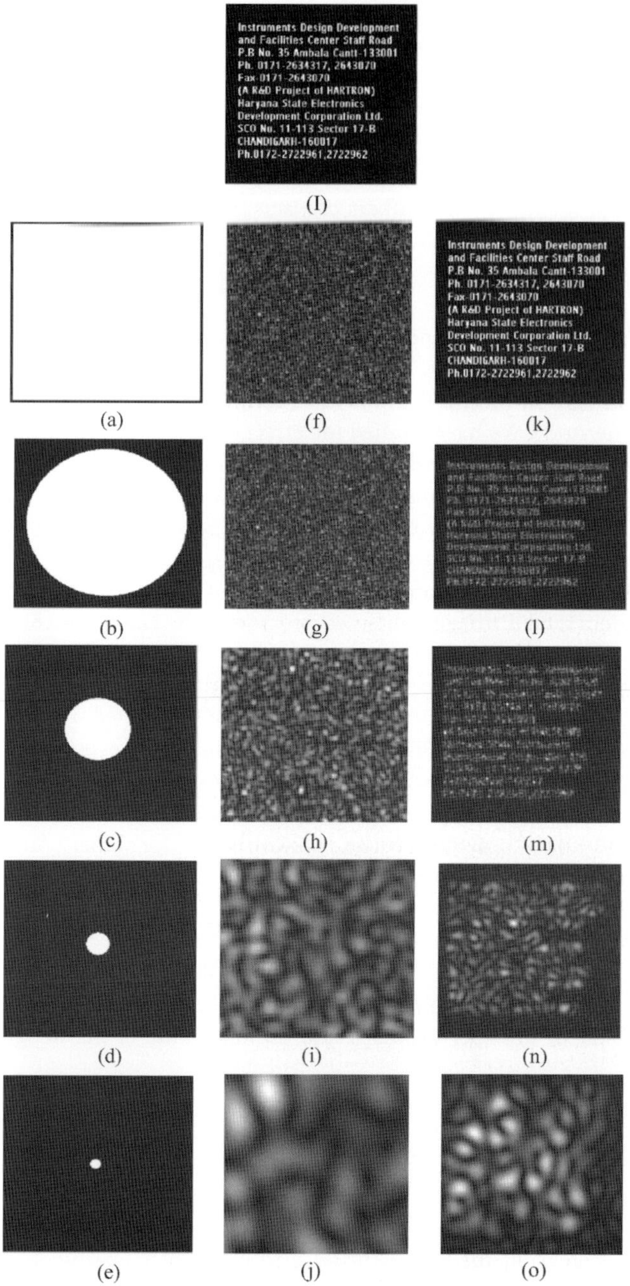

Figure 6.18 Simulation results for encryption and decryption using single aperture of different sizes. I, input image. (a–e) Apertures. (f) Encrypted image in case of full aperture. (g–j) Encrypted images by using apertures as shown in (b–e). (k) Decrypted image in case of full aperture. (l–o) Decrypted images corresponding to the apertures as shown in (b–e).

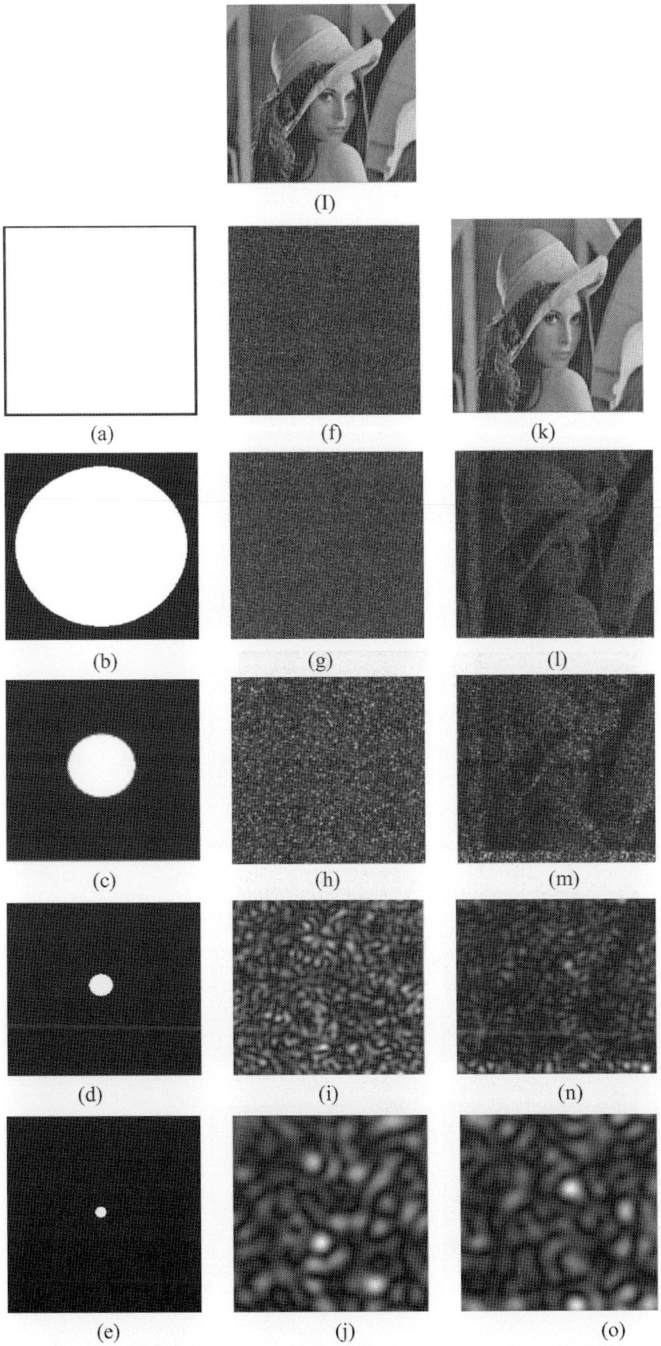

Figure 6.19 Simulation results of grayscale images for encryption and decryption using single aperture of different sizes. I, input Image. (a–e) Apertures. (f) Encrypted image in case of full aperture. (g–j) Encrypted images by using apertures as shown in (b–e). (k) Decrypted image in case of full aperture. (l–o) Decrypted images corresponding to the apertures as shown in (b–e).

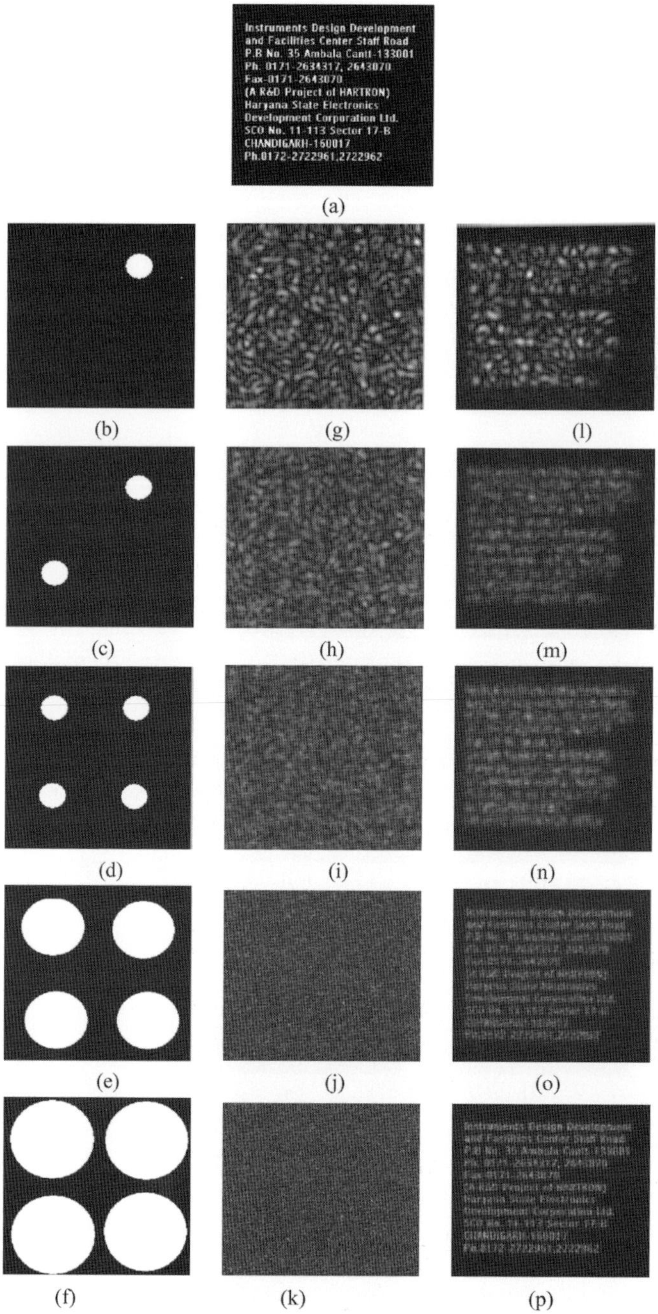

Figure 6.20 Simulation results for encryption and decryption using a single aperture of different sizes. (a) Input image. (b–f) Apertures. (g–k) Encrypted images by using the apertures as in (b–f). (l–p) Decrypted images corresponding to the encrypted images as in (g–k).

for a single aperture, double aperture, four apertures, four apertures of the size four times and eight times to that used in Figure 6.20d, respectively. From Figure 6.20 and the corresponding MSE values, it is clear that an increase in the number of apertures or the aperture size results in an improvement in the resolution of the decrypted images. So a compromise has to be made between the number of apertures to be used and the resolution of the recovered images. However, using a number of apertures facilitates the enhancement of capabilities of the recording system. In Section 6.9.5, we describe the use of an aperture system for multiplexing in encryption.

6.9.5
Multiplexing in Encryption Using Circular Apertures

In the first case, images in the form of letters A, B, C, and D are taken as objects (Figure 6.21a–d). For multiplexing, we record the images of these four objects one by one by keeping the circular apertures at different locations, as shown in Figure 6.21e–h, respectively. The multiplexed encrypted image is shown in Figure 6.21i. The decrypted images recovered from the multiplexed encrypted image are shown in Figure 6.21j–m with the apertures being kept at the previous positions. The MSEs between the decrypted image and the primary image are 0.028, 0.0339, 0.025, and 0.030, respectively. Figure 6.21o is the decrypted image if we used all four apertures simultaneously, as shown in Figure 6.21n, during the decryption. The image recovered without any aperture is shown in Figure 6.21p. The results show that the use of an aperture system for encryption suffers due to decrease in SNR as the other decrypted images with wrong keys are present while decrypting with the correct key. However, using a number of apertures facilitates the enhancement of capabilities of the recording system.

6.9.6
Multiplexing in Encryption Using Square Apertures

In this case, the images in the form of transparent letters A, B, C, and D are taken as objects (Figure 6.22a–d). Following the procedure described in Section 6.9.1 for multiplexing, we record the images of these objects one by one by keeping a pair of small square apertures at different locations, as shown in Figure 6.22e–h, respectively. The multiplexed encrypted image is shown in Figure 6.22i and decrypted images obtained from it are shown in Figure 6.22j–m. The MSE values calculated are 0.033, 0.038, 0.031, and 0.036. The combined decrypted images (Figure 6.22o and p) are obtained with a set of four apertures and without any aperture, respectively. The fringe structures observed in Figure 6.22j–m on the decrypted images is due to use of an aperture system containing a pair of apertures. From the above-mentioned results, it is clear that the use of an aperture system enhances the multiplexing capability and security of the system by a factor equal to the product of the number of apertures used.

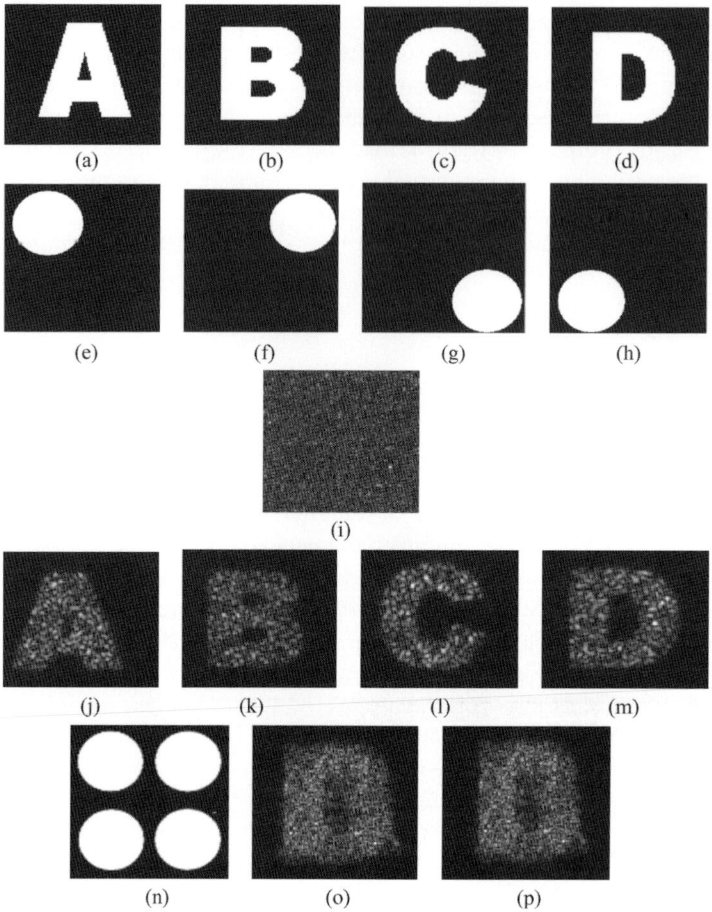

Figure 6.21 Simulation results for encryption and decryption using circular aperture at different positions. (a–d) Input images. (e–h) Aperture systems. (i) Encrypted image. (j–m) Decrypted images. (n) Set of four apertures. (o) Decrypted image with a set of four apertures. (p) Decrypted image without any aperture.

6.10
Multiplexing by In-Plane Rotation of Sandwich Phase Diffuser and Aperture Systems

In this section, the results of image multiplexing in encryption have been presented for the case when we rotate the constituent diffusers of a sandwich random phase diffuser kept in the Fourier plane of a DRPE system. Successive recording of encrypted images is done by taking the input images one by one. The results have been presented for different sets of rotation of one or both the random constituent phase masks. The use of an aperture system in addition to the rotation of one or both the RPMs enhances the multiplexing capability and security of the system by

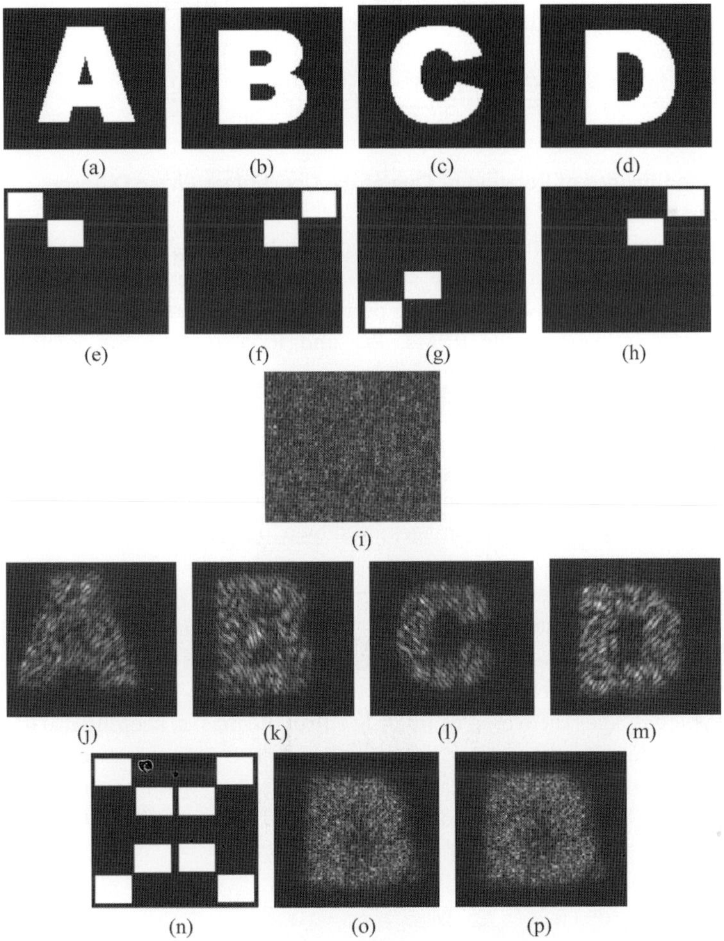

Figure 6.22 Simulation results for encryption and decryption using a pair of small size square apertures at different positions. (a–d) Input images. (e–h) Aperture systems. (i) Encrypted image. (j–m) Decrypted images. (n) Set of four apertures. (o) Decrypted image with a set of four apertures. (p) Decrypted image without any aperture.

avoiding the generation of cross talk images at the time of decryption. Simulation results are presented in support of the proposed techniques. The MSE values between the decrypted and the original image have also been calculated.

6.10.1 Methodology

For the analysis of multiplexing by use of the rotation of R_{sm}, we refer to same diagram as given earlier (Figure 6.17) except that an aperture system has been used with the R_{sm}. The investigation has been supported by the calculation of MSE and

SNR values between the original and decrypted images using Equations 6.12 and 6.13. The following are the investigations showing the effect of rotation of one or both the RPMs of the sandwich diffuser.

6.10.2
Effect on Decrypted Images of Rotation of One of the Constituent Phase Diffusers

We have studied the effects on the decrypted images of rotation of either one or both the RPMs R_{fm} and R_{sm} constituting the sandwich diffuser. The study has been carried out in two parts. In the first part, the performance of the system has been studied by rotating any one of the RPMs, R_{fm} and R_{sm}, to show the rotation sensitivity of the image decryption. A grayscale image of LENA (Figure 6.23a) of size 256 × 256 and a binary image of a set of letters **PQRS** (Figure 6.24a) of pixel size 256 × 256 have been chosen as the input images. One of the diffusers R_{sm} is rotated around the optical axis at the time of recording the decrypted images in order to determine the rotation sensitivity of the image decryption. The encrypted images obtained in case of grayscale and binary images are shown in Figures 6.23b and 6.24b, respectively.

The decrypted images in case of grayscale image are recorded (Figure 6.23c–f) for 0.00°, 0.20°, 0.40°, and 0.80°, respectively, the mean square errors being 3.470×10^{-027}, 0.345, 6.230, and 7.300, respectively. In case of a binary image, the angle of rotation is varied from 0° to 2° with an interval of 0.25°. Figure 6.24c–h shows the decrypted images recovered by rotating one of the masks, keeping it at 0°, 0.25°, 0.5°, 0.75°, 1°, 1.25° positions. The calculated MSE values between the encrypted and decrypted images is nearly zero ($\approx 4.23 \times 10^{-28}$) for the original position, that is, at 0°. The values of MSEs as calculated are, respectively, 1.26×10^3,

Figure 6.23 Encrypted and decrypted images for a grayscale object with angular rotation of the constituent diffuser R_{sm} and keeping it at various positions: input grayscale image (a); encrypted image (b); and decrypted images at 0° (original position) (c), 0.20° (d), 0.40° (e), and 0.80° (f).

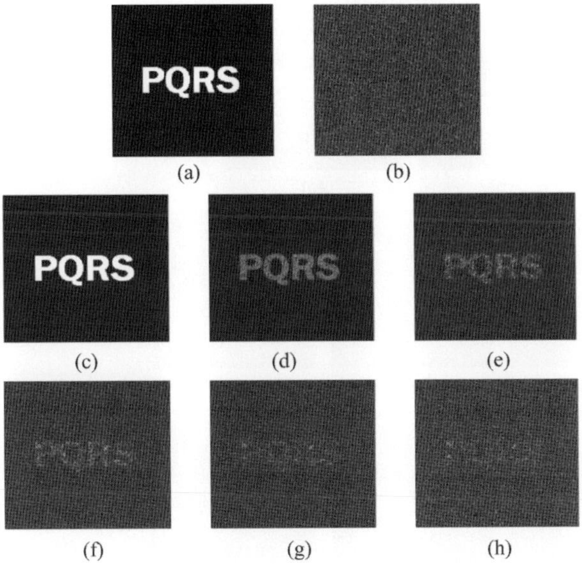

Figure 6.24 Encrypted and decrypted images of a binary object with in-plane rotation of the mask R_{sm} with respect to R_{fm}. (a) Input image. (b) Encrypted image as recorded by keeping the R_{fm} and R_{sm} at 0° position. (c–h) Decrypted images at the angles 0°, 0.25°, 0.5°, 0.75°, 1°, and 1.25°, respectively.

4.42×10^3, 5.19×10^3, 5.51×10^3, and 5.68×10^3 for 0.25°, 0.5°, 0.75°, 1°, and 1.25° rotation of the sandwich diffuser. The decrypted images and the corresponding MSEs indicate that the quality of the recovered images becomes poor with the increase of angle of rotation of the mask R_{sm}. The recovered image completely disappears if any one of the masks is rotated by 2°. The minimum limit of 2° rotation of the diffuser is found suitable, based upon the nonretrieval of the image. However, it depends on the size, as well as the grain size of the diffuser. An increase in the size of the diffuser and decrease in the grain size results in a reduction of the limit of 2° rotation. Here, it should be noted that we have rotated the diffuser only, whereas the image remains unrotated. To keep the size of the diffuser 256×256 pixels, we have compensated the area left after the rotation by filling with diffusing elements. However, it may result in the addition of noise in the decrypted images that is considered as negligible.

From Figures 6.23 and 6.24, it is clear that in case of grayscale image, the required angular rotation to get complete decrypted image is less than that in case of binary image. This is due to the fact that the binary image contains mainly the zero-order diffracted component that falls on the rotating diffuser's center where the change due to rotation is less compared to that at the outer parts of the diffuser.

In the second part, the performance of the system has been investigated for an in-plane rotation of both the constituent masks R_{fm} and R_{sm} in the opposite directions. We rotated one of the masks (R_{sm}) by 0.25° and recorded the decrypted image

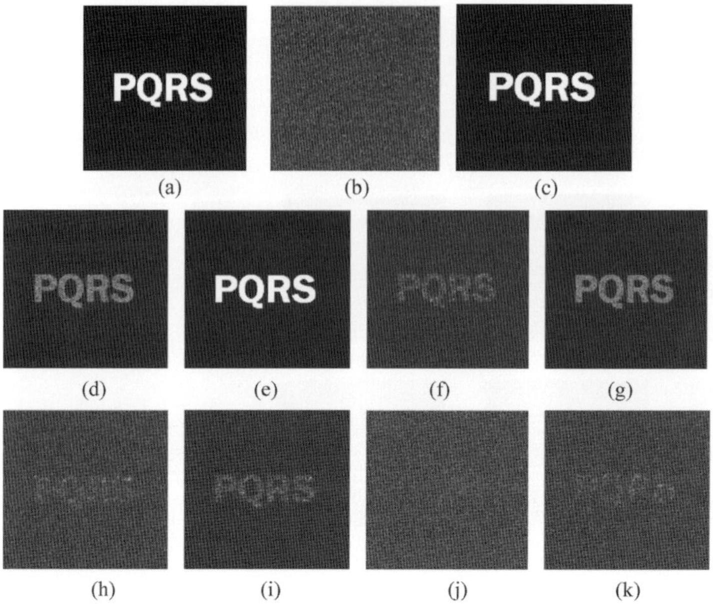

Figure 6.25 Decrypted images of a binary object due to in-plane rotation of mask R_{sm} and R_{fm}: input image (a); encrypted image (b); and decrypted image (c) with no rotation between R_{sm} and R_{fm}, (d) R_{sm} rotated by 0.25°, (e) R_{fm} and R_{sm} rotated by 0.125° in opposite directions, (f) R_{sm} rotated by 0.5°, (g) R_{fm} and R_{sm} rotated by 0.25° in opposite directions, (h) R_{sm} rotated by 0.75°, (i) R_{fm} and R_{sm} rotated by 0.375° in opposite directions, (j) R_{sm} rotated with 1°, and (k) R_{fm} and R_{sm} rotated by 0.5° in opposite directions.

(Figure 6.25b). The calculated MSE is 1.18×10^{-3}. To see the effect of rotation of both the masks in opposite directions, we rotate the masks R_{fm} and R_{sm} by an angle of 0.125°. The decrypted image recovered is shown in Figure 6.25e, the MSE value calculated being 1.33×10^{-27}. The process is repeated for different sets of rotation of the mask to get the decrypted image and the corresponding MSE value. The decrypted image in Figure 6.25f is for the rotation of mask R_{sm} by 0.5°, the MSE value being 4.37×10^{-3}. Figure 6.25g is the decrypted image when both the masks R_{fm} and R_{sm} are rotated by 0.25° in opposite directions, the MSE value being 1.19×10^{-3}. Figure 6.25h shows the decrypted image as a result of the rotation of mask R_{sm} by 0.75°, the MSE value being 5.25×10^{-3}. Figure 6.25i is decrypted image when each of the masks R_{fm} and R_{sm} are rotated by 0.375° in opposite directions keeping the total rotation of 0.75°. The MSE is 3.58×10^{-3}. Figure 6.25j shows the decrypted image when mask R_{sm} is rotated by 1°, the MSE being 5.51×10^{-3}. Figure 6.25k shows the decrypted image when both the masks R_{fm} and R_{sm} are rotated by 0.5° in opposite directions keeping the total rotation of 1°, the MSE being 4.45×10^{-3}. It is observed that the quality of decrypted images is improved as we rotate the second diffuser R_{fm} in a direction opposite to that of R_{sm}.

6.10.3 Multiplexing in Encryption Using the Rotation of the RPM R_{sm}

The multiplexing of the input images in encryption is done by using rotation of one of the masks R_{sm}. Four input objects in the form of letters P, Q, R, and S (Figure 6.26a–d) have been used, the size of each being 256×256 pixels. Following the encryption process adopted in the earlier section, we record multiple input images by using a combination of different sets of constituent random phase diffusers. Different combinations have been obtained by rotation of R_{sm} in steps of 1°. Four different combinations of masks R_{fm} and R_{sm} are considered, thus effectively making four sets of random sandwich phase diffusers by rotating R_{sm} by four different values. The multiplexed encrypted image (Figure 6.26e) is recorded with each input image by rotating one of the masks (R_{sm}) placed in the Fourier plane by different values, that is, P, Q, R, and S are recorded by keeping the R_{sm} at 1°, 2°, −1°, and −2°, respectively. The decryption is done by generating and using conjugates of the corresponding sandwich diffusers as shown in Figure 6.26f–i. The MSE values between the decrypted images (Figure 6.26f–i) and the corresponding primary input images (Figure 6.26a–d) are of the order of 10^3 in each case.

In another case, the sensitivity of the technique in decryption has been observed by rotating the masks by different angles with respect to their original orientation. The decrypted images have been recorded with a change in the angle of rotation of the

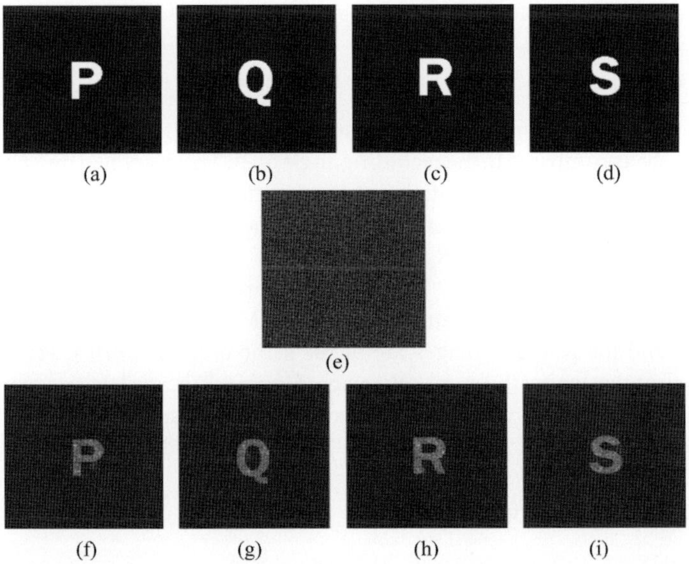

Figure 6.26 Encryption and decryption of four different objects P, Q, R, and S keeping R_{sm} at 1°, 2°, −1°, and −2°, respectively. (a–d) Input images. (e) Multiplexed encrypted image. (f–i) Decrypted images.

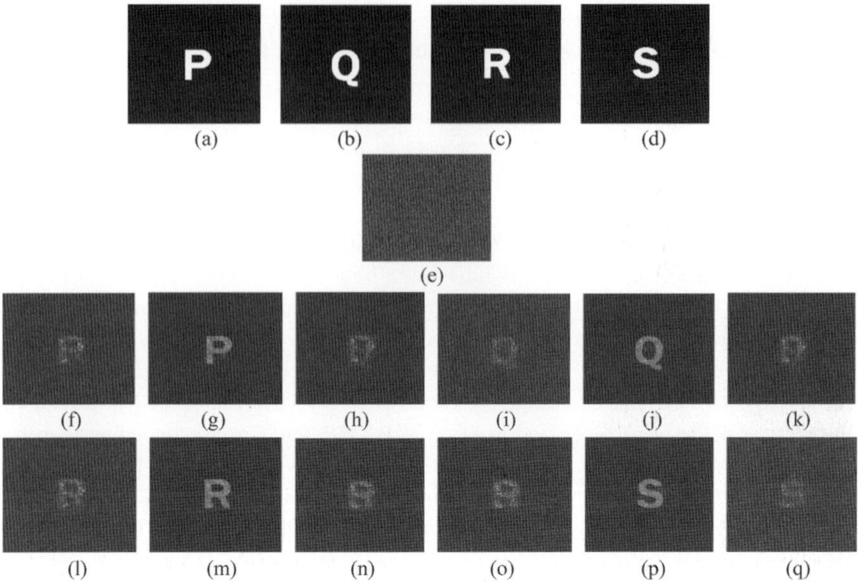

Figure 6.27 Decryption of a binary object with in-plane rotation of the mask R_{sm} with respect to R_{fm}. (a–d) Input images. (e) Multiplexed encrypted image recorded with objects P, Q, R, S with mask R_{sm} rotated by 2°, 1°, −1°, and −2°, respectively. Decrypted images with (f–h) R_{sm} at 2.5°, 2°, and 1.5°, respectively; (i–k) R_{sm} at 1.5°, 1°, and 0.5°, respectively; (l–n) R_{sm} at −0.5°, −1°, and −1.5°, respectively; and (o–q) R_{sm} at −1.5°, −2°, and −2.5°.

mask R_{sm} from its initial orientation by ±0.5°. The decrypted images shown (Figure 6.27f–h) are for different positions of the R_{sm} and keeping it at 2.5°, 2°, and 1.5°, respectively, as compared to the recording angle of 2°. The decrypted images shown in Figure 6.27i–k are for the orientation of the constituent diffuser R_{sm} at the angles 1.5°, 1°, and 0.5°, respectively, as compared to the recording angle of 1°. Figure 6.27l–n are for the orientation of R_{sm} at the angles of −0.5°, −1°, and −1.5° in place of recording angle of −1° and + 1°. Figure 6.27o–q are at the angles of −1.5°, −2°, and −2.5° in place of recording angle of −2°.The MSE values between the decrypted images and the primary images are of the same order of ($\approx 10^3$). From these results it is clear that we get the cross talk image (Figure 6.27h, i, k, l, n, and o) if we keep the angle of rotation between the successive recordings small.

6.10.4
Multiplexing by Using Set of Apertures and Angular Rotation of R_{sm}

In this study, same set of P, Q, R, and S each of pixel size 256 × 256 (Figure 6.28a–d) are chosen as primary objects. A set of apertures is kept in close contact with the sandwich diffuser in the Fourier plane. The FT of each input object is passed through

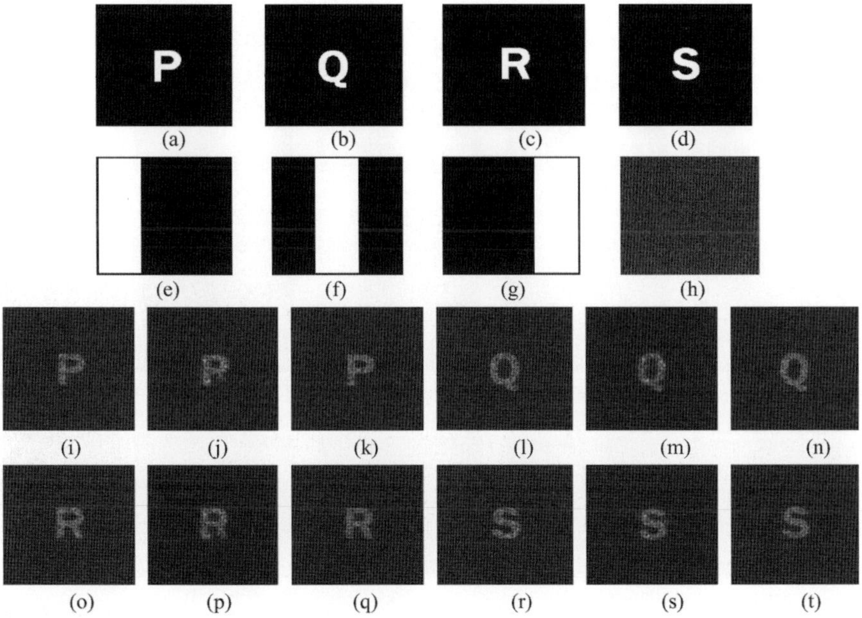

Figure 6.28 Encryption–decryption system using set of apertures with in-plane rotation of mask R_{sm}. (a–d) Input images. (e–g) Apertures. (h) Multiplexed encrypted image recorded with object P with R_{sm} at angles 2.5°, 2°, and 1.5°; Q with R_{sm} at 1.5°, 1°, and 0.5°; R with R_{sm} at −0.5°, −1°, and −1.5°; and S with R_{sm} at −1.5°, −2°, and −2.5°, and passing each through one aperture of the set of apertures (Figure e–g). Decrypted images with same apertures as used during encryption with (i–k) R_{sm} at 2.5°, 2°, and 1.5°, respectively; (l–n) R_{sm} at 1.5°, 1°, and 0.5°, respectively; (o–q) R_{sm} at −0.5°, −1°, and −1.5°, respectively; and (r–t) R_{sm} at −1.5°, −2°, and −2.5°, respectively.

all the three apertures (Figure 6.28e–g) with the corresponding three angular positions of the mask R_{sm}. Four different sets of input images are taken for multiplexing. Each input image out of P, Q, R, and S is recorded three times by keeping the R_{sm} at 2.5°, 2°, and 1.5° with P; 1.5°, 1° and, 0.5° with Q; −0.5°, −1°, and −1.5° with R; and −1.5°, −2°, and −2.5° with S, respectively. The inverse FT is calculated to get the multiplexed encrypted image (Figure 6.28h). The decrypted images (Figure 6.28i–t) are recovered from the multiplexed encrypted image with the same aperture and same position of the mask R_{sm} as they occupied at the time of recording. The MSE values between the decrypted and the primary images are of the same order ($\approx 10^{-3}$). From the results, it is clear that the cross talk images are missing. Instead of cross talk images, we get the partial decrypted images for ±0.5° orientation of the R_{sm}. So the aperture system along with the rotation of one of the constituent diffusers may be used to enhance the capability of the multiplexing without the generation of cross talk images. We have calculated the MSE and SNR values as a function of the number of multiplexed images by choosing an input image

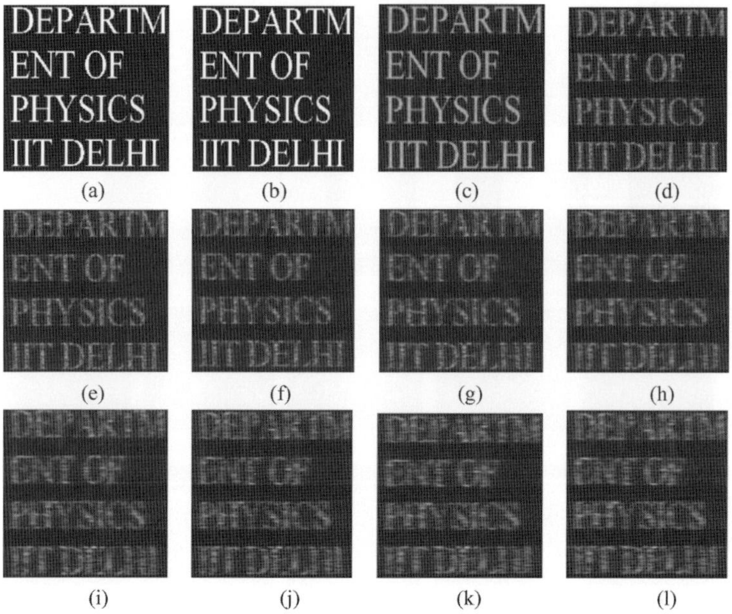

Figure 6.29 (a) Original image. Decrypted images obtained directly (b) without an aperture and (c) with one aperture. Decrypted images obtained by multiplexing with (d) 2 images and 2 apertures; (e) 3 images and 3 apertures; (f) 4 images and 4 apertures; (g) 5 images and 5 apertures; (h) 6 images and 6 apertures; (i) 7 images and 7 apertures; (j) 8 images and 8 apertures; (k) 9 images and 9 apertures; and (l) 10 images and 10 apertures.

shown in Figure 6.29a. Figure 6.29b is the decrypted image obtained when there is no aperture used at the time of recording and decryption. Figure 6.29c is the image obtained after the decryption in case of using one aperture. Figure 6.29d–l are obtained by multiplexing 2–10 similar images, respectively. Figure 6.30a shows the increase in the MSE as a function of the number of multiplexed images. Here, the MSE values are smaller than the values calculated for Figure 6.28. It is due to the fact that in case of Figure 6.29, the input images cover larger area than the images considered in Figure 6.28. From the MSE values obtained corresponding to Figures 6.28 and 6.29, it is clear that the MSE values can be reduced by taking the input images as large as possible in terms of the area covered. Figure 6.30b shows the change in SNR as a function of the number of the multiplexed images. It is clear that one should use the minimum number of apertures possible, as it affects the SNR considerably. To get the images with better resolution, the aperture size should be as large as possible so that most of the incident light may pass through it. So, a compromise has to be made between the enhancement in the capability of multiplexing by using a number of apertures and the accepted MSE and SNR values.

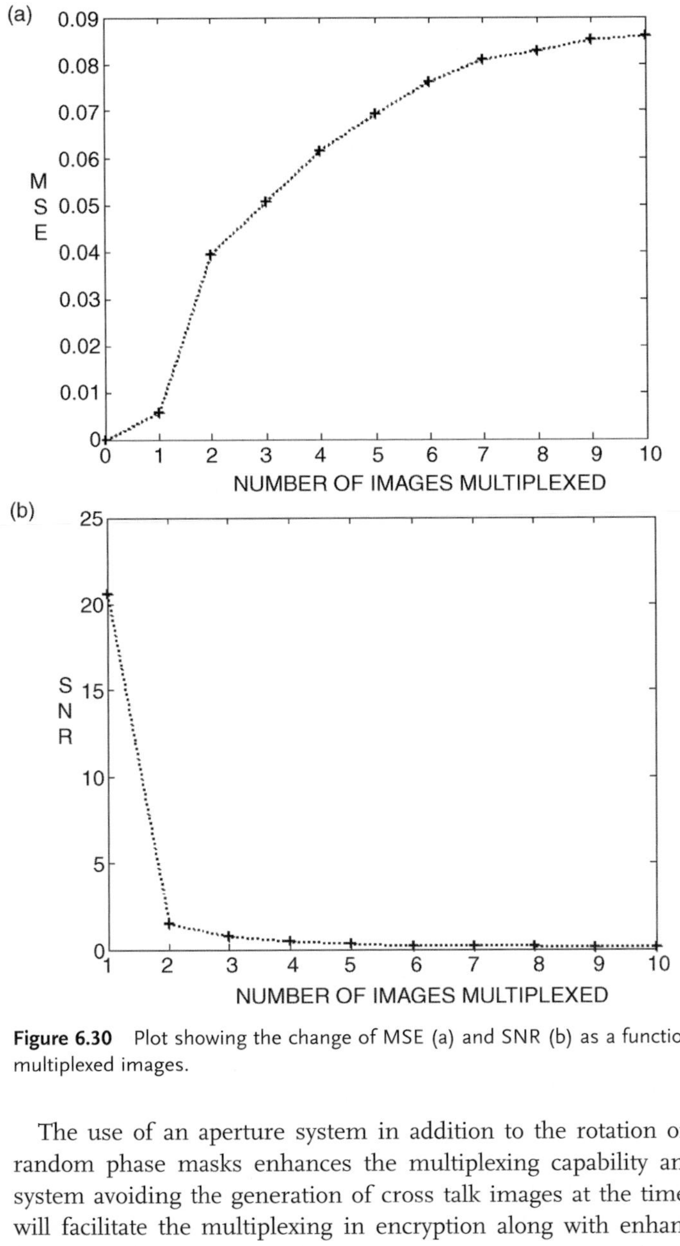

Figure 6.30 Plot showing the change of MSE (a) and SNR (b) as a function of the number of multiplexed images.

The use of an aperture system in addition to the rotation of one or both the random phase masks enhances the multiplexing capability and security of the system avoiding the generation of cross talk images at the time of decryption. It will facilitate the multiplexing in encryption along with enhanced security by a factor equal to the sum of the number of apertures used and the number of rotation steps used for the purpose. Simulation results are presented in support of the proposed techniques. The decrypted images obtained from a multiplexed encrypted image have been recorded successfully. To evaluate reliability of the technique, mean square error between the decrypted and the original images has also been calculated.

6.11 Speckles in Digital Fresnel Field Encryption

As mentioned in Section 6.1, digital holography-based techniques have certain advantages [13, 24, 31, 32, 35, 38] because the holograms can be transmitted through normal communication channels, especially after their compression. Nelleri *et al.* [154] have described a fruitful digital encryption scheme by applying it to the complex digital Fresnel field retrieved at the CCD plane from a single off-axis hologram. The quantitative complex object information can be reconstructed from the Fresnel field. Since the numerically retrieved complex wavefront is in the form of a digital complex image, one can apply the existing encryption schemes in the digital domain to secure the holographic data. A Fresnel field retrieved from an off-axis digital Fresnel hologram of a 3D object gives the appearance of a noise-like distribution. This is due to the scattering nature of the object and the resulting speckle noise, which is an inherent quality of the reconstruction process of a digital hologram. A single random Fourier plane encoding is carried out with an assumption that the noisy nature of the complex field may be equivalent to the primary random mask-bonded field in the conventional double random Fourier plane encoding.

The Fast Fourier Transform (FFT)-based implementation of the decryption process and the evaluation of the Fresnel-approximated diffraction integral for hologram reconstruction are combined for a fast and single-step decryption of the encrypted Fresnel field into complex object information. Digital encryption is helpful in handling the random phase codes, and it is the central key to the encryption and decryption processes. Apart from this, because we are encoding the Fresnel field, which ensures that the reconstruction parameters of the complex object wavefront become additional keys, security is increased. The use of a complex wave retrieval algorithm is helpful in zero-order and twin image-free object reconstruction from a single off-axis hologram. Here, we present some experimental results as a proof of the concept, based on the work of Nelleri *et al.* [154].

6.11.1 Digital Recording and Numerical Reconstruction of an Off-Axis Fresnel Hologram

As in conventional holography, both in-line and off-axis geometries are used for digital recording of holograms. The complex object wavefront is retrieved from the digital hologram by simulating the optical diffraction process of the exact reference wave from the hologram in the digital domain. Both off-axis and in-line holography have their own advantages and disadvantages. The reconstructed image quality is severely degraded by interference terms when only a single in-line hologram is used for reconstruction. Phase shifting in-line digital holography solves this problem, but one needs to record multiple holograms and more importantly ensure that the phase remains constant over the entire recording cycle, resulting in the requirement of very high precision levels of the technique. The bandwidth of off-axis holograms is wider than that of in-line holograms. The angle between the reference and object beams needs to be large enough to separate out the spatial frequencies of the zero order from the first orders. This causes the fringe width to be narrower, and the CCD camera has to have sufficient resolution to sample the fringes. This is not a serious limitation of the technique, however, with the availability of high-resolution cameras.

Off-axis holography has been recently used for high-resolution microscopic imaging. The other advantage of off-axis holography is that the problem of presence of DC and conjugate images in the reconstruction can be solved by a novel reconstruction method proposed recently by Liebling *et al.* [155]. For this, one needs to record only one hologram. If the recording device (CCD) has sufficient bandwidth, then recording of the signal generated by the optical system can be ensured by choosing an appropriate angle between the recording beams, taking into consideration the resolution of the detector. We believe that off-axis holography has definite advantages because it needs a simpler optical system and provides high data acquisition rates compared to in-line geometries with phase shifting.

Let the 3D object be illuminated by a coherent light of wavelength λ. The complex amplitude distributions in the object plane and the Fresnel plane separated by a distance d are related by a Fresnel transform:

$$\tilde{t}(x, y, d) = \frac{e^{ikd}}{i\lambda d} \iint t(x', y', 0) \times \exp\frac{i\pi}{\lambda d}\left[(x-x')^2 + (y-y')^2\right] dx' dy', \tag{6.14}$$

where (x', y') and (x, y) are the coordinates in the object plane and the CCD plane, respectively. The object wavefront is recorded as a digital off-axis hologram on a CCD using a Mach–Zehnder interferometer setup. Let the object field at the CCD plane be $\tilde{t}(x, y, d) = O_t \exp(i\vec{k}_1 \cdot \vec{r})$. Now the intensity distribution in the hologram is described by

$$I(x, y, d) = \left| O_t \exp(i\vec{k}_1 \cdot \vec{r}) + \exp(i\vec{k}_2 \cdot \vec{r}) \right|^2. \tag{6.15}$$

The resultant wavevector K of the fringe pattern is given by

$$K = \vec{k}_1 + \vec{k}_2 = (4\pi/\lambda)\sin(\theta/2). \tag{6.16}$$

The microinterference in the hologram at each point is determined by θ, which is the angle between the interfering beams. $I(x, y, d)$ is sampled by the CCD with sampling period T to give the 2D discrete hologram $I(\tilde{m}, \tilde{n})$. The Fresnel transform $\tilde{t}(m, n, d)$ of the object complex amplitude distribution can be retrieved from I by a nonlinear method proposed by Liebling *et al.* [155]. The object distribution $t(m, n, 0)$ at the object plane is obtained from $\tilde{t}(m, n, d)$ in the CCD plane by a discrete Fresnel transform using the convolution method.

6.11.2 Digital Fresnel Field Encryption

In this section, we describe the proposed encryption technique and the usefulness of speckles in the process of encrypting the digital holographic data and the direct reconstruction of a complex object wavefront from the encrypted Fresnel field. The reconstruction of a complex Fresnel field at the CCD plane from an off-axis digital Fresnel hologram is carried out by following a two-step procedure given below:

1) The off-axis hologram of the object located at a distance d from the recording plane is recorded and is described by Equation 6.15.
2) The complex wavefront at the CCD recording plane is retrieved from the off-axis hologram.

Our objective is the encryption of the Fresnel field by a single random Fourier plane encoding and then decryption leading to direct object reconstruction. The method we propose is explained in the following sections.

6.11.2.1 Digital Encryption of Fresnel Field Using Single Random Phase Encoding

Let R be a digitally produced 2D random phase code uniformly distributed in $[0, 2\pi]$ and used in the Fourier plane to encode the Fresnel field $\tilde{t}(\,)$. The encoded complex Fresnel field is given by

$$\psi(\,) = \tilde{t}(\,) \otimes \hat{R}(\,), \tag{6.17}$$

where $\hat{R}$ is the Fourier transform of R and $\otimes$ denotes the convolution operation. The digital implementation of the preceding encoding is obtained using a 2D FFT algorithm. The encrypted complex signal is given as

$$\psi(\,) = \Im^{-1}\{\Im\{\tilde{t}(\,)\} \cdot R\}, \tag{6.18}$$

where $\Im$ and $\Im^{-1}$ denote the Fourier and inverse Fourier transform, respectively. In double random phase Fourier plane encoding, a primary RPM is bonded to the input function to ensure that the encrypted function is purely white noise. The noisy appearance of the retrieved complex Fresnel field is primarily due to the scattering nature of the object and speckle noise, which is a salient trait of the reconstruction process of a hologram. In holography, the undesirable speckle noise degrades the reconstructed information, which causes reduction in the SNR. Digital holography is not an exception to this problem. The use of coherent light and the finite sampling size of the CCD pixels contribute speckles to the reconstruction. The noisy nature of the complex Fresnel field may be equivalent to the primary random mask-bonded field in the conventional double random Fourier plane encoding and can lead to the desired results, as in the case of double random phase Fourier plane encoding.

6.11.2.2 Direct Decryption of 3D Object Information from Encrypted Fresnel Field

Conventionally, after retrieving the complex Fresnel field at the CCD plane, the complex object information in a plane at a distance d can be obtained by performing a discrete Fresnel transform. There are two relevant methods of implementing a discrete Fresnel transform. One method is by direct evaluation of the Fresnel-approximated diffraction integral and the other method is called the convolution or spectral method. Both the preceding methods yield the same result if the sampling issues are taken care of properly. Using the direct method, the complex object information obtained by the discrete Fresnel transform is expressed as

$$t(\,) = \frac{1}{i\lambda d}\left[\frac{i\pi}{\lambda d}(m^2\Delta\xi^2 + n^2\Delta\eta^2)\right] \times \sum_{k=-N_x/2}^{N_x/2-1}\sum_{l=-N_y/2}^{N_y/2-1} \tilde{t}(k,l)\times$$

$$\exp\left[i\frac{\pi}{\lambda d}(k^2\Delta x^2 + l^2\Delta y^2)\right] \times \exp\left[i2\pi\left(\frac{mk}{N_x} + \frac{nl}{N_y}\right)\right]. \tag{6.19}$$

Here Δx and Δy denote the sampling periods of the CCD camera in the x- and y-directions, respectively. $\Delta\xi = (\lambda d)/(N_x\Delta x)$ and $\Delta\eta = (\lambda d)/(N_y\Delta y)$ are the reconstructed image pixel dimensions, where ζ and η are the image plane coordinates.

In digital holography, the discrete form of the preceding wave propagation is simulated using an FFT algorithm. Using the convolution method, the complex amplitude distribution of the object field may be obtained by considering the integral based on the Fresnel approximation and expressed as

$$t(\xi,\eta) = \int_{-\infty}^{\infty}\int_{-\infty}^{\infty} \tilde{t}(x,y)g(\xi - x, \eta - y)\mathrm{d}x\mathrm{d}y, \tag{6.20}$$

where the convolution kernel $g(\xi,\eta) = [\exp(ikd)]/[i\lambda d]\exp\{[ik]/[2d](\xi^2+\eta^2)\}$ is called the impulse response function. Computationally, this can be implemented to give complex object information as

$$t(\,) = \Im^{-1}\{\Im\{\tilde{t}(\,)\}\cdot\Im\{g(\,)\}\}. \tag{6.21}$$

The term $\Im\{g(\,)\}$ is the transfer function of the free-space propagation used to reconstruct the object field from the complex Fresnel field and denoted by G:

$$G(m,n) = \exp\left(\frac{i2\pi}{\lambda d}\right)\sum_{m=-N_x/2}^{N_x/2-1}\sum_{n=-N_y/2}^{N_y/2-1} \times \exp\left\{-i\pi\lambda d\left[\left(\frac{m}{M\Delta x}\right)^2 + \left(\frac{n}{N\Delta y}\right)^2\right]\right\}. \tag{6.22}$$

The decryption of the Fresnel field from the encrypted complex image represented by Equation 6.18 is achieved by exactly following the inverse algorithm with the conjugate of the random phase codes used in the Fourier plane during the encryption process:

$$\tilde{t}(\,) = \Im^{-1}\{\Im\{\psi(\,)\}\cdot R^*\}, \tag{6.23}$$

where R^* is the conjugate RPM. Accounting for the similarities between 4-f geometry of the encryption–decryption process and numerical evaluation of the diffraction integral (Equations 6.21 and 6.23), decryption of the complex object field may be directly obtained from the encrypted Fresnel field by combining the two processes in a single step, as given here:

$$t(\,) = \Im^{-1}\{\Im\{\psi(\,)\}\cdot R^*\cdot G\}. \tag{6.24}$$

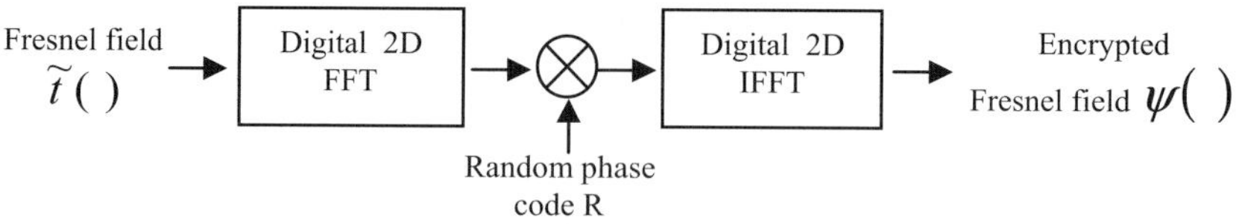

Figure 6.31 Fresnel field encryption utilizing the presence of speckles.

The true significance of digital holography is found in the property of object phase reconstruction, which contains the 3D information. The retrieved Fresnel field contains the quantitative 3D information of the object in the form of phase. Bonding a RPM with it will alter the original phase of the complex object information. A secondary process will be needed to reconstruct the exact object phase distribution. The presence of speckles in the retrieved Fresnel field is advantageous to have single random phase encoding to give white noise-like distribution in the encrypted Fresnel field. Figure 6.31 shows the encryption process. Figure 6.32 describes the numerical reconstruction of the complex object field from the retrieved Fresnel field by the convolution method. Figure 6.33 shows the single-step object reconstruction from the encrypted Fresnel field. If the transfer function G is not multiplied in the Fourier plane, the original Fresnel field is decrypted. This is shown in Figure 6.34. A separate Fresnel transform is required to reconstruct the complex object information from the decrypted Fresnel field. Correct information about the Fourier plane RPM and the transfer function can be communicated to an authentic user for the successful decryption of complex object information in a single step. In addition to the random phase key, information about the transfer function is crucial for the robust decryption of the object function. It is important to note that G is a function of the hologram reconstruction parameters, which further strengthen the performance of the key.

6.11.3 Experiment

In this section, we describe the recording and reconstruction of a hologram of a 3D object, a semipolished steel sphere of diameter 8 mm. An off-axis Fresnel digital hologram of the 3D object was recorded using a Mach–Zehnder interferometer setup (Figure 6.35). A diode-pumped Nd-YAG second harmonic laser producing light of wavelength 532 nm is used as a light source. The holograms are captured and digitized into an 8-bit gray-level image matrix of size 576×768 using a CCD camera of square pixel format and a frame grabber. The angle between the interfering beams is approximately 3°. The object is located at a distance d from the CCD recording plane. The complex wavefront retrieved at the CCD plane is propagated over a distance d using a discrete Fresnel transform algorithm by the convolution method.

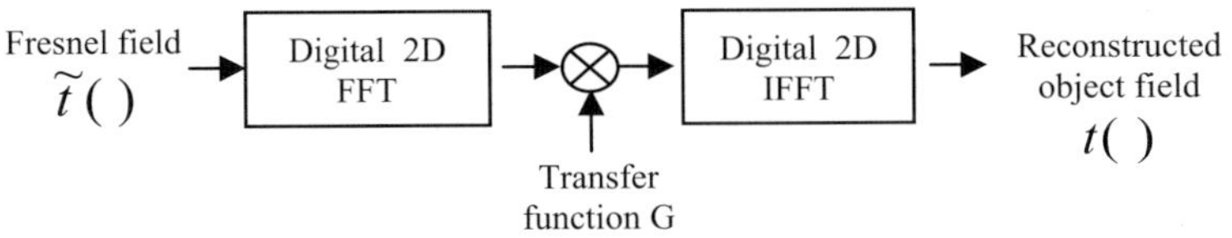

Figure 6.32 Numerical reconstruction of the object field from the Fresnel field using convolution method.

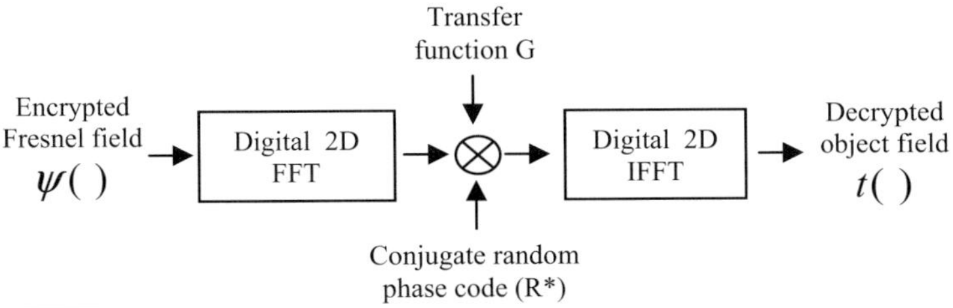

Figure 6.33 Direct decryption of the object field from the encrypted Fresnel field.

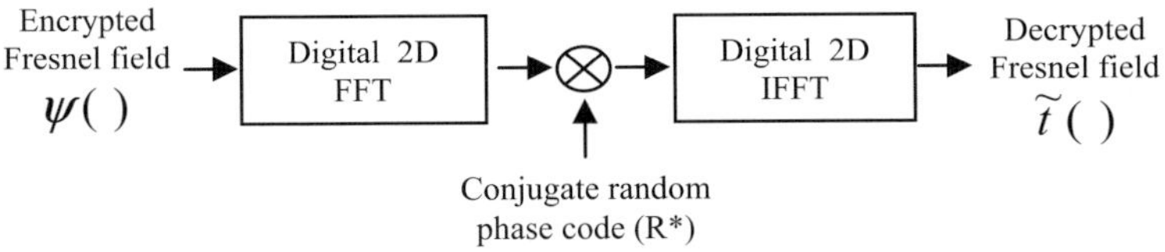

Figure 6.34 Decryption results in the Fresnel field in the absence of a transfer function.

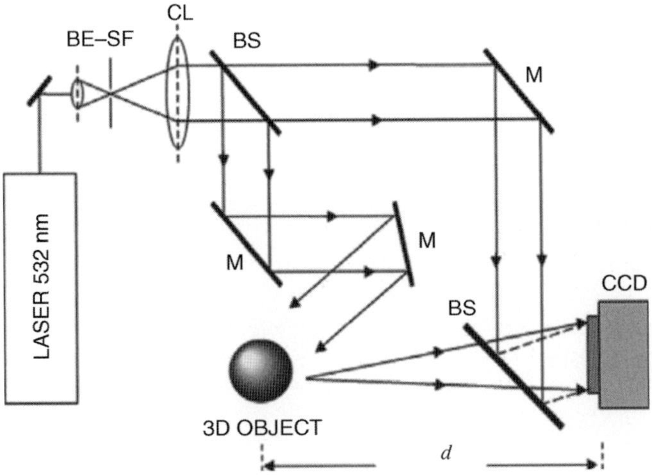

Figure 6.35 Experimental setup. BE–SF, beam expander and spatial filter; CL, collimating lens; M, mirror; BS, beam splitter.

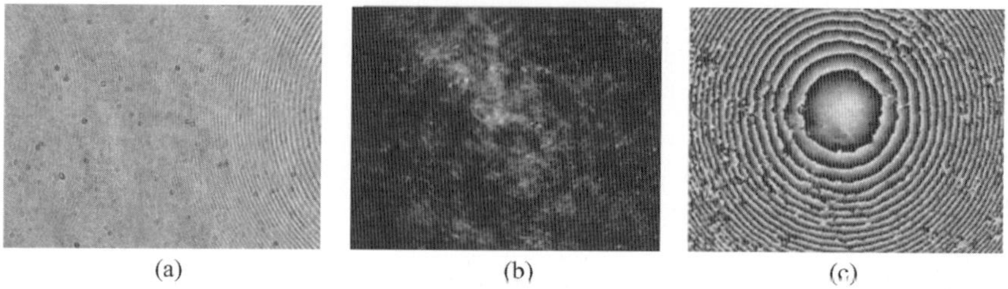

Figure 6.36 (a) Hologram of the 3D object sphere – the hologram size is 576 × 768 pixels. (b) Intensity of complex wavefront (Fresnel field) retrieved at the CCD plane at a distance 34 cm from the object. (c) Corresponding phase of the wavefront.

6.11.4 Results and Discussion

Figure 6.36a shows the digital hologram of the object at a distance 34 cm from the CCD. The intensity and phase of the complex wavefront retrieved at the CCD plane are shown in Figure 6.36b and c, respectively. The intensity of the reconstructed complex wavefront on a plane through the center of the object at a distance 34 cm from the CCD is shown in Figure 6.37a. The corresponding phase of the complex object wavefront is shown in Figure 6.37b.

6.11.4.1 Discussion of Encryption and Decryption by the Proposed Method

The retrieved complex Fresnel field at the CCD plane is encrypted by simulating the 4-f system of the optical encryption process, which is as dictated by Equation 6.18. A computer-generated random phase key uniformly distributed in $[0, 2\pi]$ serves as a Fourier plane phase mask. Figure 6.38a and b shows the intensity and phase of the encrypted signal, respectively, after digital single random phase Fourier plane encoding. In this work, the motivation for using digital random phase Fourier plane encoding is developed from numerical reconstruction of the complex object information from the Fresnel field and its FFT-based implementation of the diffraction

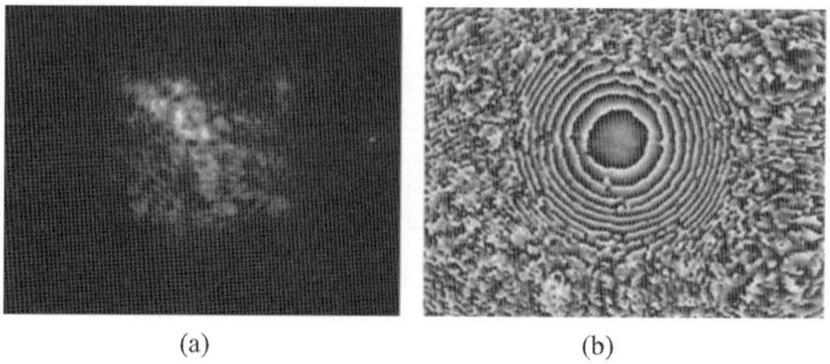

Figure 6.37 (a) Reconstructed intensity of the object wavefront in the plane through the center of the 3D object at a distance 34 cm from the CCD. (b) Reconstructed phase of the object wavefront.

(a) (b)

Figure 6.38 (a) Intensity of the encrypted Fresnel field. (b) The corresponding phase.

propagation. Moreover, one can effectively utilize optical and recording parameters of the hologram in the form of a transfer function to add to security aspects and explore the possibility of direct decryption into complex object information from the encrypted Fresnel field with lesser computational power. Figure 6.39a and b shows the intensity and phase parts, respectively, of the transfer function, which is multiplied in the inverse Fourier plane along with conjugate random phase codes, giving rise to complex object reconstruction in a single step during the decryption process. Figure 6.40a and b shows the decrypted intensity and phase of the complex object. In the conventional double random Fourier plane encoding, the Fourier spectrum of the encrypted image (FSE) is

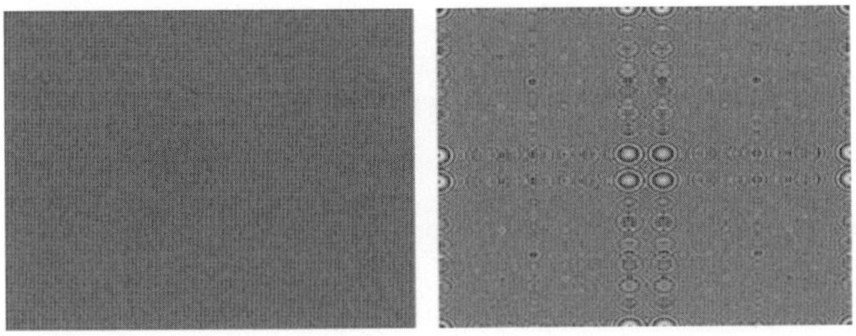

(a) (b)

Figure 6.39 (a) Intensity and (b) phase of the transfer function multiplied in the Fourier plane.

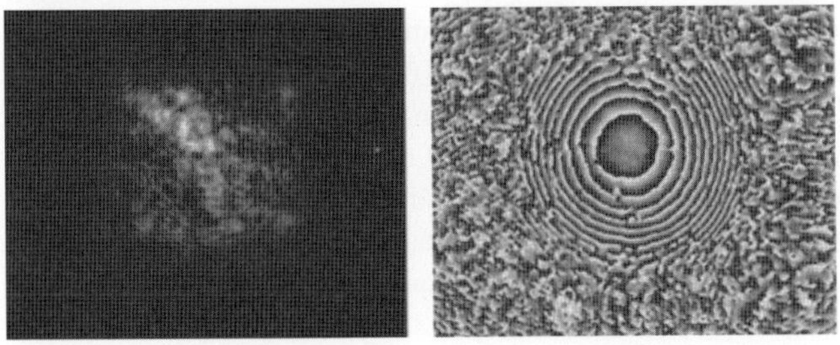

(a) (b)

Figure 6.40 Decrypted complex object wavefront from the encrypted Fresnel field. (a) Intensity. (b) The corresponding phase.

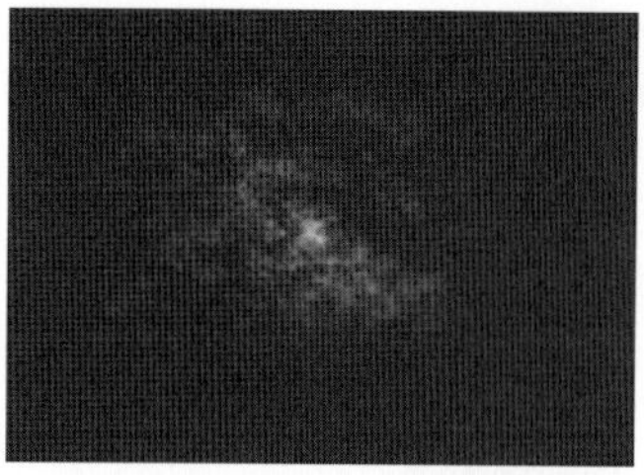

Figure 6.41 Fourier spectrum of the encrypted image (FSE).

$$\mathrm{FSE} = |\Im\{\psi(\,)\}|^2 = |\Im\{\text{input field} \times R_1\}|^2, \tag{6.25}$$

where $\psi(\,)$ is the complex encrypted image and R_1 is the first RPM bonded with the object. This is expected to be a noise, because the object is multiplied by the input RPM R_1. In the absence of R_1,

$$\mathrm{FSE} = |\Im\{\text{input field} \times R_1\}|^2. \tag{6.26}$$

In such a case, it may be possible to retrieve object information from FSE using phase retrieval techniques. The proposed method suggests that the speckle noise present in the Fresnel field plays the role of input RPM R_1 for encryption. Figure 6.41 shows the Fourier spectrum of the encrypted signal. The weak DC and higher frequencies in the figure show that the speckles in the Fresnel field effectively act as the input RPM. For a strongly light-scattering 3D object where the retrieved complex Fresnel field appears almost like a noisy complex digital image, the encrypted image tends to be a white sequence.

6.11.4.2 Some General Remarks on Digital Encryption of Holographic Information

In optical security applications, the 2D RPMs serve as essential keys for encrypting and decrypting the information. These random phase keys may be placed in the spatial and/or spatial frequency planes of a linear optical transform system such as Fourier, Fractional Fourier, or Fresnel transform in single or multiple transform stages. This changes the extent of the spatial frequency bandwidth of the encrypted complex signal at each stage. Especially in security systems based on digital holography like hybrid techniques, if one opts for encoding in the optical stage of the process, then the sufficient number of samples that are required (space–bandwidth product) to fully represent the encrypted signal and to subsequently reconstruct the signal solely depends on the resolution of the CCD. The spatial location, alignment accuracy, and so on of the phase mask used in the optical domain are very significant, as they are to be exactly repeated in the digital domain for the robust decryption of the original information. Also, it is important to note that making a pure phase modulating mask is impractical by any physical means. Requirement of a physical RPM may be avoided if the encryption is performed digitally. In digital holography, the use of a digital encryption algorithm has its own relevance, as it involves hybrid data processing.

6.12 Conclusions

In the present era, called the "information age," the security of information has become of paramount importance in many application areas, and various techniques have been developed to achieve the desired goal. Traditional methods have relied on the principle of cryptography, the science of securing information. When large amount of data are to be encoded, the requirements of using more powerful algorithm and large key length to encode may turn out to be the system's bottleneck. In view of this, scientists have been exploring various ways of using optical techniques attractive for data security applications. As the speckle-based techniques play an important role in optical information processing, they have shown great potential for useful manipulation of data and images for their security, storage, and transfer applications. This chapter attempts to describe certain speckle coding techniques for security applications. Speckle coding techniques in general and techniques based on authors' works have then been discussed. A review is first given of the double random optical encryption method and its variants.

An encryption/decryption technique has been discussed that uses a sandwich phase mask made of speckle patterns in a double random phase encoding geometry at the Fourier plane and acts as a key. The encrypted images are holographically recorded and are then decrypted by generating a conjugate of the encrypted image. If, after encryption, the two phase masks constituting a sandwich are separated, it becomes almost impossible to decrypt the image. For faithful decryption, the two constituent speckle phase masks are to be kept at the same position and orientation that they were during the encryption. This method has the advantage over conventional methods that the knowledge of position and pattern of both the RPMs is essential for successful retrieval of the information. An experimental technique using Young's fringes has been used to overcome the practical difficulty for generation of the same key in case of a sandwich diffuser. The work has been further extended by carrying out encryption with a sandwich phase mask made by using two elongated speckle patterns, as well as by using one speckle phase mask and a sinusoidal grating. The results clearly indicate that the methods are more secure compared to the DRPE. Elongated speckles also facilitate easy alignment of the constituent diffusers used in making a sandwich phase diffuser as compared to the case of using a sandwich diffuser made by using normal speckles. Simulation results are presented in support of the proposed idea. To evaluate the reliability of the technique, MSE between the decrypted and original images has been calculated.

We have also described multiplexing techniques in optical encryption of 2D images by using an aperture system along with a sandwich diffuser in the Fourier plane. It is found that any other multiplexing technique used with an aperture system increases the performance of the technique but at the cost of resolution of the images. This is due to the fact that other decrypted images with wrong keys are present while retrieving the decrypted image with a correct key. However, in some cases, it may be very useful to increase the multiplexing capability if the multiplexing is limited due to the physical and optical parameters of the recording devices. We have presented the results of the multiplexing in encryption by successive recording of encrypted images one by one. Use of aperture system may increase manifold the multiplexing capability of encryption and decryption. Results have also been presented for a case in which the encryption is done by rotating one of the constituent phase diffusers of a sandwich diffuser. By using the aforementioned techniques, one can carry out multiplexing in encryption without the generation of cross talk images. Simulation results are presented in support of the proposed idea. To evaluate the reliability of the technique, the MSE between the decrypted and original images has been calculated.

We have also demonstrated an encryption–decryption algorithm for securing digital holographic data. The FFT-based implementation of the decryption process and the evaluation of a Fresnel-approximated diffraction integral for hologram reconstruction are combined for a fast and single-step decryption of the encrypted Fresnel field into complex object information. The noisy nature of the retrieved complex Fresnel field due to the scattering nature of the object and inherent speckle noise can be utilized for single random Fourier plane encoding. The merits of using a single hologram for DC and twin image-free reconstruction in an encryption–decryption system are also demonstrated.

Acknowledgment

The authors wish to thank Mr. Pramod Kumar for his help in preparing the manuscript of this chapter.

References

1 Dainty, J.C. (ed.) (1984) *Laser Speckle and Related Phenomena*, 2nd edn, Springer-Verlag, Berlin.

2 Dainty, J.C. (1976) The statistics of speckle patterns, in *Progress in Optics*, vol. 14 (ed. E. Wolf), North Holland, pp. 1–46.

3 Erf, R.K. (ed.) (1978) *Speckle Metrology*, Academic Press, New York.

4 Francon, M. (1979) *Laser Speckles and Applications in Optics* (trans. H.H. Arsenault), Academic Press.

5 Rastogi, P.K. (ed.) (1997) *Optical Measurement Techniques and Applications*, Artech House Inc., Boston.

6 Rastogi, P.K. (ed.) (2001) *Digital Speckle Pattern Interferometry and Related Techniques*, John Wiley & Sons, Inc., New York.

7 Sirohi, R.S. (ed.) (1991) *Selected Papers on Speckle Metrology*, SPIE Milestone Series, vol. MS-35, Optical Engineering Press, Bellingham, WA.

8 Sirohi, R.S. (ed.) (1993) *Speckle Metrology*, Marcel Dekker, Inc., New York.

9 Goodman, J.W. (1984) *Statistical Optics*, John Wiley & Sons, Inc., New York.

10 Goodman, J.W. (2009) *Speckle Phenomena in Optics: Theory and Applications*, Viva Books, New Delhi.

11 Stallings, W. (1999) *Cryptography and Network Security: Principles and Practice*, 2nd edn, Prentice Hall, New Jersey.

12 Refregier, P. and Javidi, B. (1995) Optical image encryption based on input plane and Fourier plane random encoding. *Opt. Lett.*, **20**, 767–769.

13 Javidi, B. (ed.) (2005) *Optical and Digital Techniques for Information Security*, Springer.

14 Javidi, B. (ed.) (2006) *Optical Imaging Sensors and Systems for Homeland Security Applications*, Springer.

15 Matoba, O. and Javidi, B. (Sept. 2001) The keys to holographic data security. *IEEE Circuit Dev. Mag.*, **16**, 8–15.

16 Unnikrishnan, G. and Singh, K. (2001) Cryptography using optoelectronic techniques. *Proc. SPIE*, **4417**, 178–189.

17 Unnikrishnan, G. and Singh, K. (2001) Security systems using optoelectronic processors. *Asian J. Phys.*, **10**, 1–35.

18 Singh, K. and Nishchal, N.K. (2002) Fractional Fourier transforms; applications in information optics. *Proc. SPIE*, **4929**, 34–48.

19 Singh, K., Unnikrishnan, G., and Nishchal, N.K. (2002) Photorefractive optical processing for data security. *Proc. SPIE*, **4803**, 205–219.

20 Hennelly, B.M. and Sheridan, J.T. (2003) Image encryption and the fractional Fourier transform. *Optik*, **114**, 251–265.

21 Hennelly, B.M. and Sheridan, J.T. (2003) Image encryption techniques based on the fractional Fourier transform. *Proc. SPIE*, **5202**, 76–87.

22 Alfalou, A. and Brosseau, C. (2009) Optical image compression and encryption methods. *Adv. Opt. Photon.*, **1**, 589–636.

23 Matoba, O., Nomura, T., Perez-Cabre, E., Millan, M.S., and Javidi, B. (2009) Optical techniques for information security. *Proc. IEEE*, **97**, 1128–1148.

24 Tajahuerce, E., Matoba, O., and Javidi, B. (2001) Image security by digital holography, in *Smart Imaging Techniques* (ed. B. Javidi), SPIE Press, Bellingham, WA, pp. 45–67.

25 Tajahuerce, E., Lancis, J., Andres, P., Climent, V., and Javidi, B. (2005) Optoelectronic information encryption with incoherent light, in *Optical and Digital Techniques for Information Security* (ed. B. Javidi), Springer, pp. 95–112.

26 Chalasinska-Macukow, K. and Kotynski, R. (2002) Correlation methods in optical encryption and security, in *Perspectives in Modern Optics and Optical Instrumentation* (eds J. Joseph, A. Sharma, and V.K. Rastogi), Anita Publications, New Delhi, pp. 187–193.

27 Rosen, J. (2003) Security systems based on spatial correlation with meaningful output images, in *Perspective in Engineering Optics* (eds K. Singh and V.K. Rastogi), Anita Publications, New Delhi.

28 Kim, K.T., Kim, J.W., Lee, J.S., and Choi, J.U. (2004) Holographic image watermarking for secure content, in *Trust and Privacy in Digital Business* (eds S. Katsikas, J. Lopez, and G. Pernul), Springer-Verlag, pp. 219–231.

29 Kishk, S. and Javidi, B. (2005) Distortion- and noise-robust digital watermarking using input and Fourier-plane phase encoding, in *Optical and Digital Technique for Information Security* (ed. B. Javidi), Springer, pp. 37–57.

30 Marvel, L.M. (2005) Information hiding: steganography and watermarking, in *Optical and Digital Techniques for Information Security* (ed. B. Javidi), Springer, pp. 113–133.

31 Matoba, O. and Javidi, B. (2005) Secure display using encrypted digital holograms, in *Optical and Digital Techniques for Information Security* (ed. B. Javidi), Springer, pp. 155–172.

32 Naughton, T.J. and Javidi, B. (2005) Compression of digital holograms for secure three-dimensional image storage and transmission, in *Optical and Digital Technique for Information Security* (ed. B. Javidi), Springer, pp. 173–191.

33 Nomura, T. and Javidi, B. (2005) Optical image encryption using optimized keys, in *Optical and Digital Technique for Information Security* (ed. B. Javidi), Springer, pp. 193–205.

34 Nomura, T. and Javidi, B. (2005) Polarization encoding for an optical security system, in *Optical and Digital Techniques for Information Security* (ed. B. Javidi), Springer, pp. 207–220.

35 Nomura, T., Tajahuerce, E., Matoba, O., and Javidi, B. (2005) Applications of digital holography for information security, in *Optical and Digital Technique for Information Security* (ed. B. Javidi), Springer, pp. 241–269.

36 Tanida, J. and Sasaki, T. (2005) Stream cipher using optical affine transformation, in *Optical and Digital*

Techniques for Information Security (ed. B. Javidi), Springer, pp. 221–239.

37 Rosen, J. and Javidi, B. (2005) Steganography and encryption systems based on spatial correlation with meaningful output images, in *Optical and Digital Technique for Information Security* (ed. B. Javidi), Springer, pp. 59–94.

38 Nomura, T. (2006) Hybrid optical encryption of a 3D object by use of a digital holographic technique, in *Optical Imaging Sensors and Systems for Homeland Security Applications* (ed. B. Javidi), Springer, pp. 85–95.

39 Choi, H.-J., Seo, Y.-H., Yoo, J.S., and Kim, D.-W. (2006) An electronic watermarking technique for digital holograms in a DWT domain, in *Advances in Image and Video Technology* (eds L.-W. Chang, W.-N. Lie, and R. Chiang), Springer, pp. 1007–1017.

40 Shortt, A.E., Naughton, T.J., and Javidi, B. (2006) Compression of encrypted digital holograms using artificial neural networks, in *Optical Imaging Sensors and Systems for Homeland Security Applications* (ed. B. Javidi), Springer, pp. 25–45.

41 Frauel, Y., Tajahuerce, E., Matoba, O., Castro, A., and Javidi, B. (2006) 3-D object reconstruction and recognition techniques based on digital holography, in *Optical Image Sensors and Systems for Homeland Security* (ed. B. Javidi), Springer, pp. 1–23.

42 Sinzinger, S., Jahns, J., Daria, V.R., and Gluckstad, J. (2006) Planar micro optical systems for correlation and security applications, in *Optical Imaging Sensors and Systems for Homeland Security Applications* (ed. B. Javidi), Springer, pp. 339–360.

43 Cuadrado-Laborde, C. (2009) Time-variant signal encryption by dual random phase encoding setups applied to fiber optic links, in *Atomic, Molecular, and Optical Physics* (ed. L.T. Chen), Nova Science Publishers, Inc., New York, pp. 261–311.

44 Gluckstad, J. and Palima, D. (2009) Optical encryption and decryption, in *Generalized Phase Contrast: Applications in Optics and Photonics*, Springer, pp. 273–298.

45 Javidi, B. (March 1997) Processing for encryption and security system. *Opt. Photon. News*, 29–33.

46 Javidi, B. (March 1997) Securing information with optical technologies. *Phys. Today*, **50**, 27–32.

47 Javidi, B., Sergent, A., Zhang, G., and Guibert, L. (1997) Fault tolerance properties of a double phase encoding encryption technique. *Opt. Eng.*, **36**, 992–998.

48 Javidi, B. and Ahouzi, E. (1998) Optical security system with Fourier plane encoding. *Appl. Opt.*, **37**, 6247–6255.

49 Unnikrishnan, G., Joseph, J., and Singh, K. (1998) Optical encryption system that uses phase conjugation in a photorefractive crystal. *Appl. Opt.*, **37**, 8181–8186.

50 Unnikrishnan, G., Joseph, J., and Singh, K. (2000) Optical encryption by double-random phase encoding in the fractional Fourier domain. *Opt. Lett.*, **25**, 887–889.

51 Nishchal, N.K., Joseph, J., and Singh, K. (2003) Fully phase encryption using fractional Fourier transform. *Opt. Eng.*, **42**, 1583–1588.

52 Nishchal, N.K., Joseph, J., and Singh, K. (2003) Optical phase encryption by phase contrast using electrically addressed spatial light modulator. *Opt. Commun.*, **217**, 117–122.

53 Nishchal, N.K., Joseph, J., and Singh, K. (2003) Optical encryption using cascaded extended fractional Fourier transform. *Opt. Mem. Neural Net.*, **12**, 139–145.

54 Barrera, J.F., Henao, R., Tebaldi, M., Torroba, R., and Bolognini, N. (2006) Multiplexing encryption–decryption via lateral shifting of a random phase mask. *Opt. Commun.*, **259**, 532–536.

55 Barrera, J.F., Henao, R., Tebaldi, M., Torroba, R., and Bolognini, N. (2006) Multiplexing encrypted data by using polarized light. *Opt. Commun.*, **260**, 109–112.

56 Barrera, J.F., Henao, R., Tebaldi, M., Torroba, R., and Bolognini, N. (2006) Multiple image encryption using an aperture-modulated optical system. *Opt. Commun.*, **261**, 29–33.

57 Singh, M. and Kumar, A. (2007) Optical encryption and decryption using a sandwich random phase diffuser in the Fourier plane. *Opt. Eng.*, **46** (5), 055201(1–6).

58 Singh, M., Kumar, A., and Singh, K. (2008) Secure optical system that uses fully phase-based encryption and lithium niobate crystal as phase contrast filter for decryption. *Opt. Laser Technol.*, **40**, 619–624.

59 Javidi, B., Sergent, A., and Ahouzi, E. (1998) Performance of double phase encoding encryption technique using binarized encrypted images. *Opt. Eng.*, **37**, 565–569.

60 Situ, G. and Zhang, J. (2003) A cascaded iterative Fourier transform algorithm for optical security applications. *Optik*, **114**, 473–477.

61 Nomura, T., Nitanai, E., Numata, T., and Javidi, B. (2006) Design of input phase mask for the space bandwidth of the optical encryption system. *Opt. Eng.*, **45** (1), 0170061–0170065.

62 Yuan, S., Zhou, X., Li, D.-h, and Zhou, D.-f. (2007) Simultaneous transmission for an encrypted image and a double random-phase encryption key. *Appl. Opt.*, **46** 3747–3753.

63 Frauel, Y., Castro, A., Naughton, T.J., and Javidi, B. (2007) Resistance of the double random phase encryption against various attacks. *Opt. Express*, **15**, 10253–10265.

64 Barrera, J.F., Henao, R., and Torroba, R. (2005) Optical encryption method using toroidal zone plates. *Opt. Commun.*, **248**, 35–40.

65 Monaghan, D.S., Situ, G., Gopinathan, U., Naughton, T.J., and Sheridan, J.T. (2008) Role of phase key in the double random phase encoding technique: an error analysis. *Appl. Opt.*, **47**, 3808–3816.

66 Amaya, D., Tebaldi, M., Torroba, R., and Bolognini, N. (2008) Multichanneled puzzle-like encryption. *Opt. Commun.*, **281**, 3434–3439.

67 Situ, G., Monaghan, D.S., Naughton, T.J., Sheridan, J.T., Pedrini, G., and Osten, W. (2008) Collision in double random phase encoding. *Opt. Commun.*, **281**, 5122–5125.

68 Singh, M., Kumar, A., and Singh, K. (2008) Optical security system using Jigsaw transforms of the second random phase mask and the encrypted image in a double random phase encoding system. *Opt. Laser Eng.*, **46**, 763–768.

69 Singh, M., Kumar, A., and Singh, K. (2009) Encryption by using matrix-added, or matrix-multiplied input images placed in the input plane of a double random phase encoding geometry. *Opt. Laser Eng.*, **47**, 1293–1300.

70 Kumar, P., Joseph, J., and Singh, K. (2009) Impulse attack-free four random-phase mask encryption based on a 4-f optical system. *Appl. Opt.*, **48**, 2356–2363.

71 Kumar, P., Kumar, A., Joseph, J., and Singh, K. (2009) Impulse attack free double-random-phase encryption scheme with randomized lens-phase functions. *Opt. Lett.*, **34**, 331–333.

72 Barrera, J.F. and Torroba, R. (2009) Efficient encrypting procedure using amplitude and phase as independent channels to display decoy objects. *Appl. Opt.*, **48**, 3120–3128.

73 Monaghan, D.S., Gopiunathan, U., Kelly, D.P., Naughton, T.J., and Sheridan, J.T. (2009) Systematic errors of an optical encryption system due to the discrete values of a spatial light modulator. *Opt. Eng.*, **48** (2), 027001(1–7).

74 Monaghan, D.S., Situ, G., Gopinathan, U., Naughton, T.J., and Sheridan, J.T. (2009) Analysis of phase encoding for optical encryption. *Opt. Commun.*, **282**, 482–492.

75 Neto, L.G. and Sheng, Y. (1996) Optical implementation of image encryption using random phase encoding. *Opt. Eng.*, **35**, 2459–2463.

76 Javidi, B., Zhang, G., and Li, J. (1996) Experimental demonstration of the random phase encoding technique for image encryption and security verification. *Opt. Eng.*, **35**, 2506–2512.

77 Goudail, F., Bollaro, F., Javidi, B., and Refregier, P. (1998) Influence of a perturbation in a double phase-encoding system. *J. Opt. Soc. Am. A*, **15**, 2629–2638.

78 Javidi, B., Bernard, L., and Towghi, N. (1999) Noise performance of double-phase encryption compared to XOR encryption. *Opt. Eng.*, **38**, 9–19.

79 Towghi, N., Javidi, B., and Luo, Z. (1999) Fully phase encrypted image processor. *J. Opt. Soc. Am. A*, **16**, 1915–1927.

80 Javidi, B., Towghi, N., Maghzi, N., and Verrall, S.C. (2000) Error-reduction techniques and error analysis for fully phase- and amplitude-based encryption. *Appl. Opt.*, **39**, 4117–4130.

81 Matoba, O. and Javidi, B. (1999) Encrypted optical memory system using three-dimensional keys in the Fresnel domain. *Opt. Lett.*, **24**, 762–764.

82 Situ, G. and Zhang, J. (2004) Double random-phase encoding in the Fresnel domain. *Opt. Lett.*, **29**, 1584–1586.

83 Hennelly, B.M. and Sheridan, J.T. (2005) Optical encryption and the space bandwidth product. *Opt. Commun.*, **247**, 291–305.

84 Chen, L.-F. and Zhao, D. (2006) Optical color image encryption by wavelength multiplexing and lensless Fresnel transform holograms. *Opt. Express*, **14**, 8552–8560.

85 Unnikrishnan, G. and Singh, K. (2000) Double random fractional Fourier-domain encoding for optical security. *Opt. Eng.*, **39**, 2853–2859.

86 Liu, S., Yu, L., and Zhu, B. (2001) Optical image encryption by cascaded fractional Fourier transforms with random phase filtering. *Opt. Commun.*, **187**, 57–63.

87 Zhu, B., Liu, S., and Ran, Q. (2000) Optical image encryption based on multifractional Fourier transforms. *Opt. Lett.*, **25**, 1159–1161.

88 Zhang, Y., Zheng, C.-H., and Tanno, N. (2002) Optical encryption based on iterative fractional Fourier transform. *Opt. Commun.*, **202**, 277–285.

89 Kumar, P., Joseph, J., and Singh, K. (2008) Double random phase encryption with in-plane rotation of a modified Lohmann's second-type system in the anamorphic fractional Fourier domain. *Opt. Eng.*, **47** (11), 117001(1–7).

90 Liu, Z., Dai, J., Sun, X., and Liu, S. (2009) Triple image encryption scheme in fractional Fourier transform domains. *Opt. Commun.*, **282**, 518–522.

91 Zhang, S. and Karim, M.A. (1999) Color image encryption using double random phase encoding. *Microw. Opt. Technol. Lett.*, **21**, 318–323.

92 Joshi, M., Shakher, C., and Singh, K. (2008) Color image encryption and decryption for twin images in fractional Fourier domain. *Opt. Commun.*, **281**, 5713–5720.

93 Chen, L.-F. and Zhao, D. (2009) Color image encoding in dual fractional Fourier-wavelet domain with random phases. *Opt. Commun.*, **282**, 3433–3438.

94 Chang, H.T. (2001) Image encryption using separable amplitude-based virtual image and iteratively retrieved phase information. *Opt. Eng.*, **40**, 2165–2171.

95 Peng, X., Cui, Z., and Tan, T. (2002) Information encryption with virtual-optics imaging system. *Opt. Commun.*, **212**, 235–245.

96 Peng, X., Yu, L., and Cai, L. (2002) Double-lock for image encryption with virtual optical wavelength. *Opt. Express.*, **10**, 41–45.

97 Peng, X. and Zhang, P. (2006) Security of virtual-optics-based cryptosystem. *Optik*, **117**, 525–531.

98 Seo, D.-H. and Kim, S.J. (2003) Shift-tolerance property of optical security system using phase-based virtual image. *Opt. Rev.*, **10**, 175–178.

99 Wang, X. and Zhao, D. (2006) Image encryption based on anamorphic fractional Fourier transform and three-step phase-shifting interferometry. *Opt. Commun.*, **268**, 240–244.

100 Liu, Z. and Liu, S. (2007) Double image encryption based on iterative fractional Fourier transform. *Opt. Commun.*, **275**, 324–329.

101 Li, H. and Wang, Y. (2008) Double image encryption by iterative phase retrieval algorithm in fractional Fourier domain. *J. Mod. Opt.*, **55**, 3601–3609.

102 Mogensen, P.C. and Gluckstad, J. (2000) A phase-based optical encryption system with polarization encoding. *Opt. Commun.*, **173**, 177–183.

103 Unnikrishnan, G., Pohit, M., and Singh, K. (2000) A polarization encoded

optical encryption system using ferroelectric spatial light modulator. *Opt. Commun.*, **185**, 25–31.

104 Arizaga, R. and Torroba, R. (2002) Digital polarization-sensitive speckle correlation as security validation technique. *Optik*, **113**, 333–336.

105 Arizaga, R. and Torroba, R. (2003) Validation through a binary key code and a polarization sensitive digital technique. *Opt. Commun.*, **215**, 31–36.

106 Zhang, Y. and Wang, B. (2008) Optical image encryption based on interference. *Opt. Lett.*, **33**, 2443–2445.

107 Wang, B. and Zhang, Y. (2009) Double image hiding based on optical interference. *Opt. Commun.*, **282**, 3439–3443.

108 Han, Y. and Zhang, Y. (2010) Optical image encryption based on two beams' interference. *Opt. Commun.*, **283**, 1690–1692.

109 Nomura, T., Mikan, S., Morimoto, Y., and Javidi, B. (2003) Secure optical data storage with random phase key codes by use of a configuration of a joint transform correlator. *Appl. Opt.*, **42**, 1508–1514.

110 Cheng, C.-J., Lin, L.-C., Wang, C.-M., and Chen, C.-L. (2005) Optical joint transform encryption using binary phase difference key mask. *Opt. Rev.*, **12**, 367–371.

111 Rueda, E., Barrera, J.F., Henao, R., and Torroba, R. (2009) Lateral shift multiplexing with a modified random mask in a joint transform correlator encrypting architecture. *Opt. Eng.*, **48** (2), 0270061-(1–5).

112 Mogensen, P.C. and Gluckstad, J. (2001) Phase-only optical decryption of a fixed mask. *Appl. Opt.*, **40**, 1226–1235.

113 Shin, C.-M., Seo, D.-H., and Kim, S.-J. (2004) Gray-level image encryption scheme using full phase encryption and phase-encoded exclusive-OR operations. *Opt. Rev.*, **11**, 34–37.

114 Heanue, J.F., Bashaw, M.C., and Hesselink, L. (1995) Encrypted holographic data storage based on orthogonal-phase-code multiplexing. *Appl. Opt.*, **34**, 6012–6015.

115 Matoba, O. and Javidi, B. (1999) Encrypted optical storage with angular multiplexing. *Appl. Opt.*, **38**, 7288–7293.

116 Su, W.-C. and Lin, C.-H. (2004) Three-dimensional shifting selectivity of decryption phase mask in a double random phase encoding holographic memory. *Opt. Commun.*, **241**, 29–41.

117 Su, W.-C. and Lin, C.-H. (2004) Enhancement of the angular selectivity in encrypted holographic memory. *Appl. Opt.*, **43**, 2298–2304.

118 Kim, H. and Lee, Y.H. (2006) Encryption of a volume hologram by complementary input image and binary amplitude mask. *Opt. Commun.*, **258**, 9–17.

119 Carnicer, A., Montes-Usategui, M., Arcos, S., and Juvells, I. (2005) Vulnerability to chosen-cyphertext attacks of optical encryption schemes based on double random phase keys. *Opt. Lett.*, **30**, 1644–1646.

120 Peng, X., Wei, H., and Zhang, P. (2006) Chosen-plaintext attack on lensless double-random phase encoding in the Fresnel domain. *Opt. Lett.*, **31**, 3261–3263.

121 Peng, X., Zhang, P., Wei, H., and Yu, B. (2006) Known-plaintext attack on optical encryption based on double random phase keys. *Opt. Lett.*, **31**, 1044–1046.

122 Gopinathan, U., Monaghan, D.S., Naughton, T.J., and Sheridan, J.T. (2006) A known-plaintext heuristic attack on the Fourier plane encryption algorithm. *Opt. Express*, **14**, 3181–3186.

123 Monaghan, D.S., Situ, G., Gopinathan., U., Naughton, T.J., and Sheridan, J.T. (2007) Cryptosystem of optical security systems with significant output images. *Appl. Opt.*, **46**, 5257–5262.

124 Situ, G., Gopinathan, U., Monaghan, D.S., and Sheridan, J.T. (2007) Cryptanalysis of optical security systems with significant output images. *Appl. Opt.*, **46**, 5257–5262.

125 Naughton, T.J., Hennelly, B.M., and Dowling, T. (2008) Introducing secure modes of operation for optical encryption. *J. Opt. Soc. Am. A*, **25**, 2608–2617.

126 Matoba, O., Sawasaki, T., Nakajima, K., and Nitta, K. (2007) Optical identification system of three-dimensional random

phase object by use of speckle patterns in propagation distances. *J. Phys.*, **77**, 0120091–0120098.

127 Arizaga, R., Henao, R., and Torroba, R. (2003) Fully digital encryption technique, *Opt. Commun.*, **221**, 43–47.

128 Sun, C.-C., Su, W.-C., Wang, B., and Chiou, A.E.T. (2001) Lateral shifting sensitivity of a ground glass for holographic encryption and multiplexing using phase conjugate readout algorithm. *Opt. Commun.*, **191**, 209–224.

129 Jones, R. and Wykes, C. (eds) (1989) *Holographic and Speckle Interferometry*, 2nd edn, Cambridge University Press, London.

130 Kumar, A. and Singh, K. (1994) Elongated laser speckle in imaging of a rough object with slit shaped illumination region: effect of off-axis aberrations. *Optik*, **96**, 115–119.

131 Singh, M., Kumar, A., and Singh, K. (2009) Encryption and decryption using a sandwich phase diffuser made by using two speckle patterns and placed in the Fourier plane: simulation results. *Optik*, **120**, 916–922.

132 Denz, C., Muller, K.-O., Visinka, F., and Tschudi, T. (2000) A demonstration platform for phase-coded multiplexing, in *Holographic Data Storage* (eds H.J. Coufal, D. Psaltis, and G.T. Sincerbox), Springer-Verlag, pp. 419–428.

133 John, R., Joseph, J., and Singh, K. (2005) Phase-image-based content-addressable holographic data storage with security. *J. Opt. A Pure Appl Opt.*, **7**, 123–128.

134 Situ, G. and Zhang, J. (2005) Multiple-image encryption by wavelength multiplexing. *Opt. Lett.*, **30**, 1306–1308.

135 Islam, M.N. and Alam, M.S. (2006) Optical encryption and multiplexing of personal identification information using orthogonal code. *Opt. Eng.*, **45**, 098201(1–4).

136 Situ, G. and Zhang, J. (2006) Position multiplexing for multiple-image encryption. *J. Opt. A Pure Appl Opt. A*, **8**, 391–397.

137 Chang, C.-C., Liu, J.-P., Lee, H.-Y., Lin, C.-Y., Chang, T.-C., and Yau, H.-F. (2006) Decryption of a random-phase multiplexing recording system. *Opt. Commun.*, **259**, 78–81.

138 Gopinathan, U., Naughton, T.J., and Sheridan, J.T. (2006) Polarization encoding and multiplexing of two-dimensional signals: application to image encryption. *Appl. Opt.*, **45**, 5693–5700.

139 Shi, Y., Situ, G., and Zhang, J. (2007) Multiple image hiding in the Fresnel domain. *Opt. Lett.*, **32**, 1914–1916.

140 Sun, C.-C., Hsu, C.-Y., Ma, S.-H., and Su, W.C. (2007) Rotation selectivity of random phase encoding in volume holograms. *Opt. Commun.*, **276**, 62–66.

141 Hennelly, B.M., Naughton, T.J., McDonald, J., Sheridan, J.T., Unnikrishnan, G., Kelly, D.P., and Javidi, B. (2007) Spread-space spread-spectrum technique for secure multiplexing. *Opt. Lett.*, **32**, 1060–1062.

142 Inoue, A., Takayama, Y., Ishii, Y., and Kodate, K. (2008) Removable fieldstone security key used in random pattern optical multiple recording. *Jpn J. Appl. Phys.*, **47**, 5960–5963.

143 Islam, M.N., Alam, M.S., and Karim, M.A. (2008) Optical security system employing quadrature multiplexing. *Opt. Eng.*, **47**, 048201(1–5).

144 Barrera, J.F., Henao, R., Tebaldi, M., Torroba, R., and Bolognini, N. (2008) Code retrieval via undercover multiplexing. *Optik*, **119**, 139–142.

145 Hwang, H.-E., Chang, H.T., and Lie, W.-N. (2009) Multiple-image encryption and multiplexing using a modified Gerchberg–Saxton algorithm and phase modulation in Fresnel-transform domain. *Opt. Lett.*, **34**, 3917–3919.

146 Amaya, D., Tebaldi, M., Torroba, R., and Bolognini, N. (2009) Wavelength multiplexing encryption using joint transform correlator architecture. *Appl. Opt.*, **48**, 2099–2104.

147 Barrera, J.F., Henao, R., Tebaldi, M., Torroba, R., and Bologini, N. (2009) Digital encryption with undercover multiplexing by scaling the encoding mask. *Optik*, **120**, 342–346.

148 Alfalou, A. and Mansour, A. (2009) Double random phase encryption scheme to multiplex and simultaneous encode multiple images. *Appl. Opt.*, **48**, 5933–5947.

149 Xiao, Y.-L., Zhou., X., Yuan, S., Liu, Q., and Li, Y.-C. (2009) Multiple-image optical encryption: an improved encoding approach. *Appl. Opt.*, **48**, 2686–2692.

150 Barrera, J.F. and Torroba, R. (2010) One step multiplexing optical encryption. *Opt. Commun.*, **283**, 1268–1272.

151 Henao, R., Rueda, E., Barrera, J.F., and Torroba, R. (2010) Noise-free recovery of optodigital encrypted and multiplexed images. *Opt. Lett.*, **35**, 333–335.

152 Singh, M., Kumar, A., and Singh, K. (2008) Multiplexing in optical encryption by using an aperture system and a rotating sandwich random phase diffuser in the Fourier plane. *Opt. Laser Eng.*, **46**, 243–251.

153 Singh, M., Kumar, A., and Singh, K. (2009) Securing multiplexed information by in-plane rotation of random phase diffusers constituting a sandwich diffuser placed in the Fourier plane of a double random phase encoding system. *Opt. Laser Technol.*, **41**, 32–41.

154 Nelleri, A., Joseph, J., and Singh, K. (2007) Digital Fresnel field encryption for three-dimensional information security. *Opt. Eng.*, **46**, 045801(1–8).

155 Liebling, M., Blu, T., and Unser, M. (2004) Complex-wave retrieval from a single off-axis hologram. *J. Opt. Soc. Am. A*, **21**, 367–377.

Index

Advances in Speckle Metrology and Related Techniques. Edited by Guillermo H. Kaufmann

ISBN: 978-3-527-40957-0